Marine Biotechnology

Scrivener Publishing
100 Cummings Center, Suite 541J
Beverly, MA 01915-6106

Marine Nanoscience and Nanotechnology

This series focuses on various aspects of marine nanoscience and nanotechnology, covering recent developments, theory, design, studies, measurement, and real-world applications. It joins researchers, marine scientists, marine scholars, nanotechnologists, and polymer chemists who share their wide-ranging experiences and research results, thereby creating a premier interdisciplinary platform for discussing contemporary innovations, trends, and concerns, as well as challenges and solutions.

Publishers at Scrivener
Martin Scrivener (martin@scrivenerpublishing.com)
Phillip Carmical (pcarmical@scrivenerpublishing.com)

Marine Biotechnology

Studies and Applications

Edited by

Shaju S. S.
Department of Chemical Oceanography, School of Marine Sciences, Cochin University of Science and Technology, Cochin, Kerala, India

Vipin. P. M.
Central Institute of Fisheries Nautical and Engineering Training, Kochi, India

and

Visakh. P.M.
Natural Bioactive Materials Laboratory, Department of Bioengineering, Ege University, Bornova/Izmir, Turkey

Scrivener Publishing

WILEY

This edition first published 2025 by John Wiley & Sons, Inc., 111 River Street, Hoboken, NJ 07030, USA and Scrivener Publishing LLC, 100 Cummings Center, Suite 541J, Beverly, MA 01915, USA

For more information about Scrivener publications please visit www.scrivenerpublishing.com.

Wiley Global Headquarters
111 River Street, Hoboken, NJ 07030, USA

For details of our global editorial offices, customer services, and more information about Wiley products visit us at www.wiley.com.

The manufacturer's authorized representative according to the EU General Product Safety Regulation is Wiley-VCH GmbH, Boschstr. 12, 69469 Weinheim, Germany, e-mail: Product_Safety@wiley.com.

Library of Congress Cataloging-in-Publication Data

ISBN 978-1-394-30129-4

Front cover images courtesy of ChatGPT and Adobe Firefly
Cover design by Russell Richardson

Set in size of 11pt and Minion Pro by Manila Typesetting Company, Makati, Philippines

Printed in the USA

10 9 8 7 6 5 4 3 2 1

Contents

Preface

This book summarizes many of the recent research accomplishments in the field of marine biotechnology. It explores a wide range of topics, including the evolution of marine biotechnology and marine bio-discovery; biomolecular components extracted from marine bio-resources; biotechnological advances that extend the use of marine bio-resources for industrial and agricultural applications; bioprospecting of natural products from marine phytoplankton; pharmaceutical and nutraceutical applications of marine microalgae; extraction of drugs and food ingredients from marine macroalgae; and biotechnological applications of marine bacteria. Additional chapters cover marine endophytes as a natural, untapped source of bioactive compounds; nanoparticles and phytoconstituents from marine mangrove plants; the use of marine gelatin and chitosan edible nanocomposite films for food coating and packaging; and advanced antifouling alternatives derived from marine bio-resources.

Leading researchers from the marine industry, academia, government, and private research institutions across the globe have contributed to this book. Academics, researchers, scientists, marine biologists, aquaculture technologists, postdoctoral fellows, and both senior and graduate students involved in research and development will benefit from this application-oriented resource, which provides solutions to both fundamental and applied problems.

The various chapters in this book are authored by prominent experts from industry, academia, and government or private research laboratories worldwide. It presents an up-to-date record of major findings and observations in the field of marine biotechnology.

The first chapter introduces the scope, state-of-the-art, preparation methods, challenges, and new opportunities in marine biotechnology.

Chapter two discusses the therapeutic aspects of compounds derived from microalgae, framed by recent advances in their cultivation and commercialization. The chapter also covers the role of life cycle assessment in evaluating the environmental impact of microalgae-based production systems, validating cultivation methods, and designing sustainable extraction techniques aimed at enhancing the economic, environmental, and industrial implications of microalgae culture systems.

Chapter three focuses on the pharmaceutical and nutraceutical applications of marine microalgae. It covers marine microalgae-derived bioactive compounds, their pharmaceutical and nutraceutical uses, the biodiversity of marine phytoplankton, and pharmaceutical applications of marine microbiology.

The fourth chapter addresses the extraction of drugs and food ingredients from marine macroalgae. It reviews the ecological role, biochemical composition, classifications, and potential applications of marine macroalgae, particularly in the food and pharmaceutical industries.

Chapter five explores the biotechnological applications of marine bacteria. The authors highlight two main areas: the general applications of marine bacteria and detailed subtopics such as isolation, sample collection and preparation, culturing and incubation, and the use of marine bacteria in waste management, bioremediation, enzyme production, and pharmaceuticals.

The sixth chapter focuses on nanoparticles and phytoconstituents from marine mangrove plants, emphasizing their potential biochemical and medicinal applications. It discusses several bioactive substances derived from these plants and their roles in biochemistry and healthcare.

Chapter seven is dedicated to Mesopelagic Fish Protein Hydrolysates: A Source of Antihypertensive and Antidiabetic Peptides. It explains the composition and bioactivity of mesopelagic fish protein hydrolysates and their potential to produce anti-hypertensive and anti-diabetic peptides.

The eighth chapter examines the potential of mesopelagic fishes for anti-cancer and anti-microbial biotechnological applications. Topics include notable mesopelagic species with therapeutic properties, mesopelagic jellyfish, novel bioactive peptides, and the anti-cancer and antimicrobial potential of Mediterranean mesopelagic fish species.

The next chapter discusses recent advances in marine-derived drugs, particularly in cancer therapy and immune signaling.

The final chapter focuses on bioactive compounds derived from marine bivalves, covering their anticancer and antimicrobial effects. Topics include bioactive molecules, polysaccharides, and lipids from bivalves.

Finally, we express our sincere gratitude to all the contributors whose expertise and dedication made this book possible. We are deeply thankful for their commitment and the quality of their contributions. Without their support and enthusiasm, this compilation would not have been achievable. We also extend our thanks to the reviewers who provided valuable feedback on each chapter. Special thanks to Martin Scrivener and Scrivener Publishing for recognizing the growing importance of marine biotechnology and for supporting this vital project—an endeavor that few other publishers have pursued.

Dr. Shaju S. S.
Dr. Vipin. P. M.
Dr. Visakh. P.M.
August 2025

1

Marine Biotechnology: Studies and Applications—State-of-the-Art and New Challenges

Shaju S. S.[1], Vipin. P. M.[2], Visakh. P.M.[3]* and Shiv Kumari Panda[4]

[1]Department of Chemical Oceanography, School of Marine Sciences, Cochin University of Science and Technology, Cochin, Kerala, India
[2]Central Institute of Fisheries Nautical and Engineering Training (CIFNET), (Department of Fisheries, Ministry of Fisheries, Animal Husbandry and Dairying, Government of India) Kochi, India
[3]Natural Bioactive Materials Laboratory, Department of Bioengineering, Ege University, Bornova/Izmir, Turkey
[4]PI: OURIIP SEED FUND, Govt. of Odisha. Udayanath Autonomous College of Science & Technology, Adaspur, Cuttack, Odisha, India

Abstract

In this chapter, a short version of all chapters is provided covering different chapter topics such as exploring the therapeutic potential of microalgae-derived compounds, pharmaceutical and nutraceutical applications of marine microalgae, extraction of drugs and food ingredients from marine macroalgae, biotechnological applications of marine bacteria: marine biotechnology, nanoparticles and phytoconstituents from marine mangrove plant, mesopelagic fish protein hydrolysates, mesopelagic fishes: potential use for anticancer and antimicrobial biotechnological applications, marine-derived drugs: recent advances in cancer therapy and immune signaling, and marine bivalve-derived bioactive compounds: anticancer and antimicrobial effects.

Keywords: Marine biotechnology, microalgae, marine bacteria, marine mangrove plant, nanoparticles, phytoconstituents, mesopelagic fish protein, anticancer

**Corresponding author*: visagam143@gmail.com

Shaju S. S., Vipin. P. M. and Visakh. P.M. (eds.) Marine Biotechnology: Studies and Applications, (1–24) © 2025 Scrivener Publishing LLC

1.1 Exploring the Therapeutic Potential of Microalgae-Derived Compounds

Microalgae are reported to be rich in nutrients such as proteins, pigments, polyunsaturated omega-3 fatty acids, polysaccharides, vitamins, and minerals. However, their nutritional composition is reported to vary with the type of species, geographical conditions, culture conditions etc. Microalgae are reported to be a rich source of pigments such as phycobiliproteins, β-carotene, astaxanthin, luteine, and cantaxanthin. The colorant attributes, coupled with its bioactivities, boosts the applications of microalgal-based pigments in different sectors such as food, feed, biomedical, and cosmetics. Microalgal β-carotene comprises a mixture of all-*trans* and 9-*cis* isomers, which have been proven to have high bioaccessibility and potential antioxidant activity, whereas synthetic carotene has only all-*trans* isomers [1]. Natural astaxanthin is reported to be more esterified, up to 95%, than its synthetic counterpart aiding in high energy production and conferring protection to tissues from oxidative damage. Lipids, especially omega-3 polyunsaturated fatty acids (PUFAs), are reported to have important positive implications on human health. Consumption of foods rich in omega-3 PUFAs is reported to offer protection from different ailments such as inflammatory disorder, neurological disorders, and cardiovascular diseases. The content of lipids, especially fatty acids, is reported to be influenced by the type and species of microalgae and its cultivation conditions, such as temperature, light, and nutrients. It has been reported that marine microorganisms are reported to produce more PUFAs than freshwater because of the frequent exposure to the salt conditions [2]. The major microalgal polysaccharides include starch, cellulose, sugar, etc. [3].

Microalgal polysaccharides are often classified based on their function such as structure or storage or as matrix polysaccharides. The matrix-based polysaccharides that are entirely released into the environment are often referred to as extracellular polysaccharides (EPSs). In general, microalgal-based polysaccharides are heteropolymers in nature [4]. Microalgal-based EPSs are often present as a viscous mass surrounding a cell or group of cells, which are produced either as a protection mechanism or in response to fluctuating environmental conditions [5]. The microalgal-based polysaccharides are finding applications in the nutraceutical, feed and cosmetic industries as they have immense bioactivities such as antioxidant, antibacterial, antitumor, and anti-inflammatory activity [6]. Zhao *et al.* [7] reported the promising potential of UAE as a pre-treatment step in recovering carbohydrates from microalgal biomass as a renewable feedstock for bio-fermentation process. Binary mixtures of green solvents, such as water–DMSO and water–ethanol, were reported to be effective for recovering phenolic compounds and carotenoids from microalgae *via* an ultrasound-assisted extraction [8]. Sierra *et al.* [9] employed enzymatic extraction as an effective pre-treatment step for enhancing the recovery of lipids and protein. Zhang *et al.* [10] optimized enzymatic hydrolysis as an effective method for recovering lipids from the microalgae, *Scenedesmus* sp. Zhong *et al.* [11] reported a hydrothermal-based method for the bio-production of acetic acid from microalgae. Huang *et al.* [12] detailed the use of a deep eutectic solvent-assisted hydrothermal extraction method as an environment-friendly process for recovering microalgal lipids. Silva *et al.* [13] employed a biorefinery-based framework for biofuel production from microalgae using hydrothermal liquefaction. Mohit *et al.* [14] also employed LCA for evaluating the environmental benefits associated with microalgal-mediated biofuel production. Arbour *et al.* [15] employed LCA as a tool to evaluate the utilization of microalgae for waste water utilization from shrimp recirculating aquaculture firms. *Nannochloropsis* is used mainly for aquaculture, as it represents one of the major sources of EPA (eicosapentaenoic acid), an omega-3 fatty acid [16]. It is used to enrich the nutritional profile of feed for fish and shrimp enhancing their growth and health [17]. The high EPA content of *Nannochloropsis* maintains cardiovascular health, reduces inflammation, and improves the general health of aquatic organisms [18].

Chlorella is the other well-known microalga applied mainly as a dietary supplement. It becomes valuable due to its high content of chlorophyll, proteins, and also some vitamins (C, E) and minerals (zinc, magnesium) that are essential [19]. *Chlorella* is used in cosmetics and personal care products due to its feature of rejuvenating the skin [20]. It detoxifies the body by removing heavy metals, thus already proving its detoxifying effect [21].

1.2 Pharmaceutical and Nutraceutical Applications of Marine Microalgae

The benefits to the food and pharmaceutical industries have led to a growing interest in algae. Antioxidant and photoprotective characteristics are seen in two substances found in several species of marine algae. These are a type of amino acid called mycosporin and a type of organic pigment called carotenoid [22, 23]. Microalgae can survive in extreme environments and can grow in any place with light and humidity as they are amazing organisms [24]. Chemical compounds of microalgal origin called bioactive compounds, like phycocyanin, beta carotene, oleic acid, linolenic acid, cobalamin (vitamin B12), vitamin E, lutein, cyanovirin, and zeaxanthin, have exhibited anti-enzymatic, antimicrobial, anticarcinogenic, antibiotic actions, photoprotective, anti-inflammatory, antioxidant, anti-aging, and hypocholesterolemic properties [25].

Other major functions of microalgae include its capability to improve stabilization of the marine ecosystem as they are fed upon by other marine organisms. Resources, like microalgal technology, are used due to the increase in demand of microalgae as sources of bioenergy, food, and pharmaceuticals and thus meet their growing market potential [26]. Nutraceutical is a combination of two words—nutrient and pharmaceutical. These compounds have shown to be of high health benefits and supplemented through diet. Nutraceuticals of marine origin is called marine nutraceuticals and have become of great importance in recent years. Seaweed, microalgae, fish and fish by-products, crustaceans, marine fungi, and bacteria are the ones of interest as sources of ingredients that promote health or as healthy food. Microalgae do not require freshwater or arable land; they do not require pesticides or fertilizer either. They are an environment-friendly food source and are sustainable. They are also safe for the environment. They are highly efficient in accessing water and nutrients. Their photosynthetic machinery is efficient too. They have high productivity and short growth cycle [27, 28]. Microalgae have high nutritional value and are a source of pigments, proteins, and fatty acids that are required for human consumption. Nutraceutical and pharmaceutical components are produced from microalgae because of their excellent properties. Microalgal technology is an applicable resource that meets the market potential because of the increasing need for food, bioenergy, and other bioactive compounds. Microalgae have pharmaceutical applications that are dependent upon their anticancer, antioxidant, antimicrobial, and anti-inflammatory properties. The water-soluble antioxidants present in microalgae are polyphenols, phycobiliproteins, and vitamins that help in inhibiting cancer by causing the regression of premalignant lesions.

Vitamin B helps in red blood cell movement, prevents weakness, protects the heart and skin, and reduces depression. *Dunaliella* sp. microalgae have anticancerous activity against the Triple-Negative Breast Cancer (TNBC). Astaxanthin exhibits anticarcinogenic, neuroprotective, and anti-inflammatory activities, and also prevents lipid oxidation and cardiovascular diseases [29–31]. The food or food products that provide health benefits as well as helps in the prevention and treatment of diseases are nutraceuticals [32]. When essential nutrients give additional characteristics for maintenance of organisms or on normal growth, they are considered as nutraceuticals [33]. Foods that are considered nutraceuticals are not put through any testing or regulation protocols. Nutraceutical properties are shown by microalgae and provide benefits for the immune system by fighting against cancer and heart diseases [34]. The increased yield of shellfish flesh and increased possibility of safer production through aquaculture has been achieved by supplying new immunostimulants in aquafeed [35, 36]. An exogenous diet is an exclusive source when shellfish organisms are unable to synthesis carotenoids *de novo* [37, 38]. Nutraceuticals and nutritional supplements, food coloring, immune enhancers, antioxidants, and PUFAs are the factors that commercial utilization of microalgae depends on [39].

1.3 Extraction of Drugs and Food Ingredients from Marine Macroalgae

Marine macroalgae play vital roles in marine ecosystems by providing habitat and food for a diverse range of marine life, contributing to coastal protection, and influencing nutrient cycling. One of the most remarkable aspects of marine macroalgae is their complex biochemical composition, which includes a wide range of bioactive compounds with diverse properties and applications. Alginate, carrageenan, and agar, for example, are commonly used in the food industry as thickeners, stabilizers, and gelling agents. Additionally, these polysaccharides find applications in pharmaceuticals, cosmetics, and biotechnology due to their exceptional gel-forming and bioactive properties. Proteins derived from marine macroalgae are also gaining recognition for their potential nutritional value and functional properties. Marine macroalgae are rich in polyphenols, pigments, and lipids that have antioxidant, anti-inflammatory, and antimicrobial properties making them excellent candidates for pharmaceuticals, nutraceuticals, and cosmeceuticals. For example, phlorotannins, a specific type of polyphenolic compound found in brown algae, have shown promise in treating various diseases due to their antioxidant and anti-inflammatory properties. These compounds encompass a wide range of molecules, including polysaccharides, proteins, polyphenols, pigments, and lipids, each with distinct pharmacological and nutritional properties [40]. Additionally, these compounds offer nutritional benefits such as dietary fiber, proteins, essential amino acids, omega-3 fatty acids, and vitamins. As a result, marine macroalgae hold great potential as a source of functional food ingredients and dietary supplements [41]. Marine macroalgae, such as seaweeds, possess abundant bioactive compounds that offer diverse health-enhancing properties. These bioactive compounds play a significant role in the ecological and economic significance of seaweeds, and their potential applications in pharmaceuticals, nutraceuticals, cosmeceuticals, and functional foods are increasingly attracting attention [42–44]. Marine macroalgae, including seaweeds, have a wide range of pharmacological properties and health benefits due to their rich content of bioactive compounds [45–47]. Seaweeds are rich sources of vitamins, such as A, C, E, and K, as well as minerals like calcium, magnesium, iodine, and iron. These nutrients play major roles in various physiological functions, including energy metabolism, immune support, and bone health. While marine macroalgae hold immense potential for various industries, including food, pharmaceuticals, and environmental applications, they also face several challenges [48–52]. Seaweed cultivation and harvesting can have environmental impacts, including habitat alteration, nutrient runoff, and competition with native species. Sustainable practices and careful management are needed to minimize these environmental concerns and ensure the long-term health of marine ecosystems.

1.4 Biotechnological Applications of Marine Bacteria: Marine Biotechnology

The rich tapestry of marine biodiversity, encompassing the vast array of life in our oceans and seas, is crucial for the well-being and adaptability of marine ecosystems. Despite being one of the most underutilized biological resources, the marine environment is home to a staggering variety of organisms, each with unique biological systems and characteristics. Marine biodiversity patterns are influenced by the depth and characteristics of the seabed, as well as variations in water temperature, salinity, nutrient levels, and currents. The larger category include single-celled algae known as protists. These forms include diatoms and are most abundant near coasts. Occasionally, these organisms form blooms—rapid population explosions—in response to changing seasons and the availability of nutrients such as nitrogen, iron, and phosphorus. The other type of phytoplankton cells is photosynthetic bacteria. Marine microorganisms exist not only in the surface waters of the sea but also in the lower and abyssal depths from the coastal to offshore regions and from the general oceanic to specialized niches like the blue waters of coral reefs to the black smokers of hot thermal vents on the sea floor including oceans, estuaries and lagoons, mangroves and coral reefs, the deep sea, and the seafloor [53].

Bacteria exhibit varying temperature requirements for growth and activity; colder waters support less bacterial biomass and slower growth rates, whereas warmer waters support more biomass and faster growth rates. A diverse range of microorganisms, mainly bacteria and archaea, have been discovered thriving in extreme environments. These include thermophiles, halophiles, alkalophiles, psychrophiles, piezophiles, and polyextremophiles. In the marine environment, certain bacterial groups are particularly abundant, such as cyanobacteria, proteobacteria, and actinobacteria. Viruses are recognized as significant pathogens in numerous marine organisms, with the majority being bacteriophages. These viruses can exist in two main forms as follows: as virioplankton, freely circulating in the water column, or in association with macro- or microorganisms. This viral community exhibits immense genetic diversity serving as a vast reservoir of genes for prokaryotic communities. Marine bacteria can degrade various pollutants, including oil spills, heavy metals, and organic contaminants. Utilizing these bacteria to clean up contaminated water and soil helps mitigate environmental damage from human activities and restore ecosystems. Marine bacteria, particularly actinomycetes, are also significant producers of antimicrobial secondary metabolites and antibiotics. Many marine bacteria exhibit significant antibacterial activity and are considered an exciting resource for discovering new classes of therapeutics, particularly in oncology and infectious diseases. Many bacterial species are distributed across marine ecosystems worldwide. *Candidatus Pelagibacter ubique*, a marine bacterium, is recognized as being one of the most abundant organisms on Earth [54, 55]. Marine microorganisms play a significant role in marine biotechnology because of their distinct properties. They can withstand extreme temperatures and pressures and high salt conditions. These characteristics provide the ability to breakdown complex products into simple forms, recycling of nutrients, and biodegradation. The metabolic routes and physiological modifications made by marine bacteria demonstrate their adaptability to change under biological, chemical, and physical circumstances [56, 57]. Research has shown that the bioremediation potential of these marine bacteria can be enhanced, as demonstrated by Latha and Lalitha Kumari [58], who successfully transferred a catabolic plasmid from *Pseudomonas putida* containing genes for hydrocarbon degradation, into a marine bacterium, increasing its efficiency in degrading PAHs. Marine bacteria have also been found to produce amylases with remarkable pH tolerance, thermostability, and salt tolerance further expanding the potential applications of these enzymes. Protease, a versatile enzyme, is utilized in various industries, including detergent and leather manufacturing, as well as in food, health, and waste management applications. Probiotic bacteria, including lactobacilli and bifidobacteria, primarily aid in the regulation of pathogenic microorganisms by producing antibacterial proteins like bacteriocin [59] and anticancer agents. According to reports, lactobacilli dietary supplements are reducing the induction of experimental colon cancer [60].

1.5 Nanoparticles and Phytoconstituents from Marine Mangrove Plant

Mangroves are a group of tiny plants and shrubs that can withstand high salinity levels. Mangroves are rich in a variety of plants that are used for medicine, provide timber, and have exceptional CO_2 sequestration capabilities. Mangrove ecosystems are surrounded by a variety of anthropogenic agents and pollutants. Significant sources include industrial effluents, maritime activities, and wastewater runoff [61, 62]. According to Choi *et al.* [63], Doshi *et al.* [64], and Hu *et al.* [65], improper management of wastewater treatment plants and the direct or indirect disposal of untreated effluent into aquatic systems are the main causes of environmental contamination. Mangroves thus offer a favorable environment for a wide variety of micro- and macroorganisms with an abundance of bioactive substances and enzymes [66]. For survival and reproduction, everything has the special capacity to create bioactive metabolites, which serve as "chemical signals" for responding to, avoiding, or fending off environmental cues. Mangroves have been used historically as a pesticidal, insecticidal, and medicinal plant [67]. They have also been used as a source of fuel and charcoal, for building houses, furniture, boats, and fishing gear, and to produce tannins for leather and dyeing [68].

The availability of essential nutrients, such as proteins, carbs, and amino acids, which are needed for the upkeep of biological processes and bioactive chemicals, was linked to the use of mangrove plants [69].

In the tropical Pacific, mangrove plants of the rhizophora variety are widely dispersed [70]. *Rhizophora* species (e.g., *Rhizophora apiculata*, *R. mucronata*, *R. mangle*, and *R. stylosa*), belonging to the Rhizophoraceae family, contain a multitude of phytochemical substances that have great therapeutic potential. According to Lalitha *et al.* [71], mangroves offer unique ecological environments and are a rich source of bioactive compounds. Mangrove trees can yield a variety of compounds, including phenolic, flavonoid, alkaloid, terpenoid, saponin, and tannin metabolites [72]. As a result, this plant is widely used in several Indonesian regions for traditional medicine, including for its anti-inflammatory and antipyretic purposes [73]. In light of its many uses, especially in the biomedical industry, nanotechnology has emerged as a significant scientific advance [74]. Silver NPs, which can be produced *via* a range of chemical, biological, and physical processes, are among the most researched materials [75]. In biological applications, silver NPs hold great potential, especially when used as antioxidants and antibacterials [76, 77]. Mangrove species have a great potential to enhance the quality of the environment. These plants are able to withstand the harsh environmental stressors found at these places because of a variety of biological chemicals (phenol, alcohol, terpenoids, and so forth) [78].

Mangrove-mediated AgNPs, which are produced from *H. fomes* species' aqueous extracts, have been shown to be efficient against bacterial infections and mosquito vectors. TEM, FTIR, XRD, and UV–Visible spectroscopy were used in the characterization process. NPs range in size from 15 to 40 nm. Mangroves are distinct ecosystems characterized by intricate physical, chemical, and biological interactions and a high level of biodiversity. According to recent research, these conditions are capable of producing NPs with possible uses in medicine. Bioactive substances discovered in mangroves have been used to treat a wide range of illnesses. Mangrove resources produce secondary metabolites (tannins, flavonoids, alkaloids, and saponins) that have therapeutic uses. Nonetheless, compared to mangrove-synthesized NPs, the majority of studies have been on mangrove plant extracts. NPs derived from mangroves have been assessed for their potential biomedical applications (radical scavenging, antifungal, antibacterial, parasitic, antiparasitic, antidiabetic, anti-inflammatory, and cytotoxic effects), as Balakrishnan *et al.* [79] and Thatoi *et al.* [66] have shown in their study.

1.6 Mesopelagic Fish Protein Hydrolysates

The marine ecosystem is renowned for its enormous and varied oceanic environments, which have the potential to provide food and bioactive compounds. This ecosystem integrates multiple regions that contribute to the complex dynamics exhibited [80]. Present in the zone are also squid, gelatinous animals, and fish [81]. The contribution of siphonophores to mesopelagic biomass is particularly challenging to quantify given that these animals are gelatinous and delicate, breaking up when trawled. As such, fishing them (siphonophores) has no commercial value. The fish usually present at the mesopelagic zone are called mesopelagic fish or deep-sea fish. Researchers, such as Leone *et al.* [82], found that the mesopelagic zone harbors a variety of gelatinous organisms that have significant pharmaceutical and commercial potentials, which could be exploited for medicine, food, and other uses. Multiple commercial attempts to exploit mesopelagic fish species have taken place since the 1970s [83]. By and large, the fish are planktivorous in nature, small (2–15 cm), and conduct diel vertical migration (DVM) at night to the water surface (a more nutrient-rich environment) daily to feed. Hayes *et al.* conducted further *in vivo* experiments on spontaneously hypertensive rats (SHRs) using the resultant hydrolysates produced [84].

Boarfish protein hydrolysate (BPH) generated with protease AP had antihypertensive effects on the hypertensive rats. This was observed through the rapid reduction in systolic blood pressure (SBP) indicating its potential as a functional food ingredient for managing hypertension. Mesopelagic species, like the Myctophidae family (e.g., Lanternfish), Gonostomatidae family (e.g., Bristlemouths), Sternoptychidae family (e.g., Hatchetfish), Scopelarchidae family (e.g., Pearleye), and Calanidae family (e.g., *Calanus finmarchicus*), just to mention a few, are great sources of marine lipids [85]. Lipids can be extracted, and the remaining peptide fractions are by-products with bioactive characteristics during enzymatic protein hydrolysis. Matić *et al.* [86] carried out research to investigate the potential health-promoting bioactive peptides derived from *Calanus finmarchicus* through enzymatic hydrolysis with the aim of evaluating the bioactivity of the hydrolysates, particularly in terms of antioxidant activity and DPP-IV inhibition.

Mesopelagic fish protein hydrolysates contain multiple anti-hypertensive peptides that are bioavailable and stable, but their efficacy and therapeutic potential in the management of hypertension are strongly impacted by numerous factors [87]. Environmental factors, such as temperature and pH, have the potential to influence the stability of anti-hypertensive peptides [88]. Peptides might experience conformational changes or denaturation under acidic or alkaline conditions affecting their bioactivity [89]. According to Cavan *et al.* [90], Antarctic krill (*Euphausia superba*) inhibits the Southern Ocean's mesopelagic zone. They contribute significantly to the availability of nutrients for the marine food chain and to global biogeochemical cycles in the ocean while in the zone.

1.7 Mesopelagic Fishes: Potential Use for Anticancer and Antimicrobial Biotechnological Applications

The mesopelagic species *Micromesistius poutassou*, also known as blue whiting, was previously reported to have an average protein level of 16.1 g/100 g of fish, while *M. muelleri* and *B. glaciale* were found to have average protein values of 12.3 g/100 g and 14 g/100 g, respectively. According to earlier reports, total fat values for *M. muelleri* varied from 17.8 g/100 g of fish to 3.9 g/100 g of fish for *M. poutassou* [91]. The mesopelagic fish species are abundant. There are not many papers about using them to create protein-rich, bioactive hydrolysates that may be able to prevent common ailments in humans and animals. The Mediterranean krill *Meganyctiphanes norvegica* and the lanternfish *Myctophum punctatum* were found to be sources of antimicrobial and anticancer fractions rich in EPA, DHA, and eicosatetraenoic acid (ETA) in a study conducted by Lauritano *et al.* [92]. Mesopelagic fish have evolved special adaptations to withstand the high pressure at these depths, including flexible bodies and enzymes that can withstand pressure. Despite occasionally being lower than in the ocean's top layers, the mesopelagic zone has enough oxygen to support a variety of living forms. According to Sutton *et al.* [93], the zone is frequently a transitional zone where oxygen levels drop but do not drop as much as they do in the deeper bathypelagic zone. Larger fish, squid, and marine mammals feed on mesopelagic fishes, which in turn feed on smaller zooplankton and invertebrates [94–97].

Lysozymes are present in the immune systems of many marine creatures, including mesopelagic fishes. Mesopelagic fishes possess specific proteolytic enzymes that have the ability to break down bacterial toxins or microbial proteins, hence enhancing their antibacterial properties. Gram-negative bacteria are the only bacteria that create ornithine lipids (OL), which are more unique to them and are produced in greater quantities in response to phosphate being scarce [98]. *Bacillus megaterium* and *Escherichia coli* were the targets of action demonstrated by OL isolated from *Burkholderia gladioli* pv. *agaricicola* [99]. According to Li *et al.* [100], seven bile acid derivatives were identified, given nuclear magnetic resonance (NMR) characterizations, and shown to have modest anticancer efficacy in addition to selective inhibitory action against a number of human diseases. Researchers look for anticancer properties in fish extracts from mesopelagic species. To evaluate a compound's possible therapeutic effects, it is isolated and tested against cancer cell lines. The mechanisms of action, effectiveness, and safety of the chosen candidates are next investigated in further detail. The process of looking for new compounds in marine creatures is known as marine bio-prospecting. Fish belonging to this family include *Myctophum* species and *Opisthoproctus* species, which are distinguished by their transparent heads and bioluminescence, respectively. Because of their distinct biochemical makeup, which includes proteins with potential antibacterial and anticancer effects and omega-3 fatty acids, these fish are recognized for their potential [100, 101].

1.8 Marine-Derived Drugs: Recent Advances in Cancer Therapy and Immune Signaling

The utilization of marine sources is crucial in acquiring essential drugs for various types of cancers and related diseases. Persistently, marine organisms have developed the capacity to produce complex compounds that possess cytotoxic properties adapting to the challenging and hostile conditions prevalent in marine ecosystems. Regularly, challenging aquatic conditions linked to marine environments encompass elevated levels of salinity and pressure, limited oxygen availability for cellular functions, and fluctuating degrees of marine temperature in highly dynamic settings [102]. Crucial agents for treatment extracted from these drugs, such as chemically derived compounds or chemical complexes, and natural immune agents, can be revitalized by utilizing the MNP technique. Despite extensive research efforts, including innovative and medically significant clinical approaches employing novel drug agents, it is noted that mortality rates from cancer persist at high levels [103, 104]. Breakthroughs in nanomedicine have propelled innovative strategies involving nano-delivery systems and formulations, such as encapsulation techniques, to deliver anticancer drug treatments and synthetic compounds for precise targeting and eradication of diverse tumor types [105]. Additionally, brentuximab vedotin, marketed as Adcetris®, represents an antibody–drug conjugate medication employed in treating anaplastic large-cell lymphoma and relapsed or refractory Hodgkin lymphoma [11–12]. Anticancer drugs derived from marine environments originate from a diverse range of organisms, including prokaryotes like marine bacteria such as *Noctiluca scintillans*, *Lactobacilli*, algae producing secondary anticancer metabolites, mangroves containing anticancer compounds, and other lesser-explored flora for potential anticancer properties. These marine organisms together constitute more than 70% of the Earth's biosphere [108]. The depsipeptide kahalalide demonstrates promising *in vitro* efficacy against solid tumors notably displaying precision targeting in prostate cancer cell lines [103–109]. Moreover, the indole-based microbial anticancer agent trisindolal 38 has shown promising activity against melanoma (MEXF 462) and human breast (MAXF401) cancer cell lines [110]. The highly effective anticancer drug Curacin-A is synthesized from the distinctive essence extracted from Curacao samples of *Lyngbya majuscule* [111]. Pharma Mar, a pharmaceutical company, operates under joint ownership with Johnson and Johnson's drug research and innovation sector, and they have received approval for the treatment of soft tissue sarcoma. Apratoxin A, a compound from the algae *Lyngbya majuscula Harvey ex Gomont*, demonstrates cytotoxicity against adenocarcinoma [112]. Since the approval of cytarabine over the past five decades, four medications derived from marine sources have gained approval [113]. The complex network of pharmaceutical advancement, involving the development and manufacturing of marine-derived anticancer drugs, opens up innovative avenues and resources to address the treatment challenges of patients with potentially resistant cancer types [114]. Chitosan is a chemical polymeric biomaterial derived from the exoskeletons of shellfish, including crabs and shrimp. Moreover, chitosan is characterized as a highly potent, large polycationic biopolymer with biodegradable, non-toxic, and biocompatible properties [115].

Sarcodiol, a semi-synthetic derivative of sarcophine, stands as a derivative merging both semi-engineered sarcophine and a marine natural product. Interestingly, sarcodiol modulates the activation potential of STAT-3 and cyclin D1, revitalizes the tumor-suppressor protein 53 (p53) within the cell, and initiates the series of events leading to caspase-dependent cell death [116].

1.9 Marine Bivalve-Derived Bioactive Compounds: Anticancer and Antimicrobial Effects

The phylum Mollusca is one of the most expansive and diverse groups of marine animals. Within Mollusca, the class Bivalvia encompasses approximately 8,500 species [117] encompassing well-known invertebrates like clams, oysters, scallops, and mussels. Bivalves are found in variable marine habitats from shallow coastal zones to deep environments [118]. Most of the bivalves are widely consumed by the public; therefore, they have commercial importance [119].

On the other hand, marine mollusks, particularly bivalves, offer a promising yet untapped resource for discovering novel bioactive compounds.

The anticancer peptides derived from marine organisms indicate great diversity in structure, including linear, cyclic, and basic peptides especially rich in lysine amino acids. The secondary structure of peptides could be α-helix, β-sheet, cyclic, and random coil [120].

Most of the bioactive molecules were defined in species that have been used for food and medicinal importance for years. One of these species is *Mytilus coruscus*, which is consumed as a valued food and used for the treatment of reproductive disorders and to strengthen the immune system [121].

Similarly, Xu *et al.* [122] found that Mere15 inhibited the proliferation of non-small cell lung cancer and cell motility by decreasing the expression of MMP-2, MMP-9, and Snail and increasing the expression of E-cadherin. Mere15 inhibited the PI3K/Akt/mTOR pathway in NSCLC cells and suppressed xenograft growth in nude mice bearing NCI-H460 cancer cells.

The other peptide isolated from *M. meretrix* is a coelomic fluid protein, MML, which has a 40-kDa molecular weight [123]. *Sinonovacula constricta* (Lamarck, 1818), an economically and nutritionally important bivalve species, thrives in the intertidal and estuarine habitats of China, Japan, and Korea [124]. Huang *et al.* [125] isolated 12 peptides, SCH-P1–P12, by enzymatic hydrolysis from *S. constricta*. Two of these peptides, SCH-P9 and SCH-P10, inhibited the viability of human prostate cells DU-145 and PC-3 cells in a time- and dose-dependent manner more than the remaining 10 peptides. Bivalves are known for their highly developed innate immune system and their less developed acquired one [126, 127]. Antimicrobial peptides (AMPs) are determined as major components of the innate immune systems of these animals. Most marine antimicrobial peptides are thought to be obtained from mussels [128]. Polysaccharides are complex carbohydrates composed of multiple monosaccharide units linked together by glycosidic bonds [129]. Bioactive polysaccharides are known as sugar-based functional polysaccharides produced from living organisms and/or have biological effects on organisms [130]. Bioactive polysaccharides have become the focus of research as therapeutic agents and have several advantages, including biodegradability, non-toxicity, and biocompatibility making them promising candidates for treating various chronic diseases [131].

Lipids are one of the major molecular components of living beings' cells. Their function varies depending on the cells. They are classified into different classes such as free fatty acids, sterols, sterol-esters, phospholipids, and triglycerides. Fatty acids are the most studied, and more of their beneficial attributes are assessed in the antimicrobial potential of lipids. Fatty acids are reported to show a broad spectrum of antimicrobial activities against different prokaryotic and eukaryotic cells such as Gram(−) and Gram(+) bacteria, fungi, archaea, and viruses [132]. Lectins are also known for their antimicrobial activities, and they are reported to be one of the large glycoproteins from bivalve mollusks [133]. Hossain *et al.* [134] classified lectins identified from mollusks into F-type, S-type, C-type, P-type, S-type, R-type, Ficolins-, and Chitinase-like lectins based on the similarity of their amino acids on the carbohydrate recognition domain.

References

1. Sousa, V., Hijo, A.A.T., Lüdtke, F.L., Vicente, A.A., Dias, O., Geada, P., Recovery and encapsulation of Dunaliella salina β-carotene through a novel sustainable approach: Sequential application of an ionic liquid as naturally-derived solvent and emulsifier. *Food Chem.*, 458, 140232, 2024.
2. Karageorgou, D., Rova, U., Christakopoulos, P., Katapodis, P., Matsakas, L., Patel, A., Benefits of supplementation with microbial omega-3 fatty acids on human health and the current market scenario for fish-free omega-3 fatty acid. *Trends Food Sci. Technol.*, 136, 169–180, 2023.
3. Ahmad, A. and Ashraf, S.S., *Sustainable food and feed sources from microalgae: Food security and the circular bioeconomy*, p. 103185, Algal Research, UAE, 2023.
4. Caetano, P.A., do Nascimento, T.C., Fernandes, A.S., Nass, P.P., Vieira, K.R., Junior, M.R.M., Jacob-Lopes, E., Zepka, L.Q., Microalgae-based polysaccharides: Insights on production, applications, analysis, and future challenges. *Biocatal. Agric. Biotechnol.*, 45, 102491, 2022.
5. Sasaki, M., Takagi, A., Sasaki, D., Nakamura, A., Asayama, M., Characteristics and function of an extracellular polysaccharide from a green alga Parachlorella. *Carbohydr. Polym.*, 254, 117252, 2021.
6. Tounsi, L., Hentati, F., Hlima, H.B., Barkallah, M., Smaoui, S., Fendri, I., Michaud, P., Abdelkafi, S., Microalgae as feedstock for bioactive polysaccharides. *Int. J. Biol. Macromol.*, 221, 1238–1250, 2022.
7. Zhao, G., Chen, X., Wang, L., Zhou, S., Feng, H., Chen, W.N., Lau, R., Ultrasound assisted extraction of carbohydrates from microalgae as feedstock for yeast fermentation. *Bioresour. Technol.*, 128, 337–344, 2013.
8. Parniakov, O., Apicella, E., Koubaa, M., Barba, F.J., Grimi, N., Lebovka, N., Pataro, G., Ferrari, G., Vorobiev, E., Ultrasound-assisted green solvent extraction of high-added value compounds from microalgae Nannochloropsis spp. *Bioresour. Technol.*, 198, 262–267, 2015.
9. Sierra, L.S., Dixon, C.K., Wilken, L.R., Enzymatic cell disruption of the microalgae Chlamydomonas reinhardtii for lipid and protein extraction. *Algal Res.*, 25, 149–159, 2017.

10. Zhang, Y., Kong, X., Wang, Z., Sun, Y., Zhu, S., Li, L., Lv, P., Optimization of enzymatic hydrolysis for effective lipid extraction from microalgae Scenedesmus sp. *Renew. Energy*, 125, 1049–1057, 2018.
11. Zhong, H., Ma, L., Zhu, Y., Jin, B., Wang, T., Wang, Y., Jin, F., Hydrothermal conversion of microalgae and its waste residue after biofuel extraction to acetic acid with CuO as solid oxidant. *J. Supercrit. Fluids*, 157, 104717, 2020.
12. Huang, H., Liu, Y., Xu, Z., Zhang, D., Feng, M., Zhao, T., Zhang, L., Li, W., Li, X., Effect of fucoidan on kidney injury in type 2 diabetic rats based on PI3K/AKT/Nrf2. *J. Funct. Foods*, 90, 104976, 2022.
13. Silva, T.A., do Couto, E.D.A., Assemany, P.P., Costa, P.A.C., Marques, P.A., Paradela, F., dos Reis, A.J.D., Calijuri, M.L., Biofuel from wastewater-grown microalgae: A biorefinery approach using hydrothermal liquefaction and catalyst upgrading. *J. Environ. Manage.*, 368, 122091, 2024.
14. Mohit, A., Mishel, V.K., Remya, N., Life cycle assessment and technoeconomic analysis of biofuels produced from polyculture microalgae cultivated in greywater. *J. Environ. Manage.*, 356, 120711, 2024.
15. Arbour, A.J., Bhatt, P., Simsek, H., Brown, P.B., Huang, J.Y., Life cycle assessment on environmental feasibility of microalgae-based wastewater treatment for shrimp recirculating aquaculture systems. *Bioresour. Technol.*, 399, 130578, 2024.
16. Razali, W.A.W. and Pandhal, J., Outdoor pilot-scale cultivation and techno economic assessment of a novel omega-3 eicosapentaenoic acid overproducing Nannochloropsis oculata strain. *Bioresour. Technol. Rep.*, 24, 101682, 2023.
17. Chen, Y., Mitra, A., Rahimnejad, S., Chi, S., Kumar, V., Tan, B., Xie, S., Retrospect of fish meal substitution in Pacific white shrimp (Litopenaeus vannamei) feed: Alternatives, limitations and future prospects. *Rev. Aquacult.*, 16, 1, 382–409, 2024.
18. Jesionowska, M., Ovadia, J., Hockemeyer, K., Clews, A.C., Xu, Y., EPA and DHA in microalgae: Health benefits, biosynthesis, and metabolic engineering advances. *J. Am. Oil Chem. Soc.*, 100, 11, 831–842, 2023.
19. Martins, C.F., *Spirulina and Chlorella vulgaris as ingredients in diet of weaned piglets: a valorization approach using feed enzymes*, Lisboa, Portugal, 2023.
20. Ganeson, Y., Paramasivam, P., Palanisamy, K.M., Govindan, N., Maniam, G.P., LCMS and FTIR profiling of microalga Chlorella sp. for cosmetics and skin care applications. *Clean. Water*, 2, 1–10, 100028, 2024.
21. Aravind, M.K., Vignesh, N.S., Gayathri, S., Anjitha, N., Athira, K.M., Gunaseelan, S., Varalakshmi, P., Review on rewiring of microalgal strategies for the heavy metal remediation-a metal specific logistics and tactics. *Chemosphere*, 313, 137310, 2023.
22. Mouritsen, O.G., *Seaweeds: Edible, Available, and Sustainable*, University of Chicago Press, Chicago, IL, 2013, ISBN: 978-0-226-04436-1.
23. Pereira, L., *Therapeutic and Nutritional Uses of Algae*, 1st, CRC Press/Taylor & Francis Group, Boca Raton, FL, 2018.

24. Srimongkol, P., Sangtanoo, P., Songserm, P., Watsuntorn, W., Karnchanatat, A., Microalgae-based wastewater treatment for developing economic and environmental sustainability: current status and future prospects. *Front. Bioeng. Biotechnol.*, 7, 10, 904046, 2022. https://doi.org/10.3389/fbioe.2022.904046.
25. Sathasivam, R., Radhakrishnan, R., Hashem, A., Abd Allah, E.F., Microalgae metabolites: a rich source for food and medicine Saudi. *J. Biolog. Sci.*, 26, 4, 709–722, 2019. https://doi.org/10.1016/j.sjbs.2017.11.003.
26. Priyadarshini, I., Rath, B., Commercial and industrial applications of micro algae – A review, J., *Biomass Utln.*, 3, 89–100, 2012.
27. Quintana, N., Van der Kooy, F., Van de Rhee, M.D., Voshol, G.P., Verpoorte, R., Renewable energy from cyanobacteria: energy production optimization by metabolic pathway engineering. *Appl. Microbiol. Biotechnol.*, 91, 3, 471–490, 2011.
28. Rawat, I., Ranjith Kumar, R., Mutanda, T., Bux, F., Dual role of microalgae: phycoremediation of domestic wastewater and biomass production for sustainable biofuels production. *Appl. Energy*, 88, 10, 3411–3424, 2011.
29. Hamed, I., The evolution and versatility of microalgal biotechnology: a review. *Compr. Rev. Food Sci. Food Saf.*, 15, 1104e1123, 2016. https://doi.org/10.1111/1541- 4337.12227.
30. Lin, W.N., Kapupara, K., Wen, Y.T., Chen, Y.H., Pan, I., Tsai, R.K., Haematococcus pluvialis-derived astaxanthin is a potential neuroprotective agent against optic nerve ischemia. *Mar. Drugs*, 18, 1e12, 2020. https://doi.org/10.3390/md18020085.
31. Riccio, G. and Lauritano, C., Microalgae with immunomodulatory activities. *Mar. Drugs*, 18, 1e18, 2020. https://doi.org/10.3390/md18010002.
32. Kim, S.K., *Marine Nutraceuticals: Prospects and Perspectives, 2013*, CRC Press, Boca Raton, FL, 2013.
33. Wildman, R.E.C. and Taylor, C., *Wallace Handbook of Nutraceuticals and Functional Foods*, Second, CRC Press, Boca Raton, FL, 2006.
34. Bishop, W.M. and Zubeck, H.M., Evaluation of microalgae for use as nutraceuticals and nutritional supplements. *J. Nutr. Food Sci.*, 2, 5e10, 2012.
35. Robinson, C.B., Samocha, T.M., Fox, J.M., Gandy, R.L., McKee, D.A., The use of inert artificial commercial food sources as replacements of traditional live food items in the culture of larval shrimp, Farfantepenaeus aztecus. *Aquaculture*, 245, 135e147, 2005.
36. Molina Grima, E., Belarbi, E.H., Acien Fernandez, F.G., Robles Medina, A., Chisti, Y., Recovery of microalgal biomass and metabolites: process options and economics. *Biotechnol. Adv.*, 20, 491e515, 2003.
37. Linan-Cabello, M.A., Paniagua-Michel, J., Hopkins, P.M., Bioactive roles of carotenoids and retinoids in crustaceans. *Aquacult. Nutr.*, 8, 299e309, 2002.
38. Harrison, K.E., The role of nutrition in maturation, reproduction and embryonic development of decapod crustaceans: a review. *J. Shellfish Res.*, 9, 1e28, 1990.

39. Ariede, M.B., Candido, T.M., Jacome, A.L.M., Velasco, M.V.R., de Carvalho, J.C.M., Baby, A.R., Cosmetic attributes of algae- a review. *Algal Res.*, 25, 483e487, 2017. https://doi.org/10.1016/j.algal.2017.05.019.
40. Li, Z., Zhao, D., Wang, Y., Wang, J., Emerging extraction technologies for bioactive compounds from marine algae: A review. *Food Chem.*, 381, 132388, 2023.
41. Ma, H., Li, D., Li, S., Sun, Y., Emerging extraction technologies for bioactive compounds from marine algae: A review. *Food Bioprod. Process.*, 130, 144–155, 2023.
42. Vijayaraj, R., Altaff, K., Kumaran, N.S., Volatile bioactive compounds from marine macro-algae and their pharmacological properties. *Indian J. Geo-Mar. Sci. (IJMS)*, *51*, 10, 832–842, 2022.
43. Sudhakar, M.P., Kumar, B.R., Mathimani, T., Arunkumar, K., A review on bioenergy and bioactive compounds from microalgae and macroalgae-sustainable energy perspective. *J. Clean. Prod.*, *228*, 1320–1333, 2019.
44. Falkenberg, M., Nakano, E., Zambotti-Villela, L., Zatelli, G.A., Philippus, A.C., Imamura, K.B., Graminha, M.A., Bioactive compounds against neglected diseases isolated from macroalgae: a review. *J. Appl. Phycol.*, *31*, 797–823, 2019.
45. Nisar, M.F., Khadim, M., Rafiq, M., Chen, J., Yang, Y., Wan, C.C., Pharmacological properties and health benefits of eugenol: A comprehensive review. *Oxid. Med. Cell. Longevity*, 2497354, 2021. doi: 10.1155/2021/2497354. eCollection 2021.
46. Priyanka, K.R., Rajaram, R., Sivakumar, S.R., A critical review on pharmacological properties of marine macroalgae. *Biomass Convers. Biorefin.*, 1–25, 2022, https://doi.org/10.1007/s13399-022-03134-4.
47. Kizhakkekalam, V.K. and Chakraborty, K., Pharmacological properties of marine macroalgae-associated heterotrophic bacteria. *Arch. Microbiol.*, *201*, 4, 505–518, 2019.
48. Barbier, M., Charrier, B., Araujo, R., Holdt, S.L., Jacquemin, B., Rebours, C., Marine macroalgae: A sustainable biomass feedstock for biorefineries. *Biotechnol. Adv.*, 37, 3, 6–12, 2019.
49. Buschmann, A.H., Camus, C., Infante, J., Neori, A., Israel, A., Hernández-González, M.C., Critchley, A.T., Seaweed production: Overview of the global state of exploitation, farming and emerging research activity. *Eur. J. Phycol.*, 52, 4, 391–406, 2017.
50. Cai, J., He, Y., Yu, X., Tan, J., Marine macroalgae: An untapped resource for the discovery of new drugs. *Mar. Drugs*, 19, 10, 507, 2021.
51. Mac Monagail, M., Cornish, L., Morrison, L., Araújo, R., Critchley, A.T., Sustainable harvesting of wild seaweed resources. *Eur. J. Phycol.*, 52, 4, 371–390, 2017.
52. Pérez-Lloréns, J.L., Hernández, I., Vergara, J.J., Brun, F.G., Martínez-López, J., Perspectives on the use of seaweeds as a promising source of bioactive compounds for health and disease prevention. *Mar. Drugs*, 18, 2, 115, 2020.

53. Qasim, S.Z., *The Indian Ocean: Images and Realities*, pp. 57–90, Oxford and IBH, New Delhi, 1999.
54. Giovannoni, S.J., Tripp, H.J., Givan, S., Podar, M., Vergin, K.L., Baptista, D., Bibbs, L., Eads, J., Richardson, T.H., Noordewier, M., Rappe, M.S., Short, J.M., Carrington, J.C., Mathur, E.J., Genome streamlining in a cosmopolitan oceanic bacterium. *Science*, 309, 1242–1245, 2005.
55. Luna, G.M., *Biotechnological potential of marine microbes*, pp. 651-661, Springer handbook of marine biotechnology, Berlin, Germany, 2015.
56. Sivaperumal, P., Kamala, K., Rajaram, R., Bioremediation of industrial waste through enzyme producing marine microorganisms. *Adv. Food Nutr. Res.*, 80, 165–179, 2018. doi: 10.1016/bs.afnr.2016.10.006.
57. Urbanek, A.K., Rymowicz, W., Mirończuk, A.M., Degradation of plastics and plastic-degrading bacteria in cold marine habitats. *Appl. Microbiol. Biotechnol.*, 102, 18, 7669–7678, 2018. https://doi.org/10.1007/s00253-018-9195-y.
58. Latha, K. and Lalithakumari, D., Transfer and expression of a hydrocarbon degrading plasmid pHCL from Pseudomonas putida to marine bacteria. *World J. Microbiol. Biotechnol.*, 17, 523–528, 2001.
59. DeVugst, L. and Vandamme, E.J., *Bacteriocins of lactic acid bacteria. Microbiol. Genet. Appl.*, vol. 75, pp. 140174–140179, Blackie Acadamic & Profession, London, 1994.
60. Goldin, B.R. and Gorbach, S.L., Probiotics for humans, in: *Probiotics*, R. Fuller (Ed.), pp. 355–376, Chapman and Hall, London, UK, 1992.
61. Haseeba, K.P., Aboobacker, V.M., Vethamony, P., Al-Khayat, J.A., Water and sediment characteristics in the *Avicennia marina* environment of the Arabian Gulf: A review. *Mar. Pollut. Bull.*, 216, 117963, 2025.
62. Habeebula, M. and Velraj, M., Potential anti-diabetic mangroves in Kerala, India: A Review. *Int. J. Res. Ayurveda Pharm.*, 9, 194–198, 2018, DOI:10.7897/2277-4343.094142.
63. Choi, S., Johnston, M., Wang, G.S., Huang, C.P., A seasonal observation on the distribution of engineered nanoparticles in municipal wastewater treatment systems exemplified by TiO2 and ZnO. *Sci. Total Environ.*, 625, 1321–1329, 2018.
64. Doshi, B., Sillanpaa, M., Kalliola, S., A review of bio-based materials for oil spill treatment. *Water Res.*, 135, 262–277, 2018.
65. Hu, X., Ren, C., Kang, W., Mu, L., Liu, X., Li, X., Wang, T., Zhou, Q., Characterization and toxicity of nanoscale fragments in wastewater treatment plant effluent. *Sci. Total Environ.*, 626, 1332–1341, 2018.
66. Thatoi, H., Behera, B.C., Mishra, R.R., Dutta, S.K., Biodiversity, and biotechnological potential of microorganisms from mangrove ecosystems: A review. *Ann. Microbiol.*, 63, 1–19, 2013.
67. Bandaranayake, W., Traditional and medicinal uses of mangroves. *Mangroves Salt Marshes*, 2, 3, 133–148, 1998.
68. Habeebula, M. and Velraj, M., Potential anti-diabetic mangroves in Kerala, India: A Review. *Int. J. Res. Ayuverda Pharm.*, 9, 194–198, 2018.

69. Rout, P., Singh, S., Kumar, N., Basak, U.C., Nutritional and antioxidant potential of some selected edible mangrove fruits of Odisha coast. *Int. J. Adv. Sci. Res.*, 1, 349–355, 2015.
70. Kainuma, M., Kezuka, M., Inoue, T., Chan, E.W., Tangah, J., Botany, uses, chemistry and bioactivities of mangrove plants I: Rhizophora stylosa. *ISME/GLOMIS Electron. J.*, 13, 12–17, 2015.
71. Lalitha, P., Parthiban, A., Sachithanandam, V., Purvaja, R., Ramesh, R., Antibacterial and antioxidant potential of GC-MS analysis of crude ethyl acetate extract from the tropical mangrove plant Avicennia officinalis L. *S. Afr. J. Bot.*, 142, 149–155, 2021.
72. Syahidah, and Subekti, N., Phytochemical Analysis of Mangrove Leaves (Rhizophora sp.), in: *Proceedings of the 14th Pacific Rim Bio-Based Composites Symposium*, Makassar, Indonesia, 29–31 October 2018, pp. 1–7.
73. Arbiastutie, Y., Diba, F., Masriani, M., Short Communication: Ethnobotanical and ecological studies of medicinal plants in a mangrove forest in Mempawah District, West Kalimantan, Indonesia. *Biodiversitas*, 22, 3164–3170, 2021.
74. Elahi, N., Kamali, M., Baghersad, M.H., Recent biomedical applications of gold nanoparticles: A review. *Talanta*, 184, 537–556, 2018.
75. Willian, N., Syukri, S., Zulhadjri, Z., Pardi, H., Arief, S., Marine plant mediated green synthesis of silver nanoparticle using mangrove Rhizophora stylosa: Effect of variable process and their antibacterial activity. *F1000Research*, 10, 768, 2022.
76. Ali, A., Akhtar, N., Khan, B., Khan, H., Rasul, A., Khalid, N., Waseem, K., Acacia nilotica: A plant of multipurpose medicinal uses. *J. Med. Plants Res.*, 6, 1492–1496, 2012.
77. Vasanth, K., Ilango, K., Mohankumar, R., Agrawal, A., Dubey, G., Anticancer activity of Moringa oleifera mediated silver nanoparticles on human cervical carcinoma cells by apoptosis induction. *Colloids Surf. B Biointerfaces*, 117, 354–359, 2014.
78. Turkevich, J., Collodial gold part II: colour, coagulation, adhesion, alloying and catalytic properties. *Gold Bull.*, 18, 125131, 1985.
79. Bhuvaneswari, R., Xavier, R.J., Arumugam, M., Facile synthesis of multifunctional silver nanoparticles using mangrove plant Excoecaria agallocha L. for its antibacterial, antioxidant and cytotoxic effects. *J. Parasit. Dis.*, 41, 1, 180–187, 2017.
80. Najafian, L., A review of bioactive peptides as functional food ingredients: mechanisms of action and their applications in active packaging and food quality improvement. *Food Funct.*, 14, 5835–5857, 2023 https://doi.org/10.1039/D3FO00362K, 2023.
81. Drazen, J.C. and Sutton, T.T., Dining in the deep: the feeding ecology of deep-sea fishes. *Annu. Rev. Mar. Sci.*, *9*, 337–366, 2017.
82. Leone, A., Lecci, R.M., Durante, M., Meli, F., Piraino, S., The bright side of gelatinous blooms: Nutraceutical value and antioxidant properties of three Mediterranean jellyfish (Scyphozoa). *Mar. Drugs*, *13*, 8, 4654–4681, 2015.

83. Csirke, J., *Review of the state of world fishery resources: Marine fisheries*, Rome, Italy, 1997.
84. Hayes, M., Mora, L., Hussey, K., Aluko, R.E., Boarfish protein recovery using the pH-shift process and generation of protein hydrolysates with ACE-I and antihypertensive bioactivities in spontaneously hypertensive rats. *Innov. Food Sci. Emerg. Technol.*, *37*, 253–260, 2016.
85. Voronin, V.P., Artemenkov, D.V., Orlov, A.M., Murzina, S.A., Lipids and fatty acids in some mesopelagic fish species: General characteristics and peculiarities of adaptive response to deep-water habitat. *J. Mar. Sci. Eng.*, *10*, 7, 949, 2022.
86. Matić, J., Bøgwald, I., Tengstrand, E., Rønning, S.B., Afseth, N.K., Wubshet, S.G., Calanus finmarchicus as a novel source of health-promoting bioactive peptides: Enzymatic protein hydrolysis, characterization, and *in vitro* bioactivity. *Biocatal. Agric. Biotechnol.*, *52*, 102820, 2023.
87. Whitaker, R.D., Altintzoglou, T., Lian, K., Fernandez, E.N., Marine bioactive peptides in supplements and functional foods-A commercial perspective. *Curr. Pharm. Des.*, *27*, 11, 1353–1364, 2021.
88. Jogi, N., Yathisha, U.G., Bhat, I., Mamatha, B.S., Antihypertensive activity of orally consumed ACE-I inhibitory peptides. *Crit. Rev. Food Sci. Nutr.*, *62*, 32, 8986–8999, 2022.
89. Navaf, M., Sunooj, K.V., Aaliya, B., Sudheesh, C., Akhila, P.P., Mir, S.A., Khaneghah, A.M., Contemporary insights into the extraction, functional properties, and therapeutic applications of plant proteins. *J. Agric. Food Res.*, 14, 100861, 2023.
90. Cavan, E.L., Belcher, A., Atkinson, A., Hill, S.L., Kawaguchi, S., McCormack, S., Schmidt, K., The importance of Antarctic krill in biogeochemical cycles. *Nat. Commun.*, 10, 1, 40, 4742, 2019.
91. Alvheim, A.R., Kjellevold, M., Strand, E., Sanden, M., Wiech, M., Mesopelagic species and their potential contribution to food and feed security- A case study from Norway. *Foods*, 9, 344–363, 2020. doi: 10.3390/foods9030344.
92. Lauritano, C., Martínez, K.A., Battaglia, P., Granata, A., de la Cruz, M., Cautain, B., *et al.*, First evidence of anticancer and antimicrobial activity in Mediterranean mesopelagic species. *Sci. Rep.*, 10, 4929, 2020. doi: 10.1038/s41598-020- 61515-z.
93. Sutton, T.T., Clark, M.R., Dunn, D.C., Halpin, P.N., Rogers, A.D., Guinotte, J., *et al.*, A global biogeographic classification of the mesopelagic zone. *Deep-Sea Res. Pt. I Oceanogr. Res. Pap.*, 126, 85–102, 2017. http://doi.org/10.1016/j.dsr.2017.05.006.
94. Gjøsæter, J. and Kawaguchi, K., *A Review of the World Resources of Mesopelagic Fish*, FAO, USA, 1980.
95. Robinson, C., Steinberg, D.K., Anderson, T.R., Arístegui, J., Carlson, C.A., Frost, J.R., Ghiglione, J.F., Hernández-León, S., Jackson, G.A., Koppelmann, R., Quéguiner, B., Ragueneau, O., Rassoulzadegan, F., Robison, B.H., Tamburini, C., Tanaka, T., Wishner, K., Zhang, J., Mesopelagic zone

ecology and biogeochemistry - a synthesis. *Deep-sea Res. Part Ii-topical Stud. Oceanography*, 57, 1504–1518, 2010.

96. Kawabata, T., Lindsay, D.J., Kitamura, M., *et al.*, Evaluation of the bioactivities of water-soluble extracts from twelve deep-sea jellyfish species. *Fish. Sci.*, 79, 487–494, 2013. https://doi.org/10.1007/s12562-013-0612-y.
97. Dornan, T., Fielding, S., Saunders, R.A., Genner, M.J., Large mesopelagic fish biomass in the Southern Ocean resolved by acoustic properties, in: *Proc. R. Soc. B*, 2022, 28920211781, http://doi.org/10.1098/rspb.2021.1781.
98. Kim, S.K., Park, S.J., Li, X.H., Choi, Y.S., Im, D.S., Lee, J.H., Bacterial ornithine lipid, a surrogate membrane lipid under phosphate limiting conditions, plays important roles in bacterial persistence and interaction with host. *Environ. Microbiol.*, 20, 3992–4008, 2018.
99. Elshafie, H.S., Viggiani, L., Mostafa, M.S., El-Hashash, M.A., Camele, I., Bufo, S.A., Biological activity and chemical identification of ornithine lipid produced by Burkholderia gladioli pv. agaricicola ICMP 11096 using LC-MS and NMR analyses. *J. Biol. Res.*, 90, 1–8, 2017.
100. Li, H., Shinde, P.B., Lee, H.J., Yoo, E.S., Lee, C.-O., Hong, J., Choi, S.H., Jung, J.H., Bile acid derivatives from a sponge-associated bacterium Psychrobacter sp. *Arch. Pharmacal Res.*, 32, 857–862, 2009.
101. Naik, A.S., Whitaker, R.D., Albrektsen, S., Solstad, R.G., Thoresen, L., Hayes, M., Mesopelagic Fish Protein Hydrolysates and Extracts: A Source of Novel Anti-Hypertensive and Anti-Diabetic Peptides. *Front. Mar. Sci.*, 8, 719608, 2021. doi: 10.3389/fmars.2021.719608.
102. Gomes, N.G., Dasari, R., Chandra, S., Kiss, R., Kornienko, A., Marine Invertebrate Metabolites with Anticancer Activities: Solutions to the "Supply Problem". *Mar. Drugs, 14*, 5, 98, 2016. https://doi.org/10.3390/md14050098.
103. Suarez-Jimenez, G.M., Burgos-Hernandez, A., Ezquerra-Brauer, J.M., Bioactive peptides and depsipeptides with anticancer potential: sources from marine animals. *Mar. Drugs, 10*, 5, 963–986, 2012. https://doi.org/10.3390/md10050963.
104. Desbois, A.P., How might we increase success in marine-based drug discovery? *Expert Opin. Drug Discov.*, *9*, 9, 985–990, 2014. https://doi.org/10.1517/17460441.2014.927863.
105. Dias, D.A., Urban, S., Roessner, U., A historical overview of natural products in drug discovery. *Metabolites, 2*, 2, 303–336, 2012. https://doi.org/10.3390/metabo2020303.
106. van Andel, L., Rosing, H., Schellens, J.H., Beijnen, J.H., Review of Chromatographic Bioanalytical Assays for the Quantitative Determination of Marine-Derived Drugs for Cancer Treatment. MDPI journals, *Mar. Drugs, 16*, 7, 246, 2018. https://doi.org/10.3390/md16070246.
107. Mayer, A.M., Glaser, K.B., Cuevas, C., Jacobs, R.S., Kem, W., Little, R.D., McIntosh, J.M., Newman, D.J., Potts, B.C., Shuster, D.E., The odyssey of marine pharmaceuticals: a current pipeline perspective. Pub Med

publications. *Trends Pharmacol. Sci.*, *31*, 6, 255–265, 2010. https://doi.org/10.1016/j.tips.2010.02.005.

108. Alves, C., Silva, J., Pinteus, S., Gaspar, H., Alpoim, M.C., Botana, L.M., Pedrosa, R., From Marine Origin to Therapeutics: The Antitumor Potential of Marine Algae-Derived Compounds. *Front. Pharmacol.*, *9*, 777, 2018. https://doi.org/10.3389/fphar.2018.00777.
109. Ithranga Boopathy, N. and Kathiresan, K., Anticancer drugs from marine flora: an overview. *J. Oncol.*, *2010*, 214186, 2010. https://doi.org/10.1155/2010/214186.
110. Dhiman, A., Sharma, R., Singh, R.K., Target-based anticancer indole derivatives and insight into structure-activity relationship: A mechanistic review update (2018-2021). *Acta Pharm. Sinica. B*, *12*, 7, 3006–3027, 2022. https://doi.org/10.1016/j.apsb.2022.03.021.
111. Wipf, P., Reeves, J.T., Day, B.W., Chemistry and biology of curacin A. *Curr. Pharm. Des.*, *10*, 12, 1417–1437, 2004. https://doi.org/10.2174/1381612043384853.
112. Khalifa, S.A.M., Shedid, E.S., Saied, E.M., Jassbi, A.R., Jamebozorgi, F.H., Rateb, M.E., Du, M., Abdel-Daim, M.M., Kai, G.Y., Al-Hammady, M.A.M., Xiao, J., Guo, Z., El-Seedi, H.R., Cyanobacteria-From the Oceans to the Potential Biotechnological and Biomedical Applications. *Mar. Drugs*, *19*, 5, 241, 2021. https://doi.org/10.3390/md19050241.
113. Cappello, E. and Nieri, P., From Life in the Sea to the Clinic: The Marine Drugs Approved and under Clinical Trial. *Life (Basel Switzerland)*, *11*, 12, 1390, 2021. https://doi.org/10.3390/life11121390.
114. Malve, H., Exploring the ocean for new drug developments: Marine pharmacology. *J. Pharm. Bioallied. Sci.*, *8*, 2, 83–91, 2016. https://doi.org/10.4103/0975-7406.171700.
115. Jiménez-Gómez, C.P. and Cecilia, J.A., Chitosan: A Natural Biopolymer with a Wide and Varied Range of Applications. *Mol. (Basel Switzerland)*, *25*, 17, 3981, 2020. https://doi.org/10.3390/molecules25173981.
116. Mangialardi, P., Harper, R., Albertson, T.E., The pharmacotherapeutics of sarcoidosis. *Expert Rev. Clin. Pharmacol.*, *15*, 1, 51–64, 2022. https://doi.org/10.1080/17512433.2022.2032657.
117. Huber, M., *Compendium of Bivalves 2*, A full-color guide to the remaining seven families. A systematic listing of 8’500 bivalve species and 10’500 synonyms. 2015, ConchBooks.
118. Grienke, U., Silke, J., Tasdemir, D., Bioactive compounds from marine mussels and their effects on human health. *Food Chem.*, 142, 48–60, 2014.
119. Gyawali, P. and Hewitt, J., Faecal contamination in bivalve molluscan shellfish: can the application of the microbial source tracking method minimise public health risks? *Curr. Opin. Environ. Sci. Health*, 16, 14–21, 2020.
120. Xie, M., Liu, D., Yang, Y., Anti-cancer peptides: classification, mechanism of action, reconstruction and modification. *Open Biol.*, 10, 7, 200004, 2020.

121. Xu, H., Guo, T., Guo, Y.F., Zhang, J.E., Li, Y., Feng, W., Jiao, B., Characterization and protection on acute liver injury of a polysaccharide MP-I from *Mytilus coruscus*. *Glycobiology*, 18, 1, 97–103, 2008.
122. Xu, B., Guo, M., Ma, L., Farooqi, A.A., Wang, L., Qiao, G., Liu, M., Zuo, L., Ye, H., Lin, X., *et al.*, Mere15, a novel polypeptide from *Meretrix meretrix*, inhibits proliferation and metastasis of human non-small cell lung cancer cells through regulating the PI3K/Akt/mTOR signaling pathway. *Neoplasma*, 68, 1181–1189, 2021.
123. Ning, X., Zhao, J., Zhang, Y., Cao, S., Liu, M., Ling, P., Lin, X., A novel antitumor protein extracted from *Meretrix meretrix* Linnaeus induces cell death by increasing cell permeability and inhibiting tubulin polymerization. *Int. J. Oncol.*, 35, 805–812, 2009.
124. Song, J., Li, L., Liu, Z., Li, Q., Ran, P., Sequence analysis and expression of a cDNA clone encoding tropomysin in *Sinonovacula constricta*. *Mol. Bio. Rep.*, 36, 315–321, 2009.
125. Huang, F., Ding, G., Yang, Z., Yu, F., Two novel peptides derived from *Sinonovacula constricta* inhibit the proliferation and induce apoptosis of human prostate cancer cells. *Mol. Med. Rep.*, 16, 6697–6707, 2017.
126. Oh, R., Lee, M.J., Kim, Y.O., Nam, B.H., Kong, H.J., Kim, J.W., Park, J.Y., Seo, K.J., Kim, D.G., Myticusin-beta, antimicrobial peptide from the marine bivalve, *Mytilus coruscus*. *Fish Shellfish Immunol.*, 99, 342–352, 2020.
127. Tincu, J.A. and Taylor, S.W., Antimicrobial peptides from marine invertebrates. *Antimicrob. Agents Chemother.*, 48, 10, 3645–3654, 2004.
128. Cunha, S.A. and Pintado, M.E., Bioactive peptides derived from marine sources: Biological and functional properties. *Trends Food Sci. Technol.*, 119, 348–370, 2022.
129. Guo, M.Q., Hu, X., Wang, C., Ai, L., Polysaccharides: structure and solubility. *Solubility Polysaccharides*, 2, 8–21, 2017.
130. Ullah, S., Khalil, A.A., Shaukat, F., Song, Y., Sources, extraction and biomedical properties of polysaccharides. *Foods*, 8, 8, 304, 2019.
131. Colegate, S.M. and Molyneux, R.J. (Eds.), *Bioactive natural products: detection, isolation, and structural determination*, CRC press, USA, 2007.
132. Desbois, A., Potential applications of antimicrobial fatty acids in medicine, agriculture and other industries. *Recent Pat. Anticancer Drug Discov.*, 7, 2, 111–122, 2012.
133. Chikalovets, I.V., Chernikov, O.V., Pivkin, M.V., Molchanova, V.I., Litovchenko, A.P., Li, W., Lukyanov, P.A., A lectin with antifungal activity from the mussel *Crenomytilus grayanus*. *Fish Shellfish Immunol.*, 42, 2, 503–507, 2015.
134. Hossain, M.M., Rajia, S., Ohkawa, M., Yoshimoto, S., Fujii, Y., Kawsar, S.M., Ozeki, Y., Hasan, I., Physicochemical properties and antimicrobial activities of MytiLec-1, a member from the mytilectin family of mussels. *Int. J. Biol. Macromol.*, 253, 127628, 2023.

2

Exploring the Therapeutic Potential of Microalgae-Derived Compounds: Commercial Significance, Advanced Cultivation Methods, and Life Cycle Analysis

Vishnu K. Venugopalan[1]*, Lekshmi R.G. Kumar[2], Paras Nath Jha[3], Vishnuja Soman[4], A. A. Mohamed Hatha[1] and S.S. Shaju[4]

[1]*Department of Marine Biology, Microbiology and Biochemistry, Cochin University of Science and Technology, Ernakulam, Kerala, India*
[2]*Biochemistry and Nutrition Division, ICAR Central Institute of Fisheries Technology (CIFT), Kerala, India*
[3]*Fishing Technology Division, ICAR Central Institute of Fisheries Technology (CIFT), Kerala, India*
[4]*Department of Chemical Oceanography, Cochin University of Science and Technology, Ernakulam, Kerala, India*

Abstract
Microalgae are a new promising source of bioactive compounds with high therapeutic potential. The structural diversity and valuable metabolites, including proteins, lipids, polysaccharides, pigments, and antioxidants present in microalgae widen its applications in various sectors such as pharmaceutical, nutraceutical, and cosmeceutical industries. Further, owing to their advantageous characteristics, such as fast growth; the ability to be cultivated in different ecosystems, such as freshwater and marine environments; and the ability to grow on non-arable land, they are finding tremendous use as a renewable feedstock for the production of valuable bioactive compounds. Microalgae-derived bioactive molecules are reported to exert a broad range of therapeutic activities, from anti-inflammatory to antioxidant, anticancer, and antimicrobial activities. This makes them highly potential candidates for the development of new therapeutic agents. The ease in cultivation conditions have

**Corresponding author*: vishnukalladath@gmail.com

Shaju S. S., Vipin. P. M. and Visakh. P.M. (eds.) Marine Biotechnology: Studies and Applications, (25–52) © 2025 Scrivener Publishing LLC

contributed much to increase the yield and quality of these bioactive molecules. Recent innovations in this field include photobioreactors, genetic engineering, and culture condition optimization, which have increased biomass production and augmented the concentration of target compounds. Further work in the domain of omics technologies and metabolic engineering has possibly opened up large-scale production and commercialization for microalgae-based products. This review is done to understand the therapeutic aspects of compounds derived from microalgae against the backdrop of recent advances made in their cultivation and commercial scenario. Apart from this, the role of life cycle assessment in determining the environmental implications of microalgal-based production systems is also detailed. Future researches need to be aligned by validating the cultivation methods, designing sustainable extraction methods for comprehensive biomass valorization with an aim to improve the economic, environmental, and industrial implications of the microalgal culture systems.

***Keywords*:** Microalgae, bioactive molecules, pigments, antioxidants, marine compounds, nutraceutical, pharmaceutical, antimicrobial

2.1 Introduction

With the ever increase in global population, which is expected to reach 9.7 billion people by 2050, the demand for food is also increasing parallelly. Given that climate change is an important global issue, depending on agriculture alone as a food source may not be able to address this issue. Hence, identifying a sustainable and potential bioresource as a nutrient source has become the need of the hour. Of late, algae have gained immense significance as a potential bioresource in terms of nutrients. Algae is a broader term, which encompasses diverse organisms that ranges from unicellular to multicellular organisms. The former category is referred to as microalgae, which are reported to be one of the earlier forms of life on earth. Microalgae are highly diverse in nature with approximately 50,000 species reported from the freshwater and marine ecosystems [1]. Compared to their macroscopic counterparts, microalgae offer excellent advantages such as production of high biomass per unit area, the ability to grow on non-arable land, and convenience for genetic manipulation [2].

They can be very well raised in fresh water, brackish water, effluent waters, etc., making their culture easier compared to macroalgae. As a result, the microalgal culture is gaining significant momentum presently. Though there are wider varieties of microalgae available and being cultured, a few among them only have acquired commercial significance, which include *Spirulina*, *Chlorella*, *Haematococcus*, *Dunaliella*, *Botryococcus*, *Phaeodactylum*, *Porphyridium*, *Chaetoceros*, *Crypthecodinium*, *Isochrysis*, *Nannochloris*, *Nitzschia*, *Schizochytrium*, *Tetraselmis*, and *Skeletonema* [3]. In addition, some of the microalgal species, such as *Arthrospira platensis*, *Chlamydomonas reinhardtii*, *Auxenochlorella protothecoides*, *Chlorella vulgaris*, *Dunaliella bardawil*, and *Euglena gracilis*, have acquired the generally regarded as safe (GRAS) status by the FDA. Acquiring GRAS status is reported to be quite time consuming and requires passing through several safety tests [2]. Realizing its enormous potential as a source of bioactive compounds, it is presently finding wider applications in diverse sectors such as nutraceuticals, biomedical, biofuels, and cosmetics. In the present chapter, the significance of microalgae, as a source of nutrients, and its applications are detailed.

2.2 Microalgae as Source of Nutrients

Microalgae are reported to be rich in nutrients such as proteins, pigments, polyunsaturated omega-3 fatty acids, polysaccharides, pigments, vitamins, and minerals. However, their nutritional composition is reported to vary with the type of species, geographical conditions, culture conditions, etc. Figure 2.1 illustrates the major nutrients present in microalgae and their effects on human health.

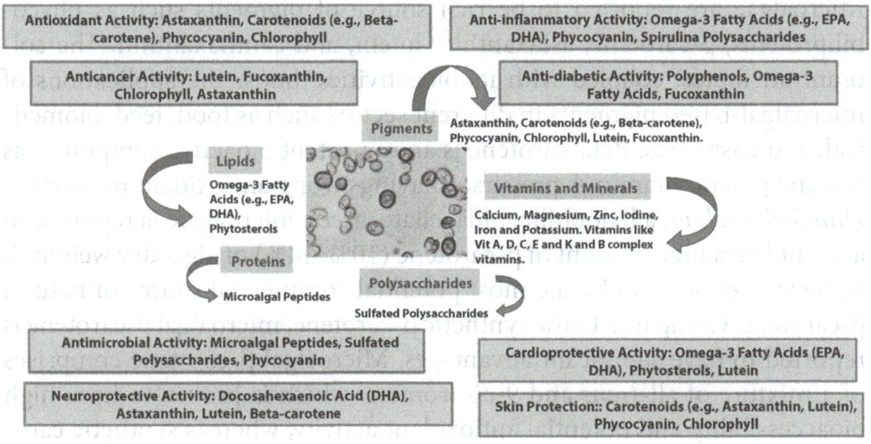

Figure 2.1 The major nutrients present in microalgae and their effects on human health.

2.2.1 Proteins

Microalgae are reported to harbor high protein content having excellent digestibility with all essential amino acids in adequate amounts. Protein content in microalgae is reported to show wider variations with contents as high as 70 g/100 g of dry weight reported for *Chlorella* species. The variations in protein content can be due to diverse parameters with environmental conditions being one of the important factors. Among environmental conditions, factors, such a salt concentration, light, pH, and temperature, are reported to significantly influence protein synthesis [4]. Because of the rich protein content, different microalgal-based products are being produced and marketed based on their protein content, such as protein concentrates, protein isolates, protein hydrolysate, and peptides, with protein content in the range of 60%–89%, 90%–95%, 70%–95%, and >95%, respectively [5]. The nutritional quality of the microalgal protein is often evaluated by their amino acid profile. It has been reported that essential amino acids, such as leucine, arginine, and lysine, constitute 7% each of the total protein content. Further, the essential amino acid content of microalgae was reported to be on par with that of eggs and soya beans. Among the non-essential amino acids, aspartic and glutamic acid were reported to the most significant contributing to 25%–30% of the total protein content [6]. Because of the complexity of microalgal cell wall, often, a cell disruption step is employed to facilitate protein extraction.

2.2.2 Pigments

Microalgae are reported to be rich source of pigments such as phycobiliproteins, β-carotene, astaxanthin, lutein, and canthaxanthin. The colorant attributes, coupled with its bioactivities, boost the applications of microalgal-based pigments in different sectors such as food, feed, biomedical, and cosmetics. Beta-carotene is an important bioactive compound as it is the potent vitamin A precursor having strong antioxidant properties. *Dunaliella salina*, a halophilic, biflagellate, green microalgae is reported to accumulate a high amount of β-carotene (10%–14%) of algal dry weight. It is, hence, reported to be the most potential commercial source of natural β-carotene. Compared to the synthetic β-carotene, microalgal β-carotene is reported to have significant advantages. Microalgal β-carotene comprises of a mixture of all-*trans* and 9-*cis* isomers, which is proven to have high bioaccessibility and potential antioxidant activity, whereas synthetic carotene has only all-*trans* isomers [7]. The natural 9-*cis* isomer assumes significance in scavenging of oxygen radicals and thereby conferring protection from oxidative damage, whereas the *trans* form is reported to have twice the activity of that of the *cis* form. Apart from *Dunaliella*, other significant microalgae notable for beta-carotene production are *Chlamydomonas*, *Chlorella*, *Muriellopsis*, *Haematococcus* spp., etc. [8]. Other than β-carotene, microalgae are also found to be promising resources for astaxanthin production. Astaxanthin, which is a natural ketocarotenoid, is reported to have superior antioxidant attributes. It is reported to have a wide range of bioactivities such as anti-oxidant, anti-aging, anti-inflammatory, anti-atherosclerotic, and anticancer [9, 10]. Natural astaxanthin is reported to be more esterified, which is up to 95% than its synthetic counterpart aiding in high energy production and conferring protection to tissues from oxidative damage. Apart from this, natural astaxanthin is non-crystalline in nature exhibiting better adsorption properties [11]. Among the several species of microalgae available, *Haematococcus pluvialis* is reported to be the major producer of astaxanthin, whereas *Chlorella sorokiniana*, *Tetraselmis* sp., and *Scenedesmus* sp. are reported as minor producers. *Haematococcus pluvialis*, which is a unicellular, eukaryotic photosynthetic, fresh water green microalga, is reported to accumulate up to 3.8%–5% of astaxanthin dry weight [12]. Canthaxanthin, which is a diketo-carotenoid pigment, acts as the precursor for astaxanthin production in *Haematococcus pluvialis*. Among the several species of microalgae, *Haematococcus pluvialis* and *Chromochloris zofingiensis* are widely studied and documented for the production of ketocarotenoids [13].

2.2.3 Lipids

Lipids, especially omega-3 polyunsaturated fatty acids (PUFAs), are reported to have important positive implications on human health. Consumption of foods rich in omega-3 PUFAs is reported to offer protection from different ailments such as inflammatory disorder, neurological disorders, and cardiovascular diseases. Fish and shellfish are often proposed as the most important dietary sources for omega-3 PUFAs. However, taking into consideration overfishing and global warming, researchers are screening for sustainable alternatives for fish-based omega-3 PUFAs. Microalgae are reported to be rich in PUFAs such as docosahexaenoic acid (DHA) and eicosapentaenoic acid (EPA) [14]. It has been reported that microalgae can accumulate a high content of PUFA, specifically EPA and DHA, in the range of 30%–50% [2]. Those microorganisms that can bioaccumulate more than 20% (w/w) of lipids in their cells are usually referred to as oleaginous microorganisms. *Thraustochytrium*, *Schizochytrium*, and *Crypthecodinium cohnii* are some of the potential oleaginous heterotrophic marine microalgae [15]. The content of lipids, especially fatty acids, is reported to be influenced by the type and species of microalgae and its cultivation conditions such as temperature, light, and nutrients. It has been reported that marine microorganisms are reported to produce more PUFAs than freshwater because of the frequent exposure to salt conditions [16].

Hence, it can be presumed that by bringing radical changes in the cultivation conditions, lipid content in microalgae can be augmented. Further, they can be proposed as a fish-free source for omega-3 PUFAs, which can open new avenues in the nutraceutical market. However, it has been reported that microalgal-based omega-3 PUFAs are still more expensive than that of fish oil. The energy intensive production systems along with the difficulties often faced in the standardization of the culture conditions are often proposed as the major constraints in microalgal-based omega-3 PUFA production, which can adversely affect the cost effectiveness and process scalability. Issues regarding the judicious selection of the appropriate microbial strain and the ideal culture conditions can be addressed to a greater extent [17].

2.2.4 Polysaccharides

Akin to macroalgae, microalgae are also reported to be a significant source of polysaccharides. In general, microalgae are reported to have 10%–25% of carbohydrates on a dry weight basis. The major microalgal polysaccharides are starch, cellulose, and sugar [18]. Microalgal polysaccharides are often classified based on their function such as structure or storage or as matrix polysaccharides. The matrix-based polysaccharides that are entirely released into the environment are often referred to as extracellular polysaccharides (EPS). In general, the microalgal-based polysaccharides are heteropolymers in nature [19]. Microalgal-based EPSs are often present as a viscous mass surrounding cells or group of cells and produced either as a protection mechanism or in response to the fluctuating environmental conditions [20]. Some of the most potent microalgal and cyanobacterial strains that produce polysaccharides are *Amphora rostrata*, *Melossira nummuloides*, *Thalassiosira* sp., *Botryococcus braunii*, *Chlamydomonas reinhardtii*, *Chlamydomonas mexicana*, *Netrium digitus*, *Arthrospira platensis*, *Spirulina platensis*, *Scenedesmus* sp., *Isochrysis* sp., *Tetraselmis* sp., *Rhodella reticulata*, *Porphyridium cruentum*, *Porphyridium aerugineum*, *Dunaliella salina*, *Anabaena* sp., *Nostoc* sp., *Haematococcus pluvialis*, and *Chlorella vulgaris* [21, 22]. Polysaccharide production and sulfation occurs in the Golgi apparatus in the case of microalgae, whereas in cyanobacteria, it is in the cytoplasm. Microalgal-based polysaccharides are finding applications in the nutraceutical, feed, and cosmetic industries as they have immense bioactivities such as antioxidant, antibacterial, anti-tumor, and anti-inflammatory activities. Apart from this, they are also reported to have potential prebiotic effects [23].

2.2.5 Vitamins and Minerals

Microalgae are reported to be a potential source for vitamins, specifically fat-soluble vitamins such as vitamin D3. Ljubic *et al.* [24] have studied the production pattern of vitamin D3 in four microalgal species, viz, *Chlorella minutissima*, *Nannochloropsis oceanica*, *Arthrospira maxima*, and *Rhodomonas salina* by exposing to artificial UVB. Out of the four species, *Nannochloropsis oceanica* was found to produce vitamin D3 with its production increased with the increased dose of UVB suggesting that this can be a novel source for vitamin D3. In a further study, it was shown that, in addition to vitamin D3, the exposure to UVB also facilitated the production of α-tocopherol and β-carotene in the case of *Dunaliella salina* and *Nannochloropsis oceanica* [25]. Santhakumaran *et al.* [26] reported that minerals, such as sodium, potassium, phosphorus, magnesium, manganese, calcium, copper, cobalt, iron, and zinc, are present in varying amounts in the microalgal biomass suggesting it as a valuable source of minerals too. Further, the lower sodium–potassium ratio of 0.17 to 1.97 also indicates its nutraceutical value.

2.3 Extraction Methods

Being rich in valuable nutrients, such as proteins, omega-3 PUFAs, and pigments, the culture of microalgae is being carried out on a larger scale globally. Followed by culture, the microalgal biomass is subjected to different extraction methods to obtain bioactive compounds out of it. The most commonly employed methods involve supercritical fluid extraction, ultrasound-assisted extraction, enzymatic extraction, microwave assisted extraction, etc.

2.3.1 Supercritical Fluid Extraction

Supercritical fluid extraction (SFE) is a green extraction technique proposed as an alternative to conventional extraction methods. The density, viscosity, and extractability of supercritical fluids can be accordingly adjusted by modulating the pressure and temperature of the extraction process. The utility of SFE in extracting lipids and pigments, such as carotenoids, and astaxanthin has been widely documented. Guedes *et al.* [27] reported that by employing a medium pressure extraction condition of 250 bar and temperature of 60°C, maximum recovery of total carotenoid was achieved. Cheng *et al.* [28] has studied the extraction efficiency of microalgal lipids by different extraction methods. It was reported that SFE has outperformed conventional solvent-based extraction methods in terms of extracting triglyceride fractions. Cheng Xiang *et al.* [29] reported a low pressure-based SFE extraction protocol for recovering astaxanthin from the microalgae, *Haematococcus pluvialis*. It was reported that the use of a co-solvent, such as ethanol, has further helped in accelerating the astaxanthin recovery from disrupted microalgae. Rodríguez-España *et al.* [30] reported *Schizochytrium* sp. as a potential microalga in terms of total lipid and docosahexaenoic acid (DHA) content. Further, the SFE-based extraction process was optimized giving 45% recovery of DHA content. It was reported that by employing pre-treatments steps, such as grinding, facilitated the efficient recovery of lipids (19.4 %), specifically DHA.

Apart from using SFE alone as an extraction method, the combined use of different extraction techniques to maximize the recovery of microalgal compounds is also being studied and documented. Feller *et al.* [31] have studied the extraction efficiency of two process, supercritical CO_2 and subcritical n-butane extraction methods, in recovering lipids from three microalgae, *Phaeodactylum tricornutum*, *Nannochloropsis oculata*, and *Porphyridium cruentum*. It was reported that the subcritical n-butane-based method was more effective in recovering omega-3 and -6 polyunsaturated fatty acids. Callejón *et al.* [32] reported that by combining two greener extraction techniques, such as supercritical fluid extraction with pressurized extraction process, the maximized recovery of microalgal lipids from *Nannochloropsis* sp. was facilitated. It was reported that the use of SFE aided in recovering the neutral saponifiable lipid fractions, whereas with pressurized ethanol, EPA-rich polar lipid fractions were recovered mostly.

2.3.2 Ultrasound-Assisted Extraction

Ultrasound-assisted extraction (UAE) is another greener extraction technique that can be employed for the extraction of valuable microalgal-based bioactive compounds such as protein, lipids, and pigments. By virtue of the cavitation process induced during UAE, the microalgal cell wall will be disrupted facilitating the leaching of the bioactives into the solvent matrix and thereby aiding in the extraction process. The technology can be easily adopted as it is economical, requires less operating cost, and lowers the temperature qualifying it as a green extraction technique. Adam *et al.* [33] reported a solvent-free UAE-based process for recovering lipids from the fresh algal biomass, *Nannochloropsis oculate.* Zhao *et al.* [34] reported the promising potential of UAE as a pre-treatment step in recovering carbohydrates from miocroalgal biomass as a renewable feedstock for bio-fermentation process. Binary mixture of green solvents, such as water–DMSO and water-–ethanol, was reported to be effective for recovering phenolic compounds and carotenoids from microalgae via an ultrasound-assisted extraction [35]. Similarly, Fernandes *et al.* [36] reported that UAE is a potential process for recovering of chlorophyll molecules from the microalgae, *Phormidium autumnale.* The potential of UAE as an effective process for recovering lipids from the microalgae, *Scenedesmus obliquus* and *Nannochloropsis* sp., has been studied and reported [37, 38].

Tavakoli *et al.* [39] has reported the use of food-grade solvents, along with UAE, as an effective method for recovering bioactive compounds, such a phycocyanin, polyphenolic compounds, and chlorophyll,. from *Spirulina platensis* for food-grade applications. Zou *et al.* [40] reported the use of buoyant bead-based UAE method as a novel extraction strategy for recovering microalgal lipids toward the production of fatty acid methyl ester and raw biodiesel. Fernandes *et al.* [41] have employed greener solvents, *viz,* ethanol and ethanol-based ionic liquids for ultrasound-assisted extraction of carotenoids and chlorophyll from the microalgae, *Scenedesmus obliquus* and *Arthrospira platensis.* Employing greener solvents for UAE-based extraction can be, hence, proposed as an effective strategy for recovering microalgal pigments. Morón-Ortiz *et al.* [42] reported that by employing green solvents, such as methyltetrahydrofuran and ethyl lactate for UAE, the extraction efficiency of lutein from freeze-dried microalgal biomass can be maximized. Further, the use of ball milling as a pre-treatment was also reported to enhance the recovery of carotenoids from microalgae.

2.3.3 Enzymatic Extraction Method

Enzymatic extraction method is another green extraction method that can be employed for microalgal biomass valorization. Often, employing enzymes is considered an effective strategy for recovering valuable bioactive compounds from microalgae via cell wall disruption and subsequent bioactive release into the solvent systems. However, the judicious selection of the appropriate enzymes, temperature, pH, medium, and concentration is considered a key parameter that can influence the extraction efficiency. Sierra *et al.* [43] employed enzymatic extraction as an effective pre-treatment step to enhance the recovery of lipids and protein. Zhang *et al.* [44] optimized enzymatic hydrolysis as an effective method to recover lipids from the microalgae, *Scenedesmus* sp. Further, employing enzymes, such as cellulase, xylanase, and pectinase, at ideal conditions was reported to give a recovery of 86.4% lipid. Qiu *et al.* [45] has reported the use of enzymatic extraction as an effective greener method for extracting lipids from wet *Nannochloropsis* sp. biomass. The utility of enzymatic extraction in recovering microalgal pigments has also been studied and documented. Zhao *et al.* [46] reported that rather than using cellulase, employing pectinase can augment the release rate of astaxanthin from the microalgae, *Haematococcus pluvialis.* He *et al.* [47] highlighted the utility of enzyme-assisted hydrolysis toward microalgal-mediated biodiesel production. Mienis *et al.* [48] highlighted the significance of enzyme-assisted disruption as a potential process for recovering lipids from the oleaginous microalgae, *Nannochloropsis* sp.

2.3.4 Other Green Extraction Methods

In addition to the abovementioned extraction process, several other greener extraction methods, such as microwave-assisted extraction and hydrothermal extraction, are being employed to recover valuable microalgal-based bioactive compounds. Biller *et al.* [49] has reported microwave irradiation as a potential pre-treatment in the disruption of microalgal cell wall for effective lipid and phytochemical extraction. Lee *et al.* [50] has developed a nitric acid-based hydrothermal pre-treatment for the extraction of lipids from microalgae. The potential of microwave-assisted process for microalgal lipid extraction toward the production of biodiesel has been also documented [51, 52]. Hu *et al.* [53] employed hydrothermal liquefaction for the production of bio-oil from microalgae. Guo *et al.* [54] combined two processes, hydrothermal liquefaction and pulsed electric field, for the simultaneous recovery of bioactive compounds, such as lipids, proteins, and amino acids, from three microalgal strains, *Auxenochlorella protothecoides*, *Chlorella vulgaris*, and *Scenedesmus almeriensis*. Zhong *et al.* [55] reported a hydrothermal-based method for the bioproduction of acetic acid from microalgae. Huang *et al.* [56] detailed the use of a deep eutectic solvent-assisted hydrothermal extraction method as an environment-friendly process for recovering microalgal lipids. Silva *et al.* [57] employed a biorefinery-based framework for biofuel production from microalgae using hydrothermal liquefaction. Marangon *et al.* [58] demonstrated that wastewater microalgae can be employed for the production of aviation fuel. Zhang *et al.* [59] has reported the use of pulsed microwave as an effective, competitive, efficient, and easy technique for augmenting lipid extraction form microalgae. Low *et al.* [60] reported the potential of a microwave-assisted extraction process for recovering lutein from microalgae.

2.4 Commercially Important Microalgae

Microalgae, due to their active growth in different aquatic environments, from fresh to marine, have been overwhelmingly noticed because of the fast growth rate and synthesis ability of a number of high-value compounds [61]. These organisms find applications in the following very important industries: health supplements, food and feed, pharmaceuticals, cosmetics, and biofuels [62].

2.4.1 *Spirulina* (*Arthrospira*)

Spirulina is the most famous microalgae; it has long been used as a dietary supplement due to its high content of proteins, vitamins, and minerals such as iron, magnesium, and essential fatty acids [63]. Animal feed is also another area in which *Spirulina* is used, specifically as a means of enhancing feed nutritional value in the case of livestock, poultry, and aquaculture [64]. It has many health benefits from being an antioxidant to having anti-inflammatory properties [65]. It helps boost the immune system maintaining gut health and maybe even helps lower blood cholesterol [66]. It also comprises a compound called phycocyanin, which has immense potential for drug production [67]. Their market mainly comes in powder, tablet, or capsule form. It is used in health food products, like smoothies, energy bars, and dietary supplements, so it has become quite a standard for health and wellness markets worldwide [68].

2.4.2 *Chlorella*

Chlorella is the other well-known microalga applied mainly as a dietary supplement. It becomes valuable due to its high content of chlorophyll, proteins, and also some vitamins (C, E) and minerals (zinc, magnesium) that are essential [69]. *Chlorella* is used in cosmetics and personal care products due to its feature of rejuvenating the skin [70]. It detoxifies the body by removing heavy metals, thus already proving its detoxifying effect [71]. It enhances the immune system and therefore promotes health in the liver and digestion [72]. Its richness in nutrients will make it an asset to any balanced diet. *Chlorella* is sold in powder form, in tablets, or as an extract.

It is a popular natural health product usually used for detoxification programs or added to health foods [73].

2.4.3 *Haematococcus pluvialis*

This is the world's richest source of astaxanthin, a very powerful natural antioxidant. Astaxanthin is utilized in nutraceuticals, cosmetics, and aquaculture [74]. It is valued for its action in protecting cells against oxidant damage and lowering inflammation [75]. Astaxanthin may also protect eye health, skin health, and cardiovascular health [76]. As an antioxidant, it reduces muscle fatigue and speeds up recovery in athletes [77]. Astaxanthin's anti-aging properties make it popular in skincare products [78]. Astaxanthin is gathering impetus, as both supplement and skincare product categories rise, with increasing use in improving the coloration and health of farmed fish and shrimp used in aquaculture [79].

2.4.4 *Dunaliella salina*

Dunaliella salina is viewed as one of the richest sources of beta-carotene, which is regarded as a provitamin A [80]. It finds applications as a dietary supplement, natural food colorant, and cosmetically for its antioxidant properties [81]. Beta-carotene obtained from *Dunaliella salina* provides protection through antioxidants to cells, promotes and supports eye health, and enhances immune function. It is also beneficial to skin health by protecting it against UV and oxidative stress [82]. *Dunaliella salina* beta-carotene finds a huge market in food and cosmetic industries. This natural pigment finds more demand compared to synthetic ones offering supplementary advantages to health [83].

2.4.5 *Nannochloropsis*

Nannochloropsis is used mainly for aquaculture, as it represents one of the major sources of EPA (eicosapentaenoic acid), an omega-3 fatty acid [84]. It is used to enrich the nutritional profile of feed for fish and shrimp enhancing their growth and health [85]. The high EPA content of *Nannochloropsis* maintains cardiovascular health, reduces inflammation, and improves the general health of aquatic organisms [86]. It plays a critical role in the larval development of marine fish. *Nannochloropsis* is of prime importance for the aquaculture feed industry—hatcheries. It is presently under research for producing biofuel due to its high lipid content [87].

2.4.6 *Isochrysis*

Isochrysis belongs to the important microalga applied to aquaculture and, more particularly, in hatcheries for both shellfish and finfish [88]. It contains a considerable amount of docosahexaenoic acid, an omega-3 fatty acid crucial for marine larvae development [89]. It feeds the growth and health of the marine species instrumental in improving survival rates at the early stages of the life cycle. With its high DHA content, it aids in the brain and eye development of marine organisms. *Isochrysis* finds its key application in hatchery feed farming to produce shellfish and finfish sustainably. It is also being researched for its value in nutraceuticals and biofuel production [90].

2.5 Cultivation Methods

2.5.1 Open Ponds

Raceway and circular ponds are low cost and suitable for large-scale production and use natural sunlight and atmospheric CO_2 as the basis for growth. Their advantages are low operational costs, simple design, and scalability. Contamination, changes in environmental factors, and less productivity compared to that of closed systems are the drawbacks [91].

2.5.2 Closed Photobioreactors

It is a controlled environment. Conditions, like light, temperature, and CO_2, are optimized in tubular, flat-panel, and column reactors. Its advantages are high productivity, better control of contamination, and consistent quality. Higher start-up costs, complicated design, and maintenance are the associated issues [92].

2.5.3 Hybrid Systems

Hybrid systems have all the benefits that both open ponds and closed photobioreactors offer. This usually entails cultivation in photobioreactors and subsequent concentration of biomass in open ponds [93]. Their advantages are higher productivity, reduced expenses, and more flexibility. Their disadvantages are they require good management to maintain a balance between the corresponding positives and negatives of the two systems.

2.5.4 Algal Turf Scrubbers

The algae will grow on inclined screens with flowing water effectively removing nutrients from wastewater and producing biomass. These systems are cost effective and environmentally beneficial but generally yield lower biomass compared to other methods. Each cultivation method can be chosen or combined based on specific production goals, whether it is

Table 2.1 The advantages and disadvantages of microalgae cultivation methods.

Sl. no.	Cultivation method	Advantage	Limitations	Reference
1.	Open ponds	**Low operational costs:** They utilize natural resources like sunlight and atmospheric CO_2, which reduces the need for artificial inputs **Simple design:** The construction and maintenance of open ponds are relatively straightforward making them accessible and easy to scale up for large-scale production **Scalability:** They can cover vast areas and produce significant quantities of biomass	**Contamination risk:** They are exposed to the environment making them more susceptible to contamination by other microorganisms, predators, or unwanted algae species **Environmental fluctuations:** The growth conditions in open ponds, such as temperature, light intensity, and pH, can vary significantly with changes in weather leading to inconsistent production **Lower productivity:** They have lower productivity due to less optimized growing conditions and higher susceptibility to contamination and environmental changes	[91]

(*Continued*)

Table 2.1 The advantages and disadvantages of microalgae cultivation methods. (*Continued*)

Sl. no.	Cultivation method	Advantage	Limitations	Reference
2.	Closed photobioreactors	**High productivity:** Closed photobioreactors allow for the precise control of growth conditions, such as light, temperature, and CO_2 levels, leading to higher biomass yields compared to open systems **Better contamination control:** Minimizes the risk of contamination by unwanted microorganisms or other algae species ensuring a more stable and controlled cultivation process **Consistent quality:** Uniform and reliable product quality, which is crucial for applications requiring specific biomass characteristics	**Higher start-up costs:** The initial investment required for building and setting up closed photobioreactors is significantly higher than for open ponds, due to the complexity of the equipment and technology involved **Complicated design:** The design and operation of photobioreactors are more complex requiring specialized knowledge and skills to manage and optimize the system effectively **Maintenance requirements:** Photobioreactors demand regular maintenance to ensure optimal functioning, which can increase operational costs and require more technical expertise	[92]

(*Continued*)

Table 2.1 The advantages and disadvantages of microalgae cultivation methods. (*Continued*)

Sl. no.	Cultivation method	Advantage	Limitations	Reference
3.	Hybrid systems	**Higher productivity:** They leverage the controlled environment of photobioreactors for initial cultivation, leading to higher biomass yields, and then use open ponds for cost-effective biomass concentration **Reduced expenses:** By combining the benefits of both systems, hybrid approaches can lower overall costs, as the expensive photobioreactor phase is followed by a more economical open pond phase **More flexibility:** Hybrid systems offer greater flexibility in managing cultivation processes allowing for the optimization of growth conditions in photobioreactors and cost-efficient biomass handling in open ponds	**Complex management:** Hybrid systems require careful management to balance the differing operational needs of photobioreactors and open ponds ensuring that the advantages of both systems are effectively utilized while minimizing their respective drawbacks **Potential for imbalance:** Without good management, the integration of the two systems could lead to inefficiencies or challenges, such as mismatches in growth rates or contamination risks during the transition from closed to open environments	[93]

(*Continued*)

Table 2.1 The advantages and disadvantages of microalgae cultivation methods. (*Continued*)

Sl. no.	Cultivation method	Advantage	Limitations	Reference
4.	Algal turf scrubbers	**Cost effective:** Algal turf scrubbers are relatively inexpensive to construct and operate making them a budget-friendly option for microalgae cultivation **Environmental benefits:** These systems are highly effective at removing nutrients from wastewater, which not only aids in water purification but also contributes to environmental sustainability **Simple design:** The design of algal turf scrubbers is straightforward involving inclined screens with flowing water, which can be easily implemented and maintained	**Lower biomass yield:** Compared to other cultivation methods, like closed photobioreactors or hybrid systems, algal turf scrubbers generally produce lower amounts of biomass, which may limit their use in applications requiring large-scale production **Space requirements:** These systems typically require a significant amount of space, especially if used for large-scale wastewater treatment or biomass production **Regular maintenance:** The screens need to be regularly cleaned and maintained to prevent clogging and to ensure optimal growth conditions for the algae	[94]

maximizing biomass, reducing costs, or targeting specific bioactive compounds [94].

The advantages and limitations of these cultivation methods are presented in Table 2.1.

2.6 Economic Importance

The microalgae market has been growing over the years in countries worldwide, mainly due to increased demand in the health, nutrition, and sustainability sectors [95]. Some heavy-duty factors have contributed to this growth. First, if one considers the health and wellness part, increasing consumer awareness about natural and healthful supplements naturally boosted the demand for microalgae-based products like *Spirulina* and *Chlorella* [96]. In aquaculture, nutritional and sustainable feeding solutions are delivered to the growing seafood sector [97]. Cosmetics take advantage of consumers' tendency to go for more natural but efficient skincare ingredients by incorporating microalgae extracts into a wide array of products [98]. Development of products and research on biofuels, with microalgae used as the ultimate source of sustainable energy solutions, are also continually going on to expand the market [99]. All these factors underline the economic importance that microalgae hold, as it has multifarious applications and potential for sustainable development.

2.7 Life Cycle Assessment

Though microalgae are gaining significance as a promising biomass, it is also reported to have serious setbacks in terms of the cultivation methods, biomass production efficiency, harvesting, economic feasibility, etc. [100]. Hence, life cycle assessment (LCA) and life cycle cost analysis (LCC) can be applied as efficient tools to validate the environmental implications and economic viability. Environmental life cycle assessment is a method employed to evaluate the environmental impact of product, process, and services [101]. Romagnoli *et al.* [102] highlighted the role of LCA and LCC in optimizing the production process and guiding the early designing of microalgal cultivation process. Xu *et al.* [103] employed LCA for analyzing the environmental and economic benefits of different microalgal cultivation systems, such as open raceway ponds, photobioreactors, and biofilm systems, toward biogas production and reported that the greenhouse emission of all the systems was positive. Xiao *et al.* [104] employed LCA to evaluate the biogas production based on two different pre-treatment methods such as hydrothermal and with solar-driven hydrothermal pretreatment. Though both the methods were reported to have good environmental and economic benefits, the solar driven method has a low cost of energy. Mohit *et al.* [105] also employed the LCA to evaluate the environmental benefits associated with microalgal-mediated biofuel production. Arbour *et al.* [106] employed the LCA as a tool to evaluate the utilization of microalgae for wastewater utilization from shrimp recirculating aquaculture firms. It can, hence, be concluded that by performing LCA, the environmental implications of microalgae-based cultivation, nutrient production, biofuel production, etc., can be ascertained.

2.8 Challenges and Future Prospects

Although microalgae have a huge potential, many challenges must be surmounted before large-scale commercialization can take place. The main problems are the high production cost, in particular that of cultivation, harvesting, and processing. Contamination is a problem that demands more solid systems, which would prevent it from happening in open-pond cultures. Growing the processes to industrial levels from successful results obtained in the laboratory is still another challenge. Gene engineering, cultivation system optimization, and improved extraction techniques are some of the future prospects that will enhance the commercial viability of microalgae and, hence, contribute to sustainable development in different industries.

2.9 Conclusion

The significance of microalgae as a renewable feedstock for the production of valuable bioactive compounds is being studied on a wider platform globally. The dependence on microalgae as a source of protein, pigments, and phenolic compounds is increasing significantly considering its sustainability. In addition, the use of microalgae as a source of lignocellulose is also increasing considerably. Fine tuning of the cultivation methods can facilitate the use of microalgae as an efficient alternative for terrestrial resources for lignocellulose toward the production of biofuel. However, the potential of microalgae as a renewable feed stock for biofuel production needs to be explored owing to the low yield. In this regard, systematic efforts should be carried out to analyze the bioconversion process and how efficient modification can be done in the process to achieve the maximization of biomass valorization. Further, the application of a systematic life cycle assessment study, along with other techno-economic evaluations, can further accelerate the application of microalgae on a commercial basis.

References

1. Richmond, A. (Ed.), *Handbook of microalgal culture: biotechnology and applied phycology*, vol. 577, Blackwell science, Oxford, 2004.
2. Torres-Tiji, Y., Fields, F.J., Mayfield, S.P., Microalgae as a future food source. *Biotechnol. Adv.*, 41, 107536, 2020.
3. Sathasivam, R., Radhakrishnan, R., Hashem, A., Abd_Allah, E.F., Microalgae metabolites: A rich source for food and medicine. *Saudi J. Biol. Sci.*, 26, 4, 709–722, 2019.
4. Guo, B., Yang, B., Silve, A., Akaberi, S., Scherer, D., Papachristou, I., Frey, W., Hornung, U., Dahmen, N., Hydrothermal liquefaction of residual microalgae biomass after pulsed electric field-assisted valuables extraction. *Algal Res.*, 43, 101650, 2019.
5. Fatima, N., Emambux, M.N., Olaimat, A.N., Stratakos, A.C., Nawaz, A., Wahyono, A., Gul, K., Park, J., Shahbaz, H.M., Recent advances in microalgae, insects, and cultured meat as sustainable alternative protein sources. *Food and Humanity*, 1,731–741, 2023.
6. Kumar, R., Hegde, A.S., Sharma, K., Parmar, P., Srivatsan, V., Microalgae as a sustainable source of edible proteins and bioactive peptides–Current trends and future prospects. *Food Res. Int.*, 157, 111338, 2022.
7. Sousa, V., Hijo, A.A.T., Lüdtke, F.L., Vicente, A.A., Dias, O., Geada, P., Recovery and encapsulation of Dunaliella salina β-carotene through a novel sustainable approach: Sequential application of an ionic liquid as naturally-derived solvent and emulsifier. *Food Chem.*, 458, 140232, 2024.
8. Pourkarimi, S., Hallajisani, A., Alizadehdakhel, A., Nouralishahi, A., Golzary, A., Factors affecting production of beta-carotene from Dunaliella salina microalgae. *Biocatal. Agric. Biotechnol.*, 29, 101771, 2020.
9. Aneesh, P.A., Ajeeshkumar, K.K., Lekshmi, R.K., Anandan, R., Ravishankar, C.N., Mathew, S., Bioactivities of astaxanthin from natural sources, augmenting its biomedical potential: A review. *Trends Food Sci. Technol.*, 125, 81–90, 2022.
10. Bule, M.H., Ahmed, I., Maqbool, F., Bilal, M., Iqbal, H.M., Microalgae as a source of high-value bioactive compounds. *Front. Biosci.*, 10, 2, 197–216, 2018.
11. Li, Y., The bioactivities of phycocyanobilin from spirulina. *J. Immunol. Res.*, 2022, 1, 4008991, 2022.
12. Patel, A.K., Tambat, V.S., Chen, C.W., Chauhan, A.S., Kumar, P., Vadrale, A.P., Huang, C.Y., Dong, C.D., Singhania, R.R., Recent advancements in astaxanthin production from microalgae: A review. *Bioresour. Technol., 364*, 128030, 2022.
13. Ye, Y., Ding, W., Huang, J., Liu, J., Characterization of a Chromochloris zofingiensis mutant with enhanced canthaxanthin accumulation. *Algal Res.*, 75, 103260, 2023.

14. Khozin-Goldberg, I., Leu, S., Boussiba, S., Microalgae as a source for VLC-PUFA production. *Lipids Plant Algae Dev.*, 86, 471–510, 2016.
15. Barta, D.G., Coman, V., Vodnar, D.C., Microalgae as sources of omega-3 polyunsaturated fatty acids: Biotechnological aspects. *Algal Res.*, 58, 102410, 2021.
16. Karageorgou, D., Rova, U., Christakopoulos, P., Katapodis, P., Matsakas, L., Patel, A., Benefits of supplementation with microbial omega-3 fatty acids on human health and the current market scenario for fish-free omega-3 fatty acid. *Trends Food Sci. Technol.*, 136, 169–180, 2023.
17. Magoni, C., Bertacchi, S., Giustra, C.M., Guzzetti, L., Cozza, R., Ferrari, M., Torelli, A., Marieschi, M., Porro, D., Branduardi, P., Labra, M., Could microalgae be a strategic choice for responding to the demand for omega-3 fatty acids? A European perspective. *Trends Food Sci. Technol.*, 121, 142–155, 2022.
18. Ahmad, A. and Ashraf, S.S., Sustainable food and feed sources from microalgae: Food security and the circular bioeconomy. *Algal Res.*, 74, 103185, 2023.
19. Caetano, P.A., do Nascimento, T.C., Fernandes, A.S., Nass, P.P., Vieira, K.R., Junior, M.R.M., Jacob-Lopes, E., Zepka, L.Q., Microalgae-based polysaccharides: Insights on production, applications, analysis, and future challenges. *Biocatal. Agric. Biotechnol.*, 45, 102491, 2022.
20. Sasaki, M., Takagi, A., Sasaki, D., Nakamura, A., Asayama, M., Characteristics and function of an extracellular polysaccharide from a green alga Parachlorella. *Carbohydr. Polym.*, 254, 117252, 2021.
21. Delattre, C., Pierre, G., Laroche, C., Michaud, P., Production, extraction and characterization of microalgal and cyanobacterial exopolysaccharides. *Biotechnol. Adv.*, 34, 7, 1159–1179, 2016.
22. Tounsi, L., Hentati, F., Hlima, H.B., Barkallah, M., Smaoui, S., Fendri, I., Michaud, P., Abdelkafi, S., Microalgae as feedstock for bioactive polysaccharides. *Int. J. Biol. Macromol.*, 221, 1238–1250, 2022.
23. Gouda, M., Tadda, M.A., Zhao, Y., Farmanullah, F., Chu, B., Li, X., He, Y., Microalgae bioactive carbohydrates as a novel sustainable and eco-friendly source of prebiotics: emerging health functionality and recent technologies for extraction and detection. *Front. Nutr.*, 9, 806692, 2022.
24. Ljubic, A., Thulesen, E.T., Jacobsen, C., Jakobsen, J., UVB exposure stimulates production of vitamin D3 in selected microalgae. *Algal Res., 59*, 102472, 2021.
25. Ljubic, A., Jacobsen, C., Holdt, S.L., Jakobsen, J., Microalgae Nannochloropsis oceanica as a future new natural source of vitamin D3. *Food Chem.*, 320, 126627, 2020.
26. Santhakumaran, P., Ayyappan, S.M., Ray, J.G., Nutraceutical applications of twenty-five species of rapid-growing green-microalgae as indicated by their antibacterial, antioxidant and mineral content. *Algal Res.*, 47, 101878, 2020.
27. Guedes, A.C., Gião, M.S., Matias, A.A., Nunes, A.V., Pintado, M.E., Duarte, C.M., Malcata, F.X., Supercritical fluid extraction of carotenoids and

chlorophylls a, b and c, from a wild strain of Scenedesmus obliquus for use in food processing. *J. Food Eng.*, 116, 2, 478–482, 2013.
28. Cheng, C.H., Du, T.B., Pi, H.C., Jang, S.M., Lin, Y.H., Lee, H.T., Comparative study of lipid extraction from microalgae by organic solvent and supercritical CO2. *Bioresour. Technol.*, 102, 21, 10151–10153, 2011.
29. Cheng, X., Qi, Z., Burdyny, T., Kong, T., Sinton, D., Low pressure supercritical CO2 extraction of astaxanthin from Haematococcus pluvialis demonstrated on a microfluidic chip. *Biores. Technol.*, 250, 481–1485, 2018.
30. Rodríguez-España, M., Mendoza-Sánchez, L.G., Magallón-Servín, P., Salgado-Cervantes, M.A., Acosta-Osorio, A.A., García, H.S., Supercritical fluid extraction of lipids rich in DHA from Schizochytrium sp. *J. Supercrit. Fluids*, 179, 105391, 2022.
31. Feller, R., Matos, AP., Mazzutti, S., Moecke, E.H., Tres, M.V., Derner, R.B., Oliveira, J.V., Junior, A.F., Polyunsaturated ω-3 and ω-6 fatty acids, total carotenoids and antioxidant activity of three marine microalgae extracts obtained by supercritical CO2 and subcritical n-butane. *J. Supercrit. Fluids*, 133, 437–443, 2018.
32. Callejón, M.J.J., Medina, A.R., Sánchez, M.D.M., Moreno, P.A.G., López, E.N., Cerdán, L.E., Grima, E.M., Supercritical fluid extraction and pressurized liquid extraction processes applied to eicosapentaenoic acid-rich polar lipid recovery from the microalga Nannochloropsis sp. *Algal Res.*, 61, 102586, 2022.
33. Adam, F., Abert-Vian, M., Peltier, G., Chemat, F., "Solvent-free" ultrasound-assisted extraction of lipids from fresh microalgae cells: a green, clean and scalable process. *Bioresour. Technol.*, 114, 457–465, 2012.
34. Zhao, G., Chen, X., Wang, L., Zhou, S., Feng, H., Chen, W.N., Lau, R., Ultrasound assisted extraction of carbohydrates from microalgae as feedstock for yeast fermentation. *Bioresour. Technol.*, 128, 337–344, 2013.
35. Parniakov, O., Apicella, E., Koubaa, M., Barba, F.J., Grimi, N., Lebovka, N., Pataro, G., Ferrari, G., Vorobiev, E., Ultrasound-assisted green solvent extraction of high-added value compounds from microalgae Nannochloropsis spp. *Bioresour. Technol.*, 198, 262–267, 2015.
36. Fernandes, A.S., Nogara, G.P., Menezes, C.R., Cichoski, A.J., Mercadante, A.Z., Jacob-Lopes, E., Zepka, L.Q., Identification of chlorophyll molecules with peroxyl radical scavenger capacity in microalgae Phormidium autumnale using ultrasound-assisted extraction. *Food Res. Int.*, 99, 1036–1041, 2017.
37. Ido, A.L., de Luna, M.D.G., Capareda, S.C., Maglinao Jr., A.L., Nam, H., Application of central composite design in the optimization of lipid yield from Scenedesmus obliquus microalgae by ultrasound-assisted solvent extraction. *Energy*, 157, 949–956, 2018.
38. Mienis, E., Vandamme, D., Foubert, I., Enzyme-assisted disruption of oleaginous microalgae to increase the extraction of lipids: Nannochloropsis as a case study. *Curr. Opin. Food Sci.*, 51, 101034, 2023.

39. Tavakoli, S., Hong, H., Wang, K., Yang, Q., Gahruie, H.H., Zhuang, S., Li, Y., Liang, Y., Tan, Y., Luo, Y., Ultrasonic-assisted food-grade solvent extraction of high-value added compounds from microalgae Spirulina platensis and evaluation of their antioxidant and antibacterial properties. *Algal Res.*, 60, 102493, 2021.
40. Zou, X., Xu, K., Chang, W., Qu, Y., Li, Y., Rapid extraction of lipid from wet microalgae biomass by a novel buoyant beads and ultrasound assisted solvent extraction method. *Algal Res.*, 58, 102431, 2021.
41. Fernandes, A.S., Caetano, P.A., Jacob-Lopes, E., Zepka, L.Q., de Rosso, V.V., Alternative green solvents associated with ultrasound-assisted extraction: A green chemistry approach for the extraction of carotenoids and chlorophylls from microalgae. *Food Chem.*, 455, 139939, 2024.
42. Morón-Ortiz, A., Mapelli-Brahm, P., León-Vaz, A., Benitez-González, A.M., León, R., Meléndez-Martínez, A.J., Ultrasound-assisted extraction of carotenoids from phytoene-accumulating Chlorella sorokiniana microalgae: Effect of milling and performance of the green biosolvents 2-methyltetrahydrofuran and ethyl lactate. *Food Chem.*, 434, 137437, 2024.
43. Sierra, L.S., Dixon, C.K., Wilken, L.R., Enzymatic cell disruption of the microalgae Chlamydomonas reinhardtii for lipid and protein extraction. *Algal Res.*, 25, 149–159, 2017.
44. Zhang, Y., Kong, X., Wang, Z., Sun, Y., Zhu, S., Li, L., Lv, P., Optimization of enzymatic hydrolysis for effective lipid extraction from microalgae Scenedesmus sp. *Renew. Energy*, 125, 1049–1057, 2018.
45. Qiu, C., He, Y., Huang, Z., Li, S., Huang, J., Wang, M., Chen, B., Lipid extraction from wet Nannochloropsis biomass via enzyme-assisted three phase partitioning. *Bioresour. Technol.*, 284, 381–390, 2019.
46. Zhao, X., Zhang, X., Liu, H., Zhu, H., Zhu, Y., Enzyme-assisted extraction of astaxanthin from Haematococcus pluvialis and its stability and antioxidant activity. *Food Sci. Biotechnol.*, 28, 1637–1647, 2019.
47. He, Y., Zhang, B., Guo, S., Guo, Z., Chen, B., Wang, M., Sustainable biodiesel production from the green microalgae Nannochloropsis: Novel integrated processes from cultivation to enzyme-assisted extraction and ethanolysis of lipids. *Energy Convers. Manage.*, 209, 112618, 2020.
48. Mienis, E., Vandamme, D., Foubert, I., Ultrasound assisted extraction of Nannochloropsis: Effects on lipid extraction efficiency and lipid stability. *Algal Res.*, 80, 103520, 2024.
49. Biller, P., Friedman, C., Ross, A.B., Hydrothermal microwave processing of microalgae as a pre-treatment and extraction technique for bio-fuels and bio-products. *Bioresour. Technol.*, 136, 188–195, 2013.
50. Lee, I., Park, J.Y., Choi, S.A., Oh, Y.K., Han, J.I., Hydrothermal nitric acid treatment for effectual lipid extraction from wet microalgae biomass. *Bioresour. Technol.*, 172, 138–142, 2014.
51. Dai, Y.M., Chen, K.T., Chen, C.C., Study of the microwave lipid extraction from microalgae for biodiesel production. *Chem. Eng. J.*, 250, 267–273, 2014.

52. Cheng, J., Huang, R., Li, T., Zhou, J., Cen, K., Biodiesel from wet microalgae: Extraction with hexane after the microwave-assisted transesterification of lipids. *Bioresour. Technol.*, 170, 69–75, 2014.
53. Hu, H.S., Wu, Y.L., Yang, M.D., Fractionation of bio-oil produced from hydrothermal liquefaction of microalgae by liquid-liquid extraction. *Biomass Bioenergy*, 108, 487–500, 2018.
54. Guo, X., Wang, Q., Wu, Y., Liu, X., Gong, Z., Comprehensive Insights into Microalgae Proteins: Nutritional Profiles and Innovative Applications as Sustainable Alternative Proteins in Health and Food Sciences. *Food Hydrocoll.*, 154, 110112, 2024.
55. Zhong, H., Ma, L., Zhu, Y., Jin, B., Wang, T., Wang, Y., Jin, F., Hydrothermal conversion of microalgae and its waste residue after biofuel extraction to acetic acid with CuO as solid oxidant. *J. Supercrit. Fluids*, 157, 104717, 2020.
56. Huang, H., Liu, Y., Xu, Z., Zhang, D., Feng, M., Zhao, T., Zhang, L., Li, W., Li, X., Effect of fucoidan on kidney injury in type 2 diabetic rats based on PI3K/AKT/Nrf2. *J. Funct. Foods*, 90, 104976, 2022.
57. Silva, T.A., do Couto, E.D.A., Assemany, P.P., Costa, P.A.C., Marques, P.A., Paradela, F., dos Reis, A.J.D., Calijuri, M.L., Biofuel from wastewater-grown microalgae: A biorefinery approach using hydrothermal liquefaction and catalyst upgrading. *J. Environ. Manage.*, 368, 122091, 2024.
58. Marangon, B.B., de Siqueira Castro, J., Calijuri, M.L., Aviation fuel based on wastewater-grown microalgae: Challenges and opportunities of hydrothermal liquefaction and hydrotreatment. *J. Environ. Manage.*, 354, 120418, 2024.
59. Zhang, Y., Soldatov, S., Papachristou, I., Nazarova, N., Link, G., Frey, W., Silve, A., Pulsed microwave pretreatment of fresh microalgae for enhanced lipid extraction. *Energy*, 248, 123555, 2022.
60. Low, K.L., Idris, A., Yusof, N.M., An optimized strategy for lutein production via microwave-assisted microalgae wet biomass extraction process. *Process Biochem.*, 121, 87–99, 2022.
61. Mahata, C., Das, P., Khan, S., Thaher, M., II, Abdul Quadir, M., Annamalai, S.N., Al Jabri, H., The potential of marine microalgae for the production of food, feed, and fuel (3F). *Fermentation*, 8, 7, 316, 2022.
62. Udayan, A., Pandey, A.K., Sharma, P., Sreekumar, N., Kumar, S., Emerging industrial applications of microalgae: challenges and future perspectives. *Syst. Microbiol. Biomanuf.*, 1, 4, 411–431, 2021.
63. Nege, A.S., Masithah, E.D., Khotib, J., Trends in the uses of spirulina microalga: a mini-review. *Sci. J. Fish. Mar.*, 12, 1, 149–166, 2020.
64. Altmann, B.A. and Rosenau, S., Spirulina as animal feed: Opportunities and challenges. *Foods*, 11, 7, 965, 2022.
65. Calella, P., Di Dio, M., Cerullo, G., Di Onofrio, V., Galle, F., Liguori, G., Antioxidant, immunomodulatory, and anti-inflammatory effects of Spirulina in disease conditions: A systematic review. *Int. J. Food Sci. Nutr.*, 73, 8, 1047–1056, 2022.

66. Sorrenti, V., Castagna, D.A., Fortinguerra, S., Buriani, A., Scapagnini, G., Willcox, D.C., Spirulina microalgae and brain health: A scoping review of experimental and clinical evidence. *Mar. Drugs*, 19, 6, 293, 2021.
67. Li, C., Gao, Y., Huan, Y., Ren, P., Zhi, J., Wu, A., Xu, J., Wei, Z., Xue, C., Tang, Q., Colon and gut microbiota greatly affect the absorption and utilization of astaxanthin derived from Haematococcus pluvialis. *Food Res. Int.*, 156, 111324, 2022.
68. Amin, M., ul Haq, A., Shahid, A., Boopathy, R., Syafiuddin, A., Spirulina as a Food of the Future, in: *Pharmaceutical and Nutraceutical Potential of Cyanobacteria*, pp. 53–83, Springer International Publishing, Cham, 2024.
69. Martins, C.F., *Spirulina and Chlorella vulgaris as ingredients in diet of weaned piglets: a valorization approach using feed enzymes*. Doctoral dissertation, Universidade de Lisboa, Portugal, 2023.
70. Ganeson, Y., Paramasivam, P., Palanisamy, K.M., Govindan, N., Maniam, G.P., LCMS and FTIR profiling of microalga Chlorella sp. for cosmetics and skin care applications. *Clean. Water*, 2, 100028, 2024.
71. Aravind, M.K., Vignesh, N.S., Gayathri, S., Anjitha, N., Athira, K.M., Gunaseelan, S., Varalakshmi, P., Review on rewiring of microalgal strategies for the heavy metal remediation-a metal specific logistics and tactics. *Chemosphere*, 313, 137310, 2023.
72. Spínola, M.P., Costa, M.M., Prates, J.A., Enhancing digestibility of Chlorella vulgaris biomass in monogastric diets: Strategies and insights. *Animals*, 13, 6, 1017, 2023.
73. Kaur, M., Bhatia, S., Gupta, U., Decker, E., Tak, Y., Bali, M., Bala, S., Microalgal bioactive metabolites as promising implements in nutraceuticals and pharmaceuticals: inspiring therapy for health benefits. *Phytochem. Rev.*, 22, 4, 903–933, 2023.
74. Elbahnaswy, S. and Elshopakey, G.E., Recent progress in practical applications of a potential carotenoid astaxanthin in aquaculture industry: a review. *Fish Physiol. Biochem.*, 50, 1, 97–126, 2024.
75. Kanwugu, O.N. and Glukhareva, T.V., Activation of Nrf2 pathway as a protective mechanism against oxidative stress-induced diseases: Potential of astaxanthin. *Arch. Biochem. Biophys.*, 741, 109601, 2023.
76. Cao, Y., Yang, L., Qiao, X., Xue, C., Xu, J., Dietary astaxanthin: an excellent carotenoid with multiple health benefits. *Crit. Rev. Food Sci. Nutr.*, 63, 18, 3019–3045, 2023.
77. Wang, Y., Chen, X., Baker, J.S., Davison, G.W., Xu, S., Zhou, Y., Bao, X., Astaxanthin promotes mitochondrial biogenesis and antioxidant capacity in chronic high-intensity interval training. *Eur. J. Nutr.*, 62, 3, 1453–1466, 2023.
78. Dutta, S., Kumar, S.J., Banerjee, R., A comprehensive review on astaxanthin sources, structure, biochemistry and applications in the cosmetic industry. *Algal Res.*, 74, 103168, 2023.
79. Kanatt, S.R., Antioxidants from the red algae Kappaphycus alvarezii, in: *Marine Antioxidants*, pp. 457–472, Academic Press, USA, 2023.

80. Cezare-Gomes, E.A., Lousada, M.E.G., Matsudo, M.C., Ferreira-Camargo, L.S., Ishii, M., Singh, A.K., Carvalho, J.C.M., Two-stage semi-continuous cultivation of Dunaliella salina for β-carotene production. *Braz. J. Chem. Eng.*, 40, 2, 367–378, 2023.
81. Tripathi, G., Dubey, P., Ahmad, S., Farooqui, A., Mishra, V., Role of Algal-derived Bioactive Compounds in Human Health. *Recent Pat. Biotechnol.*, 18, 3, 190–209, 2024.
82. Patel, A.K., Tambat, V.S., Vadrale, A.P., Kumar, P., Singhania, R.R., Chen, C.W., Dong, C.D., Marine Algae as a Source of Carotenoids: Bioactivity and Health Benefits, in: *Sustainable Marine Food and Feed Production Technologies*, pp. 1–19, CRC Press, USA, 2023.
83. Pagels, F. and Guedes, A.C., *β-Carotene from Dunaliella: Production, applications in food/feed, and recent advances*, Handbook of Food and Feed from Microalgae, 203-219, 2023.
84. Razali, W.A.W. and Pandhal, J., Outdoor pilot-scale cultivation and techno economic assessment of a novel omega-3 eicosapentaenoic acid over-producing Nannochloropsis oculata strain. *Bioresour. Technol. Rep.*, 24, 101682, 2023.
85. Chen, Y., Mitra, A., Rahimnejad, S., Chi, S., Kumar, V., Tan, B., Xie, S., Retrospect of fish meal substitution in Pacific white shrimp (Litopenaeus vannamei) feed: Alternatives, limitations and future prospects. *Rev. Aquacult.*, 16, 1, 382–409, 2024.
86. Jesionowska, M., Ovadia, J., Hockemeyer, K., Clews, A.C., Xu, Y., EPA and DHA in microalgae: Health benefits, biosynthesis, and metabolic engineering advances. *J. Am. Oil Chem. Soc.*, 100, 11, 831–842, 2023.
87. Kumar, N.A., Sridhar, S., Jayappriyan, K.R., Raja, R., Applications of microalgae in aquaculture feed, in: *Handbook of Food and Feed from Microalgae*, pp. 421–433, Academic Press, USA, 2023.
88. Xavier, B., Balla, V., Padmajarani, S., *Marine microalgae culture techniques for finfish and shellfish larval rearing*, 2023.
89. Matsui, H., Kono, S., Ishibashi, K., Ishikawa, M., Kotani, T., Assessment of microalga Tisochrysis lutea in the deceleration growth phase as an-3 HUFA source for Brachionus plicatilis rotifers in Japanese flounder larviculture, in: *Aquaculture International*, pp. 1–23, 2024.
90. Ma, M. and Hu, Q., Microalgae as feed sources and feed additives for sustainable aquaculture: prospects and challenges. *Rev. Aquacult.*, 16, 2, 818–835, 2024.
91. Razzak, S.A., Bahar, K., Islam, K.O., Haniffa, A.K., Faruque, M.O., Hossain, S.Z., Hossain, M.M., Microalgae cultivation in photobioreactors: sustainable solutions for a greener future. *Green Chem. Eng.*, 5, 4, 418–439, 2023.
92. Barboza-Rodríguez, R., Rodríguez-Jasso, R.M., Rosero-Chasoy, G., Aguado, M.L.R., Ruiz, H.A., Photobioreactor configurations in cultivating microalgae biomass for biorefinery. *Bioresour. Technol.*, 394, 130208, 2024.

93. Sarker, N.K. and Kaparaju, P., A critical review on the status and progress of microalgae cultivation in outdoor photobioreactors conducted over 35 years (1986–2021). *Energies*, 16, 7, 3105, 2023.
94. Leong, Y.K., Huang, C.Y., Chang, J.S., Pollution prevention and waste phycoremediation by algal-based wastewater treatment technologies: The applications of high-rate algal ponds (HRAPs) and algal turf scrubber (ATS). *J. Environ. Manage.*, 296, 113193, 2021.
95. Rahman, K.M. and Melville, L., Global market opportunities for food and feed products from microalgae, in: *Handbook of food and feed from microalgae*, pp. 593–602, Academic Press, Cambridge, MA, USA, 2023.
96. Abreu, A.P., Martins, R., Nunes, J., Emerging Applications of Chlorella sp. *Spirulina (Arthrospira) sp. Bioeng.*, 10, 8, 955, 2023.
97. Colombo, S.M., Roy, K., Mraz, J., Wan, A.H., Davies, S.J., Tibbetts, S.M., Turchini, G.M., Towards achieving circularity and sustainability in feeds for farmed blue foods. *Rev. Aquacult.*, 15, 3, 1115–1141, 2023.
98. Goyal, N. and Jerold, F., Biocosmetics: technological advances and future outlook. *Environ. Sci. Pollut. Res.*, 30, 10, 25148–25169, 2023.
99. Olabi, A.G., Shehata, N., Sayed, E.T., Rodriguez, C., Anyanwu, R.C., Russell, C., Abdelkareem, M.A., Role of microalgae in achieving sustainable development goals and circular economy. *Sci. Total Environ.*, 854, 158689, 2023.
100. Huang, R., He, Y., Yao, X., Yu, Y., Song, W., Yang, W., Cheng, J., Disintegration of wet microalgae biomass with deep-eutectic-solvent-assisted hydrothermal treatment for sustainable lipid extraction. *Green Chem.*, 24, 4, 1615–1626, 2022.
101. Klöpffer, W., The Hitch Hiker's Guide to LCA-An orientation in LCA methodology and application. *Int. J. Life Cycle Assess.*, 11, 2, 142–142, 2006.
102. Romagnoli, F., Spaccini, F., Boggia, A., Paoli, R., Feofilovs, M., Ieviņa, B., Rocchi, L., Microalgae cultivation in a biogas plant: Environmental and economic assessment through a life cycle approach. *Biomass Bioenergy*, 182, 107116, 2024.
103. Xu, Y., Wei, C., Liu, D., Li, J., Tian, B., Li, Z., Xu, L., Life-cycle and economic assessments of microalgae biogas production in suspension and biofilm cultivation systems. *Bioresour. Technol.*, 395, 130381, 2024.
104. Xiao, C., Fu, Q., Liao, Q., Huang, Y., Xia, A., Chen, H., Zhu, X., Life cycle and economic assessments of biogas production from microalgae biomass with hydrothermal pretreatment via anaerobic digestion. *Renew. Energy*, 151, 70–78, 2020.
105. Mohit, A., Mishel, V.K., Remya, N., Life cycle assessment and technoeconomic analysis of biofuels produced from polyculture microalgae cultivated in greywater. *J. Environ. Manage.*, 356, 120711, 2024.
106. Arbour, A.J., Bhatt, P., Simsek, H., Brown, P.B., Huang, J.Y., Life cycle assessment on environmental feasibility of microalgae-based wastewater treatment for shrimp recirculating aquaculture systems. *Bioresour. Technol.*, 399, 130578, 2024.

3

Pharmaceutical and Nutraceutical Applications of Marine Microalgae

Nayomi John*, Parvathy Raj, Jinsa Jose, Swathy T., Fathima Ferin Haneef, Gokuljith S. and Rukkiya Sunil

Department of Microbiology, Mar Athanasius College (Autonomous), Kothamangalam, Kerala, India

Abstract
This chapter focuses mainly on the review of marine microalgae and marine phytoplanktons. Recent research accomplishments in the fields of pharmaceutical application of marine microalgae (phytoplanktons) and nutraceutical application of marine microalgae (phytoplanktons) have also been discussed in detail.

***Keywords*:** Marine microalgae, marine phytoplanktons, pharmaceutical application, marine nanoalgae, nutraceutical application

3.1 Introduction

Approximately 70% of planet Earth is occupied by its seas and oceans, which provide the Earth with its largest habitat and have abundant diversity in biology and chemistry. Less than 0.01% of this enormous area has been thoroughly examined, and only approximately 5% of the seafloor has been explored. This has been exploited, but due to its technical limitations, exploration is significantly affected. Numerous creatures, including sponges, macro- and microalgae, and corals, have yielded a wide range of chemicals with unique chemical structures [1, 2]. A significant worldwide issue facing the 21st century is the lack of resources to satisfy the growing population's needs for food and energy [3]. The challenge of nutriment scarcity will be faced by 12% of the population, which is expected to reach

**Corresponding author*: nayomijohn@macollege.in

Shaju S. S., Vipin. P. M. and Visakh. P.M. (eds.) Marine Biotechnology: Studies and Applications, (53–110)

approximately 9 billion by the year 2030 [4]. The highest population of hungry and malnourished people are in the Asian countries and Central African Republic according to the global hunger index (GHI) 2019 report [5]. Food crops lack necessary nutrition in our present scenario. Food starvation is mitigated by the use of alternative food sources that is sustainable. Generation of wastewater and emissions of carbon dioxide are another major issue [6, 7]. In recent years, the general public's concerns about their general health have prompted a hunt for novel functional meals. The race to a successful life has contributed significantly to a decrease in the quality of life, and moreover, people have become slaves to lifestyle diseases such as diabetes, obesity, and allergies. Nutritious balanced diet is of utmost importance for the prevention of such diseases. Our oceans and seas are rich sources of chemicals called bioactive compounds. Extraction of these chemicals and their addition into conventional foods can yield functional foods [8].

Enhanced nutritional value, decreased disease risk, and improved health may be beneficial for human function. Foods of marine origin, such as seaweed, can be included in our daily diet as they are high in bioactive compounds. Conventional food can be improved in the following ways:

i) Harmful ingredients can be changed chemically or eliminated.
ii) Effective concentration of probiotic microorganisms or ingredients that promote health can be added.
iii) Food ingredients that promote health can be added, or the concentration can be increased.
iv) Food ingredients that promote health can be made more stable, and its bioavailability can be increased [8].

Microalgae are eukaryotic single-celled (unicellular) organisms that can feed either phototrophically or heterotrophically [9]. It produces biomass by the process of photosynthesis in the presence of sunlight by fixation of atmospheric CO_2 [7, 10]. Materials that can be used as health products, supplements, feed, and food are accumulated, and they include carbohydrates, lipids, amino acids, proteins, natural pigments, and carotenoids [11]. Humans use many species of marine macroalgae as part of their diet or as raw materials in industries like food, pharmaceutical, cosmetics, and paint. Thus, they are of great economic importance. Antioxidant and anticancer activity are the major beneficial activities shown by seaweeds or marine microalgae. Minerals, including potassium, calcium, iron, magnesium, and iodine, are abundant in them. Vitamins B1, B2, B6, C, and niacin are also present. Their main composition consists of essential amino acids, proteins, and minimal fat [1]. Interest in algae is rising as a result of the advantages enjoyed by the food and pharmaceutical industries. Two compounds present in a variety of marine algae species have antioxidant and photoprotective properties. These include a class of chemical pigment known as carotenoid and an amino acid known as mycosporin [1, 12]. To meet the world's food needs, food output must double [13, 14]. Over 10.8% of the world's population lack access to enough food for sustenance [15]. Microalgae are remarkable organisms that can grow anywhere there is light and humidity and can survive in harsh conditions [16]. Phycocyanin, beta-carotene, oleic acid, linolenic acid, cobalamin (vitamin B12), vitamin E, lutein, cyanovirin, and zeaxanthin are examples of chemical compounds of microalgal origin that are known as bioactive compounds. These compounds have been shown to have anti-enzymatic, antimicrobial, anticarcinogenic, antibiotic, photoprotective, anti-inflammatory, antioxidant, anti-aging, and hypocholesterolemic qualities [17, 18]. As a result, they lower and prevent illnesses. Bioactive nutrients (bioproducts) can include vitamins, enzymes, pigments, sterols, proteins, fatty acids, and alkaloids. They are physiologically active compounds having functional qualities that may enhance health [19]. Marine microalgae produce bioactive substances, such as carotenes, chlorophylls, antioxidants, and phycobiliproteins, as a defense mechanism that allows them to thrive under harsh environmental and phototrophic circumstances [20].

Microalgae produce a variety of beneficial bioproducts such as astaxanthin, beta-carotene, eicosapentaenoic acid (EPA), docosahexaenoic acid (DHA), natural colors, bioactive and functional pigments, polysaccharides, algal extracts, and antioxidants. Japan began growing *Chlorella* for commercial use as a nutraceutical in 1960. Because of their high protein concentration, microalgae serve as a source of "single-cell protein." Their rapid growth allows them to create significant amounts of biomass [21]. Among the many goods derived from microalgae are chemicals, algal oils, beta-carotene, omega-3 fatty acids, and nutraceuticals. The increasing demand for these compounds cleared the path for sustainable microalgal technology. Different kinds of secondary metabolites can be produced by the algae under stress [22–24]. The most commercially available biochemical product to date is beta-carotene, which is halophilic and generated by the green algae *Dunaliala salina* [25]. Significant advancements in sustainable cultivation technologies and strength development lower high production costs [26]. Other major functions of microalgae include its capability to improve stabilization of the marine ecosystem as they are fed upon by other marine organisms. Resources, like microalgal technology, are used due to the increase in demand of microalgae as sources of bioenergy, food, and pharmaceuticals and thus meet their growing market potential [27]. Among the several types of microalgae, green algae, blue algae, and red algae, the green algae have been used commercially for their proteins, lipids, carbohydrates, and enzymes [28]. The incorporation of seaweeds as medicines in traditional Asian cultures rather than food and herbs is seen in our modern-day approach. They are employed in the treatment and prevention of fatal illnesses. Additionally, they satisfy the present need for sustainable and healthful food [29, 30].

Nutraceutical is a combination of two words—nutrient and pharmaceutical. These compounds have been shown to be of high health benefits and can be supplemented through diet. Nutraceuticals of marine origin are called marine nutraceuticals and have become of great importance in recent years. Fish and fish byproducts, crustaceans, seaweed, microalgae, marine fungi, and bacteria are all of interest as sources of health-promoting substances or as nutritious foods. They include minerals, vitamins, pigments, lipids, chitin, polysaccharides, proteins, and peptides [8]. Details about marine nutraceuticals will be dealt in Section 3.5 of this chapter. Production of pharmaceutical and organic chemicals, which were done by conventional method of using fossil fuels is now being replaced with renewable bioresource—microalgae. They can also be used for supplements as they contain vitamins, carotenoids, and other biomolecules. The drawbacks of such large-scale production of microalgal biomass are the harvesting methods that require high energy, techno-economic analysis that lack in various aspects, biorefinery routes that are optimized, and variation in composition. Newer strategies have been brought for bioactive compound production enhancement. Microalgal bioproducts are becoming more and more in demand on the global market. Safety issues related with the usage of microalgal biomass is also a key challenge in this process.

Freshwater and arable land are not necessary for microalgae and neither are fertilizer or pesticides. They are an environment-friendly food source and are sustainable. They are also safe for the environment. They are highly efficient in accessing water and nutrients. Their photosynthetic machinery is efficient too. They have high productivity and short growth cycle [31, 32]. Natural matrices provide more stability to bioactive compounds than to extracts. So, the whole biomass of microalgae can be used without extraction [33, 34]. Mechanical or thermal treatments are used to break cell walls, and microalgae pigments are extracted using solvents [35, 36]. Recent years have seen the use of innovative, environmentally friendly, and effective selective extraction technologies to extract bioactive compounds from microalgae, including enzyme-assisted extraction, microwave-assisted extraction, ultrasound-assisted extraction, super critical fluid extraction, and pressurized liquid extraction [37–39]. To make microalgal biomass economically viable, technologies must be developed, and all produced compounds must be used optimally [40, 41]. Efficient production of high-value compounds is possible as photosynthetic capability of microalgae allows efficient utilization of sunlight compared to other higher plants [42]. The protein content generated by microalgae is significantly higher than that of conventional protein sources such as meat (43% DW protein), soy (37% DW protein), yeast (39% DW protein), peanuts (26% DW protein), and dried skim milk (36% DW protein) [43, 44]. Compared to proteins generated from microalgae, animal proteins require 100 times more water to extract [45]. These autotrophic microalgae have health benefits and exhibit properties of nutraceuticals. They benefit our immune system to fight off cancer cells and also heart diseases [46].

3.2 Marine Microalgae

3.2.1 Introduction

Algae is a group of microorganisms with different biochemical and physiological characteristics. Microalgae is a unicellular photosynthetic organism and includes both prokaryotes and eukaryotes. Several phyla, including Cyanophyta, Chlorophyta, Rhodophyta, Haptophyta, Streotophyta, and Heterokondophyta, are found in microalgae. With the aid of water, carbon dioxide, and solar energy, they create complex organic substances like primary and secondary metabolites. By adjusting to various environmental stressors, they can continue to develop from freshwater to extremely salinized environments. They once inhabited damp, arid sands and black dirt [47]. In addition to producing a wide range of chemical compounds with numerous features, they also lead to the low-cost manufacturing of metabolites and the discovery of new metabolites [48]. The physiological states of microalgae include versatility and plasticity. Microalgae are considered to be a food source for the marine ecosystem, so they stabilize the marine ecosystem. Red, blue, and green algae are examples of microalgae. The primary uses of green algae, a kind of algae that belong to the Chlorophyta division, are for its protein, lipid, carbohydrates, and enzymes [28]. Microalgae are rich in nutrients and contain proteins, fatty acids, and pigments that are necessary for human consumption. Nutraceutical and pharmaceutical components are produced from microalgae because of their excellent properties. The growing demand for food, bioenergy, and other bioactive chemicals makes microalgal technology a useful resource that is able to meet market potential [27]. The active compound in microalgae are polysaccharides. They repair damaged skin, prevent blemishes, and inhibit the inflammation process and are used in producing moisturizers and gelling agents. Certain physiologically active substances, including those used in biofuels, meals, feeds, and medications, can be found in microalgae [49]. Additionally, microalgae contain particular bioactive chemicals that support the skin's healing process and preserve its moisture content [50, 51]. Microalgae have a special photosynthetic aptitude that enables them to manufacture a variety of bioactive substances because they are able to use sunlight more effectively than other larger plants [42].

Microalgae produce a variety of components, including chemicals, nutraceuticals, beta-carotene, omega-3 fatty acids, and algal oil. Antimicrobial, anti-inflammatory, antioxidant, anticarcinogenic, anti-lipidemic, antihypertensive, and antidiabetic properties are among the bioactivities exhibited by bioactive compounds derived from microalgae. These bioactivities raise the demand for microalgae in the production of new products in the food, pharmaceutical, and nutraceutical industries. Microalgae do not compete for the resources needed to produce food crops and can be grown in brackish or marine water as well as in non-arable soil [52, 53].

3.2.2 Bioactive Compounds from Microalgae

Antioxidants, proteins, pigments, and polyunsaturated fatty acids are the substances that microalgae create. Secondary metabolites are organic substances that have an indirect role in the growth and development of living things. Microalgae produce secondary metabolites through different metabolic pathways [54]. High-value molecules (HMV) are the secondary metabolites of microalgae.

3.2.2.1 Pigments

The photosynthetic system of microalgae consists of colorful chemical substances called pigments. The pigments in algae give them their color. The three main types of pigments are phycobiliproteins (up to 8% of dry weight), carotenoids (0.1%–0.2% of dry weight), and chlorophyll (0.5%–1.0% of dry weight) [55–58].

3.2.2.1.1 Carotenoids

The group of algal pigments with the most commercial value is the carotenoids. They are isoprenoid molecules and some of the earliest pigments found in nature ranging in color from yellow to red. Carotenes (hydrocarbons like alpha-carotene, beta-carotene, and lycopene) and xanthophylls (oxygenated molecules like astaxanthin, lutein, and canthaxanthin) are the two main categories of carotenoids [34, 59, 60]. The two additional categories of carotenoids are primary and secondary categories. Secondary carotenoids are overproduced by some species of microalgae of the group Chlorophyta [61, 62]. Astaxanthins that are produced by microalgae have commercial applications. Astaxanthins have certain advantages compared to other carotenoids. Astaxanthins give shrimp, salmonoids, crayfish, and lobsters their pinkish red flesh [59, 63]. Lutein is a yellow main carotenoid found in microalgal species including *Chlorella* and *Scenedesmus* [37].

3.2.2.1.2 Chlorophylls

Chlorophylls are a useful bioactive chemical found in all higher plants that are isolated from microalgal biomass. Because they are green pigments that are found in nature everywhere, they are in charge of the photosynthetic process [64, 65]. Microalgae have many forms of chlorophyll, including a, b, c, d, and f. Two types of chlorophyll (a, b) are found in *Chlorella*, and under optimal conditions they are produced more [66, 67]. Under stress conditions, the chlorophyll content in microalgal biomass is reduced [57].

3.2.2.1.3 Phycobiliproteins

The water-soluble, deep-colored proteins present in Rhodophyta and *Cyanobacteria* are phycobiliproteins. During photosynthesis, the light captured by phycobiliproteins are passed on to chlorophylls. During stress condition, such as phosphorus, sulfur, and nitrogen starvation, the phycobiliproteins found in microalgae are degraded [68, 69].

3.2.2.2 PUFAs or Polyunsaturated Fatty Acids

The usage of PUFA, which are derived from fish oil, is restricted by their disagreeable taste and odor. Microalgae are the source of PUFAs such as eicosapentanoic acids (EPA) and docosahexanoic acids (DHA). *Nannochloropsis oculate* and *Tetraselmis* sp. consume the fatty acid-rich microalgae strains. EPA and DHA are the two primary omega-3 fatty acids found in marine oils, and they have a variety of biological actions.

3.2.2.3 Proteins

Microalgae have the potential to be cellular factories for protein synthesis. Microalgae are utilized as feed stocks for animals and poultry as well as for human consumption. *Chlorella*, *Arthrospira*, *Spirulina*, and *D. salina* are among the microalgae strains that have been shown to have a high protein content [70, 71]. Microalgal peptides have anti-inflammatory and anti-cancer properties [72, 73]. Antioxidant, anti-inflammatory, anticancer, hepatoprotective, neuroprotective, and immunomodulatory properties are shown by phycobiliproteins from marine cyanobacteria and red algae [72, 74]. Microalgae abundantly produce a biomolecule called MAAs (mycosporine-like amino acids), which shows anti-photoaging property on the skin [72, 75].

3.2.2.4 Vitamins

Provitamin A,vitamin B (B1, B2, B3, B5, B8, B9, and B12), vitamin C [11, 76], and vitamin K [77] are the different types of vitamins produced by microalgae. The accumulation of vitamins in microalgae is dependent on a number of parameters, including species kinds, climate changes, and physiological or nutritional factors [78]. Light intensity and photoperiodism have an impact on vitamin production. *Chlorella* sp. and *D. teritiolecta* manufacture riboflavin (B2), biotin (B7), and cobalamin (B12) [45].

3.2.3 Use of Bioactive Compounds

Marine microalgae are a rich source of bioactive compounds because they develop quickly and are simple to grow [79]. Marine microalgae produce bioactive compounds that have important applications in the field of health, nutrition, energy, and environment [80]. Polysaccharide is one of the bioactive substances made by marine microalgae, and it has a variety of uses, such as thickening and gelling agents in the food sector. Polysaccharides enhance the texture and quality of food ingredients [81]. Microalgal-derived polysaccharides are particularly crucial for soil treatment because they serve as a nutrient carrier and soil conditioner, which benefits fertile soil. Additionally, polysaccharides boost the soil's ability to store water [81, 82] and promote crop development [83].

Proteins are also produced from marine microalgae. Bioactive peptides are obtained by hydrolyzing high levels of protein content. The length is usually 10–20 amino acids, and proteins play an important role in the cosmetic and pharmaceutical industry. Since most species have a higher protein content, and extraction techniques are less costly, microalgal peptides are employed as an alternative to conventional food protein sources. Because of their superior efficiency, selectivity, and tolerance, microalgal bioactive peptides are a great option for the pharmaceutical and cosmetic industries [84].

3.2.4 Microalgae as a Pharmaceutical Source

Pharmaceutical uses for microalgae rely on their anti-inflammatory, antioxidant, anticancer, and antibacterial qualities. Polyphenols, phycobiliproteins, and vitamins are water-soluble antioxidants found in microalgae that aid in the prevention of cancer by causing premalignant lesions to regress [85]. Microalgal metabolites are helpful against inflammation, chronic conditions, and cardiovascular diseases, and they also reduce oxidative damage. The pharmaceutical products from microalgae with immune-stimulating activity help in reducing brain damage and improve the immune defense system, and also acts as a provitamin source [11]. Wounds, peptic ulcers, and constipation can be treated with the help of microalgal biomolecules. Tumorigenic cells in humans are prevented with them [86]. *C. vulgaris* produce a pharmaceutical product, 1,3 β-glycan, with immunostimulatory activity and reduce the levels of cholesterols [4]. Detoxification of body from ROS is also done with the aid of microalgae-based pharmaceutical products [4]. Anticarcinogenic activity is shown by vitamin B [11, 45]. Vitamin B prevents weakness, protects the skin and heart, eases depression, and aids in the flow of red blood cells. The microalgae *Dunaliella* sp. exhibit anticancer properties against Triple-Negative Breast Cancer (TNBC). In addition to preventing lipid oxidation and cardiovascular disorders, astaxanthin has anti-inflammatory, neuroprotective, and anticarcinogenic properties [11, 87, 88].

3.2.5 Cosmeceutical

Sustainability in the production of cosmetic products is provided by primary (carbohydrates, lipids, and proteins) and secondary (pigments and vitamins) metabolites generated by microalgal biomass that have anti-aging and antioxidant qualities [89]. Microalgae, such as *Haematococcus* sp., *Dunaliella* sp., and *Chlorella* sp., are typically used in the development of cosmetic products [90, 91]. The colorant in the cosmetic product is microalgal carotenoid [11]. The anti-tyrosinase activity exhibited by a few microalgae inhibits the tyrosine enzyme and reduces the skin darkling [89]. The *Chlorella* sp. and *Dunaliella* sp. of microalgae can be used as cosmetic products because of their medicinal properties [92]. Cosmeceutical properties are shown by the microalgal species like *Chlorogloeopsis* sp., *Chlamydocapsa* sp., *Nannochloropsis* sp., *Chlorococcum* sp., and *P. tricornutum* [89].

These properties prevent hair loss and photooxidation, moisturize the hair and skin like sunscreen, reduce skin sagging and wrinkle formation, and also increase collagen stimulation and skin whitening [89, 93].

3.2.6 Nutraceuticals

The food or food products that provide health benefits as well as helps in prevention and treatment of diseases are nutraceuticals [94]. When essential nutrients give additional characteristics for maintenance of organisms or on normal growth, they are considered nutraceuticals [95]. Foods that are considered nutraceuticals are not put through any testing or regulation protocols. Nutraceutical properties are shown by microalgae and provide benefits for the immune system by fighting against cancer and heart diseases [46]. Because shellfish have an underdeveloped immune system, nutraceuticals are used in aquaculture to boost the growth of shellfish under regulated and severe settings. The fatty acids, provitamin compounds, and nutritious proteins that microalgae provide improve the color of fish and crustacean flesh, particularly that of Salmonids and Penacidae. Commercial and present prospective nutraceuticals use microalgae such as *Chlamydomonas*, *Scenedesmus*, *Chlorella*, *Dunaliella*, and *Porphyridium*. Certain microalgae accumulate essential elements utilized in nutraceuticals, including potassium, zinc, phosphorus, cobalt, calcium, sodium, nitrogen, magnesium, and molybdenum.

3.2.6.1 Nutraceutical and Aquaculture

Aquaculture operations and industry use microalgae as a source of biomass and biomolecules for feed. Aquatic species cultivated in hatcheries or other regulated environments are deficient in nutrients. Nutraceuticals then provide these nutrients for the organisms' healthy growth and improved characteristics. By adding new immunostimulants to aquafeed, shellfish flesh yields have grown, and the likelihood of safer aquaculture production has increased [96, 97]. When shellfish organisms cannot synthesize carotenoids from scratch, they must rely only on their exogenous food [98, 99]. Commercial use of microalgae is dependent on the use of nutraceuticals and nutritional supplements, food coloring, immune enhancers, antioxidants, and polyunsaturated fatty acids [89]. When given carotenoid pigments in their food, aquatic species that are subjected to harsh breeding conditions, such as goldfish, shrimp, lobsters, seabream, and koi carp, get their distinctive muscle color. Astaxanthin and canthaxanthin are carotenoids that give growth and stimulatory effects in farmed fish and shrimp, as well as benefits to animal health and welfare and larval development [100, 101].

3.2.6.2 Marine Nutraceutical

The compounds derived from sea are referred to as marine nutraceuticals. Marine lipids, both microbial and animal derived, marine natural pigments, bioactive marine peptides, enzymes, and vitamins, as well as marine probiotics, are examples of marine nutraceuticals.

Marine Lipids of Animal Origin

The main composition of marine lipid includes phospholipid, triacyl glycerol, wax esters, and their metabolic products and small amounts of glycolipids, sulfolipid, glycerol esters, and hydrocarbons. Fish and other aquatic creatures, such as crustaceans, are the source of marine lipids.

Lipids of Microbial Origin

Lipids also originate from marine microalgae. These microalgal-derived lipids have a variety of uses in aquaculture larval nutrition, mostly for enriching live feed. They have antiviral, anti-inflammatory, anti-allergic, and therapeutic qualities. The presence of various substances, like HUFA and PUFA, provide marine lipids with various properties.

Marine Natural Pigments

Various bioactive compounds are provided by marine microalgae and macroalgae. Natural pigments are an important and an attracting feature of marine algae. Pigmentation and food by photosynthesis are provided by natural pigments of marine algae. Natural pigments are an important marine nutraceutical, which provide certain health benefits.

Vitamins

A large amount of alpha-carotene is present in marine microalgae. Microalgae, such as *I. galbana*, *P. cruentum*, *Arthrospira*, and *Tetraselmis* contain alpha-carotene, which has potent antioxidant properties, as well as vitamins A, K, E, and C.

3.2.7 Conclusion

The sustainable source that fulfills the demand of society for nutritional food is microalgae. Microalgae fulfill the needs of various sustainable industries. Marine microalgae produce a variety of primary and secondary metabolites. Several marine microalgae species, such as *D. salina*, *Arthrospira*, and *Haemotococcus* sp., are utilized to separate these bioactive substances. The metabolites that microalgae generate include proteins, PUFAs, and carotenoids. These metabolites are used for the treatment of inflammation, infection, and cancer; so, they improve the lifestyle of humans. Nutraceutical properties are reported in polysaccharides, pigments, and lipids produced by microalgae.

3.3 Marine Phytoplankton

3.3.1 Introduction

The small photosynthesis organisms, known as marine phytoplankton, are found in practically every kind of marine habitat. The marine world is made up of creatures, including corals, sponges, microalgae, and macroalgae, that have been shown to include marine nutraceuticals like lipid, polysaccharides, chitin, marine natural pigments, marine probiotics, and other associated goods.

From fresh water to extremely high salinity, they can endure a variety of environmental stressors that prolong their growth. The large-scale commercial production of microalgae has generated a lot of interest, and various scientific disciplines are investigating algal biomass. One of the primary sources of chemicals with high biological value is a vast aquatic habitat with an abundance of microalgae. B-carotene, astaxanthin, docosahexaenoic acid (DHA), eicosapentaenoic acid (EPA), bioactive and functional pigments, natural colors, polysaccharides, antioxidants, and algal extract are among the beneficial bioproducts that microalgae create. The market for nutraceuticals in the food supplement industry is growing quickly. Numerous bioactive substances found in macro- and microalgae have therapeutic effects on gastrointestinal disorders, neurological illnesses, and lifestyle disorders. Proteins, peptides, polysaccharides, and vitamins provide information about the role algae play in strengthening the immune system and enhancing the nutritional composition of the body. Furthermore, there is a growing interest in phyco-compounds, such as polysaccharides and polyphenols, for use in cosmeceutical products to shield the skin from UV ray damage. This review has covered the action mechanism of algae and how they can be incorporated into the diet to help manage and prevent chronic diseases like cancer and lung and heart diseases. Ongoing studies in the food supplement industry offer valuable insights into algae, which is also known as a superfood and a new-generation nutritional supplement [102].

3.3.2 Biodiversity of Marine Phytoplankton

Microalgae are unicellular, photosynthetic creatures that typically occur in freshwater, marine habitats, and wet soils. The foundation of the marine food chain, planktonic microalgae, often known as phytoplanktons, generate almost half of the planet's molecular oxygen [2]. Algae are considered to have been among the first living things to perform photosynthesis and contributed to the formation of the present Earth's atmosphere. Since microalgae are primary producers, and hence provide the food base for other marine organisms, they are crucial to the flow of energy throughout the trophic chain [2]. While they can be grown on unproductive land or along coastlines, microalgae, as opposed to higher plants, do not compete with the production of food and feed. Because of these factors, microalgal biomass is garnering interest worldwide to meet the need for the so-called bioeconomy [103]. The diversity of microalgae is vast and represents an intact resource. The scientific literature indicates the existence of 200,000 to several million species of microalgae when compared to approximately 250,000 species of higher plants [104]. The microalgae have a basic morphology and are typically unicellular, with a modest size (generally 5–50 μm). As a result, most species are not seen as individual cells or specimens; instead, they only become noticeable when they form enormous colonies, particularly when they appear as brown, red, black, or green patches on the water's surface. Algal farming can be done along coastlines between 45°N and 30°N, especially in areas south of the Mediterranean, with warmer climates and year-round temperatures that are never below 15°C [105].

3.3.3 Uses of Microalgae

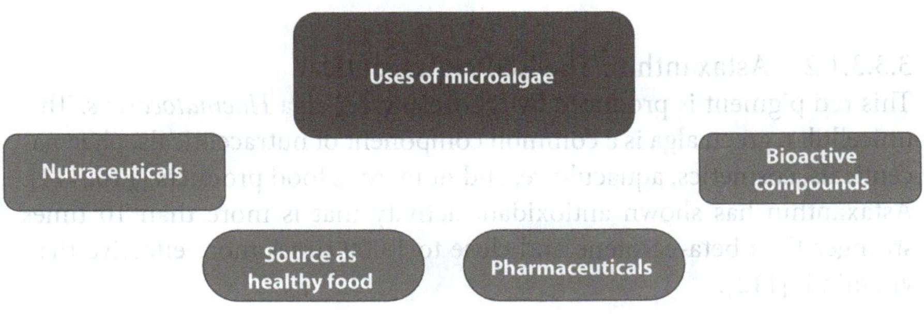

Microalgae have a wide range of industrial applications, in food industries, wastewater purification, and pharmaceutical formulations [106]. Microalgae can also be used for high-value food, health food for humans, polysaccharides, food and fodder additives, cosmetics, antioxidants, anti-inflammatory objects, dyes and feed for aquaculture, and preparation of biofilms [21].

3.3.3.1 As Nutraceutical

Nutraceuticals play a positive role in enhancing health and wellness in the consumers. Hence, health-promoting nutrients are derived from food or food products to enhance the prevention or treatment of a disease and/or disorder [46].

A varied group of autotrophic organisms, microalgae can fix CO_2 from the atmosphere, grow rapidly, utilize light energy efficiently, and produce more biomass per surface than any other natural source. Microalgae's potential to produce a wide range of substances gives them nutraceutical qualities and increases their potential for industrial use [46].

3.3.3.1.1 *Spirulina*: The Original Blue Nutraceutical

Spirulina is probably the best documented example of microalgae as a food, being used by ancient cultures such as the Aztecs in Mexico and the Kanembu tribes in Lake Chad, Africa [107]. Several medicinal businesses produce *Spirulina*. To meet the current demand, this algae is widely cultivated, mostly in specially made raceway ponds and photo bioreactors, and is regarded as a super food. In comparison to carrots, *Spirulina* has up to 10 times more beta-carotene per unit mass and 60%–70% protein by weight, including various amino acids [108]. B vitamins, phycocyanin, chlorophyll, vitamin omega-6 fatty acids, and other minerals are among the many nutrients found in *Spirulina* [109].

3.3.3.1.2 Astaxanthin: The Red Nutraceutical

This red pigment is produced by the freshwater alga *Haematococcus*. This unicellular, green alga is a common component of nutraceuticals, pharmaceuticals, cosmetics, aquaculture, and numerous food products [110, 111]. Astaxanthin has shown antioxidant activity that is more than 10 times stronger than beta-carotene and close to 1,000 times more effective than vitamin E [112].

3.3.3.2 *Source as Healthy Food*

Microalgae, commonly referred to as phytoplankton, are microscopic marine organisms that live in the ocean's benthic and littoral habitats. They consist of diatoms (phylum Ochrophyta, class Bacillariophyceae), dinoflagellates (phylum Myzozoa, class Dinophyceae), green and yellow-brown flagellates (chlorophyta, prasinophyta, prymnesiophyta, and others), and blue-green algae (phylum Cyanobacteria, class Cyanophyceae) [109]. PUFAs and pigments, like carotenoids and chlorophylls, are abundant in microalgae. The pigments of interest in blue-green algae are phycocyanins, phycoerythrin (phycobilins), and chlorophyll a. These compounds have biological properties like immunosuppressive, antifungal, antibacterial, and anticancer properties. In cold, nutrient-rich seas, diatoms—photosynthetic organisms—dominate the phytoplankton. They generate antioxidants like fucoxanthin and chlorophyll, as well as PUFAs like eicosapentaenoic acid (EPA), docosahexaenoic acid (DHA), and other omega-3 FAs [109].

3.3.3.3 *Bioactive Compounds and Their Potential Biological Activities and Roles*

Several researchers have demonstrated in literature that bioactive compounds isolated from various microalgae species possess several pharmaceutical applications, like anti-inflammatory, antimicrobial, and antioxidant activities, and have the capability to improve health and lessen the risk of degenerative diseases [113].

3.3.3.3.1 Antioxidant Activity

In usual physiological and environmental conditions, homeostasis is maintained between producing and degrading reactive oxygen species (ROS) through an antioxidant defense system comprising various antioxidants and antioxidative enzymes. However, disturbances in the favorable circumstances or by factors such as exposure to X-rays, gamma rays or UV rays can overwhelm the antioxidant defense system resulting in an imbalance that generates oxidative stress. Uncertainty in ROS homeostasis can lead to the activation of free radicals' chain reactions ultimately disturbing or damaging cell membranes, cells, tissues, organelles, and generating potentially toxic compounds that cause various diseases' initiation such as cancer, kidney failure, arthritis, Alzheimer's, hastened aging, and Parkinson's [114].

3.3.3.3.2 Anti-Inflammatory Activity

Inflammation is one of the most vital innate defense mechanisms whose long-lasting effect can cause dysfunctions and abnormalities in metabolic pathways leading to various types of renal, cardiovascular, neurodegenerative, cancerous and skin diseases. Microalgae-based bioactive metabolites, such as pigments, modified polysaccharides, lipids, unsaturated fatty acids, vitamins, phenolic compounds, and terpenoids, can have anti-inflammatory properties when used as dietary supplements [115].

3.3.3.3.3 Antimicrobial Activity

In humans, resistance developed against persistent uses of antibiotics led to the necessity to find out innovative antimicrobial compounds. It has been reported in the literature that various bioactive compounds synthesized in microalgae, such as carbohydrates, lipids, fatty acids, sulfur-containing heterocyclic compounds, terpenoids, phenolic acids, and sterols possess antibacterial, antiviral, and antifungal properties. Therefore, microalgae can be an inventive basis for antimicrobial compound production. The first antibacterial compound isolated from microalgae is chlorellin in the 1940s [116].

3.3.3.3.4 Anticarcinogenic Activity

The abnormal and uncontrolled growth of the cells with their ability to invade or expand in other body parts is known as cancer, and it is a persistently growing threat to human health. Cancer is the second major cause of deaths worldwide. It has been reported by the International Agency for Research on Cancer (IARC) that 10 million deaths are caused by cancer, and the number of new cases is expected to increase in the upcoming years. Despite their sometimes-deadly side effects and resistance, chemotherapy and radiation therapy remain the primary methods for treating cancer. Treating a number of illnesses with natural or plant-derived products is a cutting-edge strategy in the field of medication development research. Microalgae are easily cultivated and offer an environmentally friendly approach to drug development. They are also a reliable source of valuable bioactive compounds with medicinal applications [117–120].

3.3.3.4 Microalgae in the Pharmaceutical and Cosmetic Industry

A broad range of nutritional supplements and industrial biomaterials may be derived from algae. For instance, chlorophyll's chelating ability makes it a valuable commodity for the pharmaceutical business, as it may be used to heal liver ulcers. Additionally, published research attributes anticancer, antimutagenic, and antioxidant properties to chlorophyll and its derivatives [121]. Agar, for example, is used in the manufacture of tablets and capsules, surgical lubricants, and various types of emulsions. It is also used as a disintegrating agent and as an excipient in medicinal dragées. In the cosmetics industry, it is used in the production of creams, lotions, lip balms, and deodorants [122]. Seaweed extracts have been used in the cosmetics industry for a long time, and a number of products made from these organisms are available on the market, specifically for skin care (anti-aging creams, regenerative or rejuvenating items). Additional cosmetic uses for microalgae include sun protection and hair care products. An extract of *Chlorella vulgaris* is one example of a commercially accessible substance that promotes tissue regeneration and collagen synthesis in the skin, hence reducing the appearance of aging indications [121]. However, alginates derived from brown macroalgae (Ochrophyta, Phaeophyceae) work well as laxatives because they expand in volume and make intestinal transit easier by absorbing a lot of water as it moves through the colon. These same substances are employed in dentistry to prepare dental molds. Burns are treated using alginate-impregnated bandages because this phycocolloid promotes healing and makes the healing process less uncomfortable. In cases of intoxication from consumption, alginates can also be utilized as neutralizing agents for specific heavy or radioactive elements [121]. Antiviral, antifungal, antibacterial, and anticarcinogenic compounds are not the only uses for microalgae. Numerous novel options have been explored. One recent example is the reported use of microalgae to produce an HIV vaccine. To prevent the release of transgenic material into the environment, a faulty cell approach was employed. The algae can therefore be utilized to make edible vaccinations. The foundation for using algae as a possible vaccine source was also laid by the discovery that some algal components, such as sulfated polysaccharides and bioactive peptides, prevent HIV from entering the body. Smallpox, polio, tuberculosis, and other illnesses, like antihypertensive and anticoagulant disorders, can all be treated with the same method. However, there are still several obstacles for it [123].

3.4 Pharmaceutical Applications of Marine Microbiology

Microalgae is extensively used in the pharmaceutical, nutraceutical, and other industries. Due to the high yield and large diversity of microalgae, they exhibit wide varieties of biological activities because of the production of primary and secondary metabolites. So, they can be used to develop different pharmaceutical products. Secondary metabolites were previously isolated from plants and applied to the medical industry. But later, due to the seasonal variation of plants, microalgae were taken into consideration. The bioactive compound from the microalgae is similar to that of the aquatic community. In addition, they are capable of producing different chemical entities. The microalgae show antiviral, antibacterial, antifungal properties, etc.

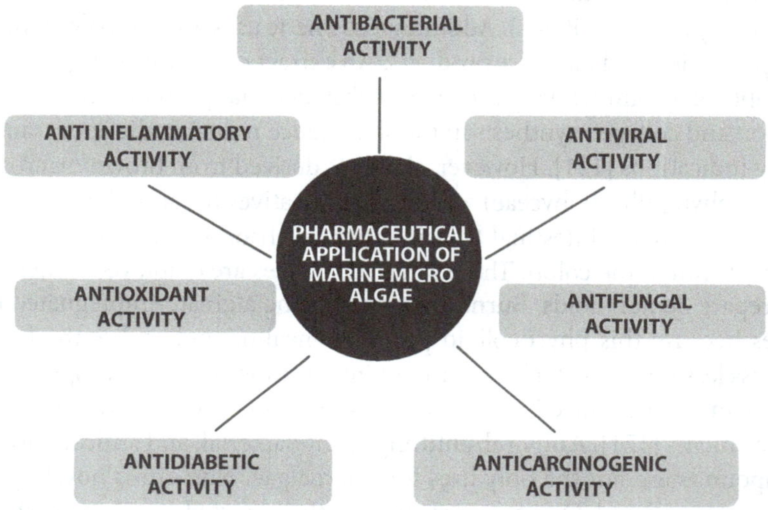

Antibacterial Activity of Microalgae

Antibiotics was discovered by a microbiologist named Alexander Fleming. Antibiotics control the growth or rupture the cell and kill the organism. Due to their overuse in the last decade, there has been a decline in the effectiveness and treatment of bacterial infections. Today, drug resistance is a major problem, and finding a novel antibacterial substance is crucial to combating bacterial diseases [12]. These infections have a great impact on public health and can also be deadly. The resistance mechanism of antibacterials should be monitored [124]. Microalgal cells are utilized as synthetic antibiotics for the composition of food and feed. Additionally, it is administered as a subtherapeutic medication during animal reproduction. Gram-positive and -negative bacteria are inhibited by chlorellin, the first antibacterial chemical identified from the microalgae *Chlorella* [116]. Fatty acids with 10 carbon atoms found in organic extracts extracted from the microalgae *Scenedesmus costatum* prevent aquaculture microorganisms from acting as antibacterial agents. Thus, the degree of unsaturation and length of the chain demonstrate the fatty acids' capacity to infiltrate bacteria in microalgae [125].

Antiviral Activity of Microalgae

There is now a need for an antiviral active chemical that is not exposed to the viral pool because of drug-resistant mutant strains and the development of viruses. Despite the development of numerous antiviral medications, they are unsuccessful. It was identified that the microalgae have several antiviral compounds. According to certain research, the following microalgae exhibit antiviral activity against the viral *hemorrhagic septicema* virus: *Porphyridium pupureum* (Rhodophydata), *Chlorella vulgaris var. autotrophica* (Chlorophyta), *Isochrysis galbana* (Haptophya), and *Dunaliella tertiolecta* (Chlorophyta). Both *P. purpureum* and *C. vulgaris var. autotrophica* extracts have the capacity to suppress the African swine fever virus. Damonte *et al.* [126] identified that the cell walls of brown algae have sulfated polysaccharides. They demonstrated a wide range of antiviral activity against African swine fever disease (AFSV), HIV-1, and the viral *hemorrhagic septicemia* virus (VHSV). This prevents both dengue virus (DENV) [128] and *human papillomavirus* (HPV) infection [127]. The *encephalomyocarditis* virus is inhibited by the homopolysaccharide of galactose with uronic acid and sulfate groups found in the microalgae *Gyeodinium impudicun* [129]. Encephalomyocarditis virus is an acute form of virus, and piglets can die suddenly from the acute version of this disease.

Antifungal Activity of Microalgae

There are various pathologies caused by fungi that are problematic because of their therapy. The main property of fungi that causes the pathogenicity is its dimorphism. According to dimorphism, when the temperature of the human host rises, the algae can exist in mycelial form and transform into fungi. A well-advised introduction of antifungal alga occurred during a deadly fungal infection. A study was conducted by Katircioglu *et al.* for the development of infection of three different yeasts by using 10 microalgal strains. The strains were gathered from Turkey's freshwater reservoir to test the microalgae's antifungal qualities. It was determined that the antifungal properties of *Chlorella* and *Oscillatoria* had been achieved [130]. The effectiveness of methyl lactate and butanoic acid against *C. albicans* was debated using pressured liquid extracts of *H. pluvialis*, which exhibits antifungal properties [131, 132]. *S. quadricauda* and *C. pyrenoidosa* were studied in relation to eight distinct fungi. This indicates that the antifungal properties of these microalgae extend to all of the fungi under investigation [133].

Anticarcinogenic Activity of Microalgae

Cancer is a disease in which the cells grow abnormally and uncontrollably by invading or expanding to other body parts. It is a serious danger to the medical field. According to reports, it is the second leading cause of death worldwide. The International Agency for Research on Cancer (IARC) reports that over 10 million people have died from cancer [117]. There will be increase in the rate of mortality in the upcoming future. Mutations, inactivation of tumor-suppressor genes, and overexpression of growth-promoting oncogenes are the primary genetic causes of cancer. The oncogenes that are overstimulated include EGFR, ERBB2, CDK4, KRAS, and BCL2. The main treatments against cancer are radiotherapy and chemotherapy. Microalgae have different biologically active metabolites, which include fucoxanthin, astaxanthin, omega-3 fatty acid, sulfated polysaccharides, and peptides which have antitumor properties. Microalgae are also a rich source of carotenoids. The carotenoids consist of astaxanthin, violaxanthin, lutein, and beta-carotene. Beta-carotene [134] suppresses the M2 macrophages, which help in the metastasis and progression of tumors. When animals with colon cancer are given beta-carotene twice a week for 11 weeks *in vitro*, the migration and intrusion of HCT116 (colon cancer cells) are decreased, which slows the growth of the tumor [135]. Metalloproteinase in LoVo colon cancer cells is likewise inhibited by beta-carotene. In HeLa and MCF7 cell lines, C-*phycocyanin*, which is isolated from *Spirulina platensis*, exhibits anticancer properties [136]. Antitumor chemicals that function as anticancer drugs have been identified in *Chaetomorpha* sp. [137]. *Borophycin* and *cryptophycin* 1, which are isolated from Nostoc, exhibit antitumorigenic properties against human tumor cells, including KB and LoVo cell lines [138]. Methanolic extract, which is extracted from *H. pulvialis* also showed anticancerous activity as they suppress invasion of breast cancer cells and also promote apoptosis. Both polyketide and alkaloid neurotoxins, such as brevetoxin and saxitoxin, have anticancer properties [139]. *Cymopolia barbata*, a marine green algae, yielded the following two novel chemicals that can prevent mutagenicity: *cymobarbatol* and 4-isocymobarbatol.

Antidiabetic Activity of Microalgae

Diabetes is associated with high blood pressure, heart problems, kidney failure, and neurological disorders. It was estimated that 463 million people had diabetes in 2019. The number is expected to increase by roughly 578 million by 2030 and 700 million by 2045 [140]. The condition called diabetes mellitus has extremely increased the mortality rate and morbidity in many countries. Microalgae have various bioactive compounds that help in new therapies and hypoglycemic agents in the case of diabetes [141]. These bioactive compounds are screened and then evaluated to develop antidiabetic drugs. Hypoglycemic effects have been found in *Chlorella sorokiniana* [142], *Chlorella pyrenoidosa* [143], and *Chlorella vulgaris* [144]. In a study done on streptozotocin-induced mice, there was an increased hypoglycemic effect of insulin, which is supplied exogenously due to the liquid extract of *Chlorella vulgaris* [145]. When water-soluble *Spirulina* extract was added to diabetic rats, a hypoglycemic effect was also observed [146]. Additionally, research has shown that *Nannochlorpsis gaditana* possesses antioxidant and antidiabetic qualities. This stops oxidative damage and changes the metabolic processes associated with diabetes [147]. A study also conducted on *Spirulina* showed that it had antihyperglycemic and antihyperlipidemic effects [148]. Palinurin is a sesquiterpene found in isopropanol extracts of *Ircinia* dendroides that inhibits the glycogen synthase kinase 3β (GSK-3β) enzyme, which is present when diabetes first appears [149]. The isolated red algae chemical 3,4-dibromo-5-(2-bromo-3,4-dihydroxy-6-(isopropoxymethyl)benzyl)benzene-1,2-diol (HPN) aids in the dose-dependent decrease in blood lipids, insulin, HbA1c, and glucose levels. HPN helps in the development of novel antidiabetic drugs, even though antidiabetic drugs are not still available on the market.

Antioxidant Activity of Microalgae

An antioxidant defense system that produces and breaks down reactive oxygen species and contains a variety of antioxidants and antioxidative enzymes maintains homeostasis. The antioxidant defense system is disrupted by exposure to X-rays, gamma rays, or ultraviolet radiation, which results in oxidative stress. The unreliability in the ROS homeostasis causes the activation of free radical chain reaction that damages the cell membrane, cells, organelles, and tissues. This results in the production of numerous harmful substances and causes illnesses like cancer, kidney failure, arthritis, Alzheimer's, Parkinson's, and more [114]. Numerous secondary metabolites that are antioxidants and beneficial to health can be produced by microalgae. Phenolic acids are abundant in *Neochloris oleobundans* and *Wilmottia murravi*. Lipid-soluble substances are abundant in *Anisancylus obliquus*. Moreover, *Phormidium* sp. contains a lot of ascorbic acid. *Prymnesium* simplex has bioactive compounds having antioxidant properties [150]. *Porphyra* has mycrosporine amino acids, which are effective biochemical agents for blocking UV radiation [151]. The antioxidant-sulfated polysaccharides found in *Porphyra haitanensis, Chlorella vulgaris*, and *Scenedesmus quadricauda* are utilized in food, medicine, and stabilizers [152]. *Messastrum gracile, ParaChlorella kessleri, Chlorella vulgaris*, and *AuxenoChlorella pyrenoidosa* possess high ferric-reducing antioxidant capability, which is also a characteristic of *Desmodesmus subspicatus*.

Table 3.1 The health benefits and mode of action of various microalgae.

Health beneficial application	Microalgae	Bioactive compound	Mode of action	Reference
Antioxidant activity	*Chlorella* vulgaris	Phenolics and flavanoids	Radicle scavenging activity	[158]
	Dunaliella salina	Beta-carotene	Inhibits ROS production	[159]
	Haematococcus pluvialis	Astaxanthin	Antioxidant activity	[160]
	Scenedesmus sp.	Lutein	Maintains homeostasis	[161]
	Porphyridium sp.	Sulfated polysaccharides	Inhibit ROS formation	[162]
	Chlorella vulgaris	Phenolics	Maintain homeostasis	[163]
	Nannochloropsis oculata	Peptides	Antioxidant activity	[164]
	Spirulina maxima	Phenolic compounds	Radicle scavenging activity	[165]
	Nannochloropsis sp.	Phenolics	Radicle scavenging activity	[166]
	Gymnodinium mikimotoi	Monogalactosyl diacylglycerol	Inhibits ROS production	[167]
	Stephanodiscus sp.	Digalactosyl diacylglycerol	Enhances cell differentiation	[168, 169]
	Scytosiphon lomentaria	Protein extract	Inhibits ROS production	[170]
Anti-inflammatory activity	*Spirulina platensis*	Phycocyanin	Suppresses IL-6, 8 and increases TGF-b1 production, inhibits COX-2 expression	[171, 172]
	Chlorella vulgaris	Protein hydrolysates	Modulate cytokines	[173]
	Chlorella vulgaris	Sterols	Anti-inflammatory activity	[174]
	Anabaena cylindrical	Vitamin K	Prevent from toxic agents/pollutants	[175]
	Chlorococcum sp. and *Scenedesmus obliquus*	Linolenic acid	Anti-inflammatory activities and reduces acne	[176]

(Continued)

Table 3.1 The health benefits and mode of action of various microalgae. (*Continued*)

Health beneficial application	Microalgae	Bioactive compound	Mode of action	Reference
	Chlorella vulgaris and *Chlorococcum* sp.	Canthaxanthin	Anti-inflammatory actions	[177]
	Chlorella zofigiensis and *Dunaliella salina*	Astaxanthin	Anti-inflammatory properties	[178]
	Chlorella sp., *Scenedesmus* sp., and *Spirulina* sp.	Phycobiliproteins, hormone-like bioactive peptides	Inhibits inflammation	[179]
	Nannochloropsis oculata	Docosapentaenoic acid (DPA)	Inhibits pro-inflammatory prostaglandin E2, NO, TNF-a, IL-6	[180]
	Chlorella stigmatophora	Sulfated polysaccharide	Immunosuppressive activity	[181]
Antibacterial activity	*Skeletonema costatum*	Unsaturated, saturated long-chain fatty acids	Acts on *Vibrio* sp.	[182]
	Euglena viridis	Organic extracts	Acts on *Pseudomonas, Aeromonas, Edwardsiella, Vibrio,* and *Escherchia coli*	[183]
	Dictyosphaerium pulchellum	Methanolic extracts	–	[184]
	Chlorococcum sp.	Aqueous extracts	–	[184]
	Chlorella vulgaris	Methanolic and hexanolic extracts	Acts on *Staphylococcus aureus, Staphylococcus epidermidis, Bacillus subtilis, Escherichia coli,* and *Salmonella typhi*	[185]
	Chlorella pyrenoidosa	Saturated and unsaturated fatty acids	Acts on *Propionibacterium acnes*	[186]

(*Continued*)

Table 3.1 The health benefits and mode of action of various microalgae. (*Continued*)

Health beneficial application	Microalgae	Bioactive compound	Mode of action	Reference
	Chlorella humicola	Pigments (carotenoid, chlorophyll)	Acts on *Bacillus subtilis*, *Staphylococcus aureus*, *Escherichia coli*, *Pseudomonas aeruginosa*, *Salmonella typhimurium*, *Klebsiella pneumonia*, and *Vibrio cholera*	[187]
	Scenedesmus obliquus	Long-chain fatty acids	Acts on *Staphylococcus aureus*, *Escherichia coli*, *Pseudomonas aeruginosa*, and *Salmonella* sp.	[188]
	Rhodella reticulata	Extracellular sulfated polysaccharides	Acts on *Staphylococcus aureus*, *Streptococcus pyogenes*, *Bacillus cereus*, and *Salmonella typhimurium*	[189]
Antiviral activity	*Navicula directa*	Polysaccharide	Inhibits hyaluronidase	[190]
	Chlorella autotrophica	Sulfated polysaccharides	Inhibit replication	[191]
	Cochlodinium polykrikoide	Extracellular sulfated polysaccharides	Inhibition of cytopathic effect	[192]
	Porphyridium cruentum	Highly sulfated polysaccharide	HSV-1, HSV-2, and vaccina	[152, 193]
	Navicula directa	Sulfated polysaccharide	HSV-1, HSV-2, and influenza-A	[190, 194]
	Gyrodinium impudicum	p-KG03 exopolysaccharide	Inhibition of cytopathic effect	[195]
Antifungal activity	*Chlamydomonas reinhardtii*	Methanolic extracts	Act on *Candida kefir*, *Aspergillus niger*, and *Aspergillus fumigatus*	[185]
	Amphidinium sp.	Karatungiols	Act on *Aspergillus niger* and *Trichomonas foetus*	[196]

(*Continued*)

Table 3.1 The health benefits and mode of action of various microalgae. (*Continued*)

Health beneficial application	Microalgae	Bioactive compound	Mode of action	Reference
	Prorocentrum lima	Polyether compounds	–	[184]
	Haematococcus pluvialis	Butanoic acid and methyl lactate	Act on *Candida albicans*	[197]
Anticancerous activity	Botryococcus braunii and Microcystis aeruginosa	Lipids	Cytotoxic activity against prostate, SHSY-5Y neuroblastoma, and AGS gastric adenocarcinoma cell line	[198]
	Chlorella pyrenoidosa	Polypeptide CPAP	Induces apoptosis	[199]
	Chlorella vulgaris	Polypeptide	Antiproliferative activity and arrests cell cycle at G1 phase	[200, 201]
		Sterols	Stabilize phospholipid bilayers	[202]
	Pavlova lutheria	Phytosterols	Inhibit colon cancer development	[203]
	Chaetoceros calcitrans and Dunaliella tertiolecta	Vitamin D	Inhibits prostrate cell cancer growth	[204]
	Chlorella pyrenoidosa	Mycosporine-like amino acid (MAA)	Inhibits cancerous cell growth	[205]
	Chlorella stigmatophora	Polysaccharides	Tumoricidal activity	[206]
	Chlorella sp.	Chlorophyll	Cytotoxic activity toward tumoral cells	[207]
	Spirulina limacinum, *Spirulina* bacillaris, and *Chlorella* cohnii	Phenolics	Cytotoxic special effects against human hepatocellular liver carcinoma cells	[208]

(*Continued*)

Table 3.1 The health benefits and mode of action of various microalgae. (*Continued*)

Health beneficial application	Microalgae	Bioactive compound	Mode of action	Reference
	Scenedesmus sp., *Chlorella* pyrenoidosa, and Chlorococcum sp.	Exopolysaccharides	Inhibit human colon cancer cell line proliferation	[209]
	Dunaliella tertiolecta	Violaxanthin	Antineoplastic effects	[210]
	Isochrysis galbana	(1 ? 3, 1 ? 6)-b-D-glucan-sulfated exopolysaccharide	Cytotoxic against lymphoma cells	[211]
	Arthrospira platensis	Extracellular polysaccharide	Cytotoxic against kidney and colon cancer cell line	[212]
Antihypertensive and antihyperlipidemic activity	Nannochloropsis oculata	Protein extract	Inhibits angiotensin I-converting enzyme	[164]
	Chlorella vulgaris	Polypeptide	Inhibits angiotensin I-converting enzyme	[213]
	Nanochloropsis sp.	Sterols	Reduce blood cholesterol levels in hyper- and normocholesterolemia	[174]
	Chlorella, Spirulina, and Pavlova	Vitamin B	Reduces cholesterol	[214]
	Chlorella sp. and Dunaliella sp.	Glycoprotein	Inhibits angiotensin I activities and reduces LDL-cholesterol	[215, 216]
	Chlorococcum sp., Dunaliella primolecta, and *Spirulina* sp.	c-Linoleic acid	Reduces blood pressure and prevents heart diseases	[217, 218]
	Chlorella vulgaris and Nannochloropsis sp.	Eicosapentaenoic acid	Regulates blood pressure, reduces blood clotting, and prevents heart diseases	[219]
	Chlorella sorokiniana and Chlorococcum sp.	Beta-carotene	Declines low-density lipoprotein	[76]

(*Continued*)

Table 3.1 The health benefits and mode of action of various microalgae. (*Continued*)

Health beneficial application	Microalgae	Bioactive compound	Mode of action	Reference
	Spirulina platensis and *Chlorella* vulgaris	Protein extracts and hydrolysates	Inhibit activity of angiotensin I-converting enzyme and acetylcholinesterase	[220]
Antidiabetic activity	Porphyridium sp.	Liquid extract	Inhibits a-glucosidase	[221]
	Nannochloropsis sp. and Pavlova salina	Polyunsaturated fatty acids	Antihyperglycemic, reduce blood glucose	[222]
	Dunaliella salina and Isochrysis sp.	Fucoxanthin and zeaxanthin	Suppresses hyperglycemia, prevents diabetic diseases	[223]
	Chlorococcum sp. and *Chlorella* fusca	Lutein	Prevents diabetic retinopathy	[224, 225]
	Chlorella vulgaris and Haematococcus pluvialis	Astaxanthin	Reduces diabetes	[226]
	Cosmospora sp. SF5060	Aquastatin A	Hypogluycemic activity	[227]
	Phaeodactylum tricornutum	Fucoxanthin	Suppresses hyperglycemia	[228]

Anti-Inflammatory Activity of Microalgae

Inflammation is an innate defense mechanism that can cause dysfunction and abnormalities in the metabolic pathway, which cause cardiovascular, renal, cancerous, neurodegenerative, and skin diseases. The bioactive metabolites present in microalgae include pigments, lipids, modified polysaccharides, vitamins, unsaturated fatty acid, and terpenoids, which have anti-inflammatory properties. Studies conducted on rats suffering from the autoimmune disease acephalomyelitis showed that the introduction of phycocyanin caused downregulation of IFN-γ and IL-6 expression [153]. The microalgae *Spirulina platensis* have a high content of phyacocyanin [154]. Astaxanthin produced from *Haematococcus* decreases the levels of proinflammatory cytokines, improves phagocytosis, and reduces oxidative damage [155]. Polysaccharides extracted from *Chlorella stigmatophora* and *Phaeodactylum tricorutum* have immunosuppressive and immunostimulatory effects [156]. The bioactive metabolites obtained from microalgae can reduce inflammation by inhibiting the expression of genes that promote inflammation, delaying the synthesis of cytokines and eicosanoids, changing the activity of enzymes like phospholipase A2, cyclooxygenase-2 (COX2), and nitric oxide synthase (NOS), varying cellular activity, and disrupting signaling pathways [157]. The Table 3.1 shows the health benefits and mode of action of different microalgae [278].

3.5 Nutraceutical Application of Marine Microalgae (Phytoplanktons)

A nutraceutical is a broad term used to describe products derived from food sources that offer extra health benefits in addition to their basic nutritional value. The term combines "nutrition" and "pharmaceutical" reflecting the idea that these substances have a role in maintaining health and preventing or treating disease. Nutraceuticals can be part of a regular diet or taken as dietary supplements. They include a variety of products, such as vitamins, minerals, herbal products, and other botanicals, amino acids, and certain components of foods, such as phytochemicals and probiotics. Generally, nutraceuticals play a positive role in enhancing health and wellness in the consumers. Hence, health-promoting nutrients that are derived from food or food products to enhance the prevention or treatment of a disease and/or disorder [46] are also considered nutraceuticals.

Microalgae as Nutraceuticals

Microalgae are a diverse group of photosynthetic organisms capable of rapid growth and efficient utilization of light energy. They are highly effective at fixing atmospheric CO_2 and producing significant biomass making them valuable for various applications. These organisms can synthesize a wide range of compounds giving them substantial nutraceutical properties with high industrial potential [46]. Various types of microalgae, such as *Chlamydomonas*, *Scenedesmus*, *Synechococcus*, *Dunaliella*, *Haematococcus*, *Chlorella*, and *Porphyridium*, are currently used or have potential for use in commercial nutraceuticals. Microalgae can accumulate essential elements, like potassium, zinc, iodine, selenium, and others, which can be incorporated into nutraceutical products. Some microalgae produce secondary carotenoids, such as canthaxanthin and astaxanthin, which are stored in lipid globules within their chloroplasts. They are also rich in essential amino acids and omega-6 and -3 fatty acids (e.g., arachidonic acid, DHA, and EPA) [229]. The market for algae-derived nutraceuticals is lucrative due to the high concentration of health-promoting molecules and nutrients found in these organisms. Microalgae can harness solar energy to convert wastewater, CO_2, and nutrients into biomass rich in lipids, sugars, proteins, and other organic compounds. This biomass is valuable for producing a variety of nutraceutical products [236]. Microalgae offer a rich array of nutrients that support human and animal health. They are found globally in diverse environments highlighting their vast potential as nutraceutical sources. Their unique properties and ability to produce beneficial compounds have significant advantages for human health, including critical nutrients like phycobilins, phycoerythrins, carotenoids, micronutrients, amino acids, and carbohydrates.

Nutraceutical Applications of Marine Microalgae

Marine microalgae have significant potential in the nutraceutical industry due to their rich nutrient profiles and bioactive compounds. The following are some key nutraceutical applications of marine microalgae:

- Omega-3 fatty acids
- Antioxidants
- Protein supplements
- Vitamins and minerals
- Phycobiliproteins
- Polysaccharides
- Carotenoids
- Detoxification
- Prebiotics

Omega-3 Fatty Acid

There were significant efforts that have focused on utilizing algal oils rich in long-chain polyunsaturated fatty acids (LCPUFAs) as nutraceuticals. Among these, omega-3 LCPUFAs, such as docosahexaenoic acid (DHA) and eicosapentaenoic acid (EPA), have gained prominence [230]. DHA, in particular, is an omega-3 LCPUFA with 22 carbon atoms and six methylene-interrupted *cis*-double bonds (22:6). It is a major fatty acid in neurological tissues making up 20%–25% of the total fatty acids in the gray matter of the human brain and 50%–60% in the retina rod outer segments [230]. DHA is also abundant in heart muscle tissues and sperm cells. The limited storage of n−3 fatty acids, like DHA in mammalian adipose tissues, indicates that an external dietary source is necessary [231]. Thus, it is essential to obtain adequate supplies of DHA from dietary sources. Fish and fish oils have long been recognized as the primary sources of LCPUFAs due to their high DHA and EPA content. Microalgae serve as an alternative source since, like humans, fish acquire much of their LCPUFAs from their diet, primarily from the ocean's primary producers [232]. A significant advantage of microalgae over fish oils is the absence of an unpleasant odor. These fatty acids, characterized by methylene-interrupted double bonds, are crucial for normal cell function. The understanding of these fatty acids' roles in obesity and cardiovascular diseases has led to their inclusion as nutraceuticals [233, 234]. For humans and animals, ensuring an adequate dietary intake of DHA is important because its precursor, linolenic acid, is insufficient [235]. Docosahexaenoic acid is also considered essential during infancy, with breast milk being a natural source for infants. Several algae, especially diatoms, such as *Nitzschia* sp., *Navicula* sp., and *Phaeodactylum*, have been suggested for EPA production. EPA, a 20-carbon LCPUFA with five double bonds (20:5), plays a critical role in cellular and tissue metabolism regulating membrane fluidity, electron and oxygen transport, and thermal adaptation [236]. The PUFA family, particularly eicosapentaenoic acid (EPA, C20:5D5,8,11,14,17, 20:5 ω-3), is well studied and considered effective. EPA's five double bonds include one at the third carbon from the methyl (ω) end [237]. EPA forms complex lipid molecules and is a key precursor to eicosanoids, hormone-like substances, such as prostaglandins, thromboxanes, and leukotrienes, which are vital for regulating development and health.

Antioxidants

Marine microalgae are rich in antioxidants, including carotenoids (e.g., astaxanthin, beta-carotene) and phycobiliproteins (e.g., phycocyanin, phycoerythrin). These antioxidants protect cells from oxidative damage potentially reducing the risk of chronic diseases such as cancer and cardiovascular diseases.

Beta-Carotene

Dunaliella salina, a microalgae thriving in hypersaline coastal waters, is the richest natural source of the carotenoid beta-carotene. It can be cultivated on a large scale in open raceway ponds. The production of beta-carotene by *D. salina* increases significantly under high salinity and nutrient deprivation conditions turning the green cells red due to carotenoid accumulation. Beta-carotene, which comprises 95% of the dry weight of these pigments, is stored in the chloroplast's interthylakoid spaces [238]. Beta-carotene is utilized in nutraceuticals as a non-toxic precursor to vitamin A, found in multivitamin and specialty formulations [98]. Due to its tolerance to high salinity (up to 100 g NaCl/L), *Dunaliella* strains do not compete with agricultural or domestic water sources. *Dunaliella* produces several carotenoid pigments, primarily beta-carotene, which can make up 14% of its dry weight, along with smaller amounts of alpha-carotene, lutein, and lycopene [239]. As an antioxidant, beta-carotene can neutralize reactive oxygen species involved in aging [240, 241]. Research indicates that beta-carotene may help prevent cancers in various organs due to its antioxidant properties. It has also shown positive effects on intracellular communication, immune response [242], and protection against many types of cancers [243]. Supplements of *Dunaliella* have demonstrated hepatoprotective effects and reduced liver lesions [244]. Despite these benefits, over 90% of commercial beta-carotene is produced synthetically. However, naturally derived beta-carotene from *Dunaliella* has higher bioavailability compared to its synthetic counterpart [245]. Naturally produced beta-carotene also contains higher levels of antioxidant enzymes like catalase, peroxidase, and superoxide dismutase. Consuming *Dunaliella* supplements is safe, as evidenced by a study where rats fed with a 10% *Dunaliella* diet showed no significant negative effects underscoring its safety for human consumption [246].

Astaxanthin

Astaxanthin is a red pigment produced by the freshwater alga *Haematococcus*, a unicellular green alga commonly used in nutraceuticals, pharmaceuticals, cosmetics, aquaculture, and various food products [110, 111]. This pigment has demonstrated innovative anti-inflammatory and antioxidant benefits in human nutrition. *Haematococcus*, the largest natural source of astaxanthin, contains 1.5%–3% astaxanthin of its dry weight [86, 247]. Astaxanthin's antioxidant activity is over 10 times stronger than beta-carotene and nearly 1,000 times more effective than vitamin E [112, 248]. Research has shown that this pigment can reduce arterial blood pressure, plasma triglyceride levels, and nonesterified fatty acids [96]. Astaxanthin's metabolic functions in humans include protecting polyunsaturated fatty acids (PUFAs) from oxidation, shielding against ultraviolet radiation, and enhancing vision, immune response, pigmentation, and reproductive behavior [249]. Studies have highlighted astaxanthin's antioxidative [250], anticancer [251], anti-inflammatory, and antibacterial properties. There are no reports of negative effects from using *Haematococcus* as a dietary supplement. In rat studies, no adverse effects were observed at consumption levels of 5–18 g/kg/day [252]. Human studies found no significant differences over 8 weeks, and doses of 20 mg/day for 4 weeks showed no adverse effects on blood biochemistry or hematology [253]. Since 1999, *Haematococcus pluvialis* has been considered safe by the FDA and approved as a new dietary ingredient in the United States [254].

Protein Supplements

Microalgae, such as *Spirulina* and *Chlorella*, are excellent sources of high-quality protein containing all essential amino acids. They are beneficial for muscle repair, immune function, and as a dietary supplement for vegetarians and vegans.

Spirulina

Spirulina is a well-documented example of microalgae used as food. This cyanobacterium uses photosynthesis to convert sunlight into proteins, fatty acids, carbohydrates, and nearly all essential nutrients. Its blue color comes from the pigment phycocyanin (PC), which is a light-harvesting pigment found in *cyanobacteria* and certain algae genera, *Rhodophyta* and *Cryptophyta*. The blue hue of cyanobacteria is due to PC, and the phycobiliproteins in this cyanophyte, including allophycocyanin (APC) and c-phycocyanin (CPC), have applications such as nutraceuticals, colorants in food, and fluorescent labels. These proteins contain a tetrapyrrolic pigment called phycobilin, which is covalently attached to their structure. *Spirulina* is known as a supplementary food rich in nutrients such as B vitamins, phycocyanin, chlorophyll, vitamin E, omega-6 fatty acids, and numerous minerals [107, 255]. Depending on the culture conditions, this cyanophyte can accumulate approximately 60%–70% protein by weight, including many amino acids and beta-carotene. Commercially available *Spirulina* is high in protein, essential amino acids, gamma-linolenic acid, alpha-linolenic acid, linoleic acid (LA), stearidonic acid, EPA, DHA, and arachidonic acid (AA). It is also a good source of potassium, calcium, chromium, copper, iron, magnesium, manganese, phosphorus, selenium, sodium, and zinc. Commercial *Spirulina* is marketed to promote weight loss, managing diabetes, high blood pressure, and hypertension [256]. There are reports of its antiviral and anticancer properties [257]. *Spirulina* is used for weight loss and to control diabetes, high blood pressure, and hypertension. It has been reported to help lower cholesterol by increasing HDL levels promoting cardiovascular health [258]. The antioxidant and anti-inflammatory properties of C-phycocyanin in *Spirulina* have been demonstrated in mice and humans [259]. Research has also investigated *Spirulina*'s potential to improve digestion, food absorption, and enhance the immune system to fight infections [260]. Despite its benefits as a nutraceutical, questions remain about the assimilation, digestion, and bioconversion capacity of *Spirulina*'s nutrients and the antagonistic effects of other nutrients in *Spirulina* extracts or meals. Efficient processing and engineering techniques are needed to make its nutrients more digestible for humans. Potential side effects of *Spirulina* consumption include allergic reactions, diarrhea, nausea, and vomiting.

Chlorella

Chlorella, a green cellular microalga, is commonly marketed as a health food, dietary supplement, and nutraceutical [261]. In the Far East, it has been used traditionally as an alternative medicine for centuries. In regions, like China and the Orient, *Chlorella* is viewed similarly to a nutraceutical. Due to its nutrient content and health benefits, *Chlorella* is considered a significant functional food and nutraceutical. It comprises 55%–60% protein, 1%–4% chlorophyll, 9%–18% dietary fiber, along with various minerals and vitamins [262]. Studies show that *Chlorella* contains all essential amino acids necessary for the nutrition of heterotrophic organisms. Its ability to detoxify metals and pesticides is linked to the porphyrin rings in chlorophyll or the glutathione pathway induced by vitamin B12. Additionally, *Chlorella* is rich in lutein, which is associated with the prevention and treatment of macular degeneration [263]. It can also help lower cholesterol and reduce blood pressure significantly decreasing low-density lipoprotein and cholesterol levels. However, like other microalgae, *Chlorella* can have side effects when consumed as a nutraceutical [264]. These side effects can vary between commercial brands and include gastrointestinal issues, nausea, and vomiting. *Chlorella* is classified as a weak allergen.

Vitamins and Minerals

Marine microalgae are rich in essential vitamins (e.g., B vitamins, vitamin C, and E) and minerals (e.g., iron, magnesium, calcium, and zinc). For instance, *Chlorella* is known for its high vitamin B12 content, which is particularly important for vegans [84].

Phycobiliproteins

Phycobiliproteins found in microalgae, like *Porphyridium*, have antioxidant, anti-inflammatory, and immune-boosting properties. These compounds are useful in developing supplements to enhance immune function and reduce inflammation [265].

Polysaccharides

Marine microalgae produce various bioactive polysaccharides, such as sulfated polysaccharides, which have antiviral, antibacterial, and anti-inflammatory properties. These compounds can be incorporated into nutraceutical products to support immune health and combat infections. Marine microalgae are recognized as a rich source of bioactive compounds [266]. The polysaccharides they produce, whether sulfated or not, have shown potential in various pharmacological applications. These molecules have been noted for their immunomodulatory, anti-inflammatory, antitumor, prebiotic, anticoagulant, antibacterial, hypoglycemic, antioxidant, and skin tissue regeneration-enhancing properties [152]. Microalgae polysaccharides have demonstrated significant utility in the food industry through their roles as emulsifying agents, gelling agents, flocculants, stabilizers, thickeners, and even edible protective films [267–269]. Furthermore, prebiotics have been shown to influence gut microbiota beneficially [270, 271]. Nonetheless, the application of microalgae polysaccharides in the food sector is still in its early stages, primarily due to the strong competition from polysaccharides derived from terrestrial plants, macroalgae, and other microorganisms [272].

3.6 Conclusion

Marine microalgae are microscopic organisms found in the ocean that perform photosynthesis. They are crucial to marine ecosystems and global biogeochemical cycles. Microalgae are photoautotrophic organisms with different phyla such as Cyanophyta, Rhodophyta, Haptophyta, Streptophyta, and Heterokondophyta. Primary and secondary metabolites produced in microalgae are complex organic compounds accumulated by microalgae with the help of H_2O, CO_2, and energy from sunlight. They are adapted to different environmental stress conditions that extend their growth ranging from freshwater to extreme salinity [273]. Microalgae are autotrophic organisms that can grow in a wide range of habitats. To survive under different conditions, different microalgae produce different types of secondary metabolites, which have beneficial therapeutic activities like anti-cancerous and anti-microbial. These secondary metabolites of microalgae are referred to as "high-value molecules" (HMV). Microalgae have the ability to co-produce other compounds, like pigments, proteins, polyunsaturated fatty acids, and antioxidants, and are attracting interest as therapeutic agents for a variety of health conditions as well as additional uses in the food, cosmetics, energy, and pharmaceutical industries [274]. In the pharmaceutical industry, marine microalgae produce a variety of antioxidants, such as carotenoids (e.g., astaxanthin), polyphenols, and vitamins. These compounds help in combating oxidative stress and may be useful in preventing chronic diseases like cancer, cardiovascular diseases, and neurodegenerative disorders.

Several species of marine microalgae produce bioactive molecules with anticancer properties. For instance, fucoxanthin from brown algae exhibits significant antiproliferate effects on cancer cells by inducing apoptosis and inhibiting cell cycle progression. Marine microalgae contain compounds, such as omega-3 fatty acids (e.g., EPA and DHA), that possess anti-inflammatory properties. These can be used in the treatment of inflammatory diseases, including arthritis, inflammatory bowel disease, and other chronic inflammatory conditions [275]. Microalgae produce a range of antimicrobial compounds effective against bacteria, viruses, fungi, and protozoa. For example, sulfated polysaccharides from red algae have shown antiviral activity against viruses, such as the herpes simplex virus (HSV) and human immunodeficiency virus (HIV). Bioactive compounds from marine microalgae, such as phycocyanin from *Spirulina*, have neuroprotective effects. These compounds can potentially be used in the treatment of neurodegenerative diseases, like Alzheimer's and Parkinson's disease, by reducing oxidative stress and inflammation in neural cells [255]. Certain microalgae produce compounds that can help manage lipid levels in the blood. For example, the omega-3 fatty acids found in microalgae can reduce triglyceride levels and may have a beneficial effect on overall lipid profiles, thus reducing the risk of cardiovascular diseases [276].

Marine microalgae can act as immune modulators enhancing or suppressing the immune response as needed. Compounds, like β-glucans and other polysaccharides, can stimulate the immune system potentially aiding in the defense against infections and in cancer immunotherapy. Marine microalgae are a source of several antiviral compounds. For example, the sulfated polysaccharides from red algae have shown inhibitory effects on HIV, the herpes simplex virus, and influenza. These compounds interfere with viral replication and attachment providing a basis for developing new antiviral medications. Compounds derived from marine microalgae, such as polyphenols and fatty acids, exhibit antifungal properties. These substances can be used to develop treatments for fungal infections, including those caused by *Candida* species and dermatophytes, which affect skin, nails, and mucous membranes. Certain marine microalgae produce metabolites with antiparasitic activity. For example, extracts from microalgae, like *Spirulina*, have been investigated for their potential to combat parasites, such as *Plasmodium*, the causative agent of malaria. These compounds could be developed into new antiparasitic drugs. Bioactive substances from marine microalgae, such as alginates and carrageenans, have been utilized in wound healing applications. These polysaccharides possess properties that promote tissue regeneration, reduce inflammation, and provide a moist environment conducive to healing. They are used in dressings and other wound care products. Marine microalgae produce compounds that have analgesic properties. For example, certain extracts from algae have shown the potential to relieve pain by interacting with pain receptors and reducing inflammation. This could lead to the development of new, natural pain relief medications. Marine microalgae are a source of anticoagulant compounds, such as sulfated polysaccharides, which can prevent blood clot formation. Additionally, omega-3 fatty acids from microalgae help maintain cardiovascular health by reducing blood pressure, lowering triglycerides, and preventing arrhythmias. Certain microalgae produce enzyme inhibitors that can be used in the treatment of various diseases. For instance, inhibitors of angiotensin-converting enzyme (ACE) from marine algae are of interest for managing hypertension. Additionally, inhibitors of digestive enzymes, such as lipase, can aid in weight management and diabetes control.

Marine microalgae-derived compounds can enhance the bioavailability and efficacy of other drugs. For example, certain polysaccharides can act as carriers for drug delivery improving the stability and controlled release of active pharmaceutical ingredients. Marine microalgae produce compounds, such as mycosporine like amino acids (MAAs), which absorb UV radiation and protect against UV-induced damage. These compounds are being explored for use in sunscreens and other photoprotective skincare products. Marine microalgae offer a diverse array of bioactive compounds and nutrients that can significantly contribute to human health and wellness making them valuable assets in the nutraceutical industry. The demand for natural and sustainable nutraceuticals is driving research and development in the cultivation and processing of marine microalgae. Advances in biotechnology are enhancing the efficiency of microalgae production and extraction methods making these potent nutraceuticals more accessible and cost effective [95]. Marine microalgae are excellent sources of high-quality proteins and essential amino acids. For instance, *Spirulina* and *Chlorella* are known for their high protein content. Many microalgae, such as *Schizochytrium* and *Crypthecodinium*, are rich in omega-3 fatty acids (DHA and EPA), which are crucial for cardiovascular health and cognitive function. Microalgae are abundant in vitamins (A, B, C, E, and K) and minerals (iron, magnesium, calcium, and iodine) making them valuable for preventing deficiencies and supporting overall health. Many marine microalgae are available as supplements in powder, capsule, or tablet forms providing concentrated doses of their beneficial compounds. Microalgae are incorporated into various food products, like smoothies, snack bars, and health drinks, to enhance their nutritional value. Due to their anti-aging and skin-protective properties, microalgae extracts are used in skincare and cosmetic products.

Apart from the pharmaceutical and nutraceutical industries, marine microalgae are seen in various other industries, such as food, cosmetics, biofuels, and environmental management. In the food industry, marine microalgae, such as *Spirulina*, *Chlorella*, and *Haematococcus pluvialis*, are popular as dietary supplements for their high protein content, vitamins, minerals, and antioxidants. Algae, like *Schizochytrium* and *Crypthecodinium*, are used as vegetarian sources of omega-3 fatty acids (DHA and EPA), which are essential for brain and cardiovascular health. Algae are added to food, like pasta, bread, snacks, and beverages, to enhance their nutritional value. Algae-derived pigments, such as chlorophyll, phycocyanin, and astaxanthin, are used as natural food colorants and antioxidants [259]. In cosmetics, algae extracts are incorporated into skincare products for their antioxidant, anti-inflammatory, and moisturizing properties. Certain compounds in algae offer natural UV protection and are used in sunscreen formulations. Algae are used in shampoos and conditioners to nourish the scalp and hair. Algae serve as a primary feed in aquaculture to improve the growth and health of fish and shellfish [277]. Algae are added to livestock feed to enhance the nutritional quality of meat and dairy products. In the biofuel industry, algae are considered a promising source of biofuels due to their high lipid content and rapid growth rates. Research focuses on converting algal biomass into bio-diesel, bio-ethanol, and biogas. Microalgae cultivation can be integrated with carbon capture technologies helping to reduce greenhouse gas emissions.

Algae are used to treat wastewater by removing nutrients, heavy metals, and other pollutants, thereby improving water quality. Algae cultivation captures and utilizes carbon dioxide helping mitigate the effects of climate change. In the present scenario, people are very health cautious and prefer to consume organic foodstuffs, which are free from antibiotics, pesticides, hormones, and other contaminants. The ban on the usage of antibiotics, pesticides, and hormones in the aquaculture industry improved the farmed fish quality, but still, it needs certain value additions to enhance health benefits of consumers. Similarly, to mitigate the stress in culture condition (present intensive farming practices), in addition to feed, certain compounds are desired by cultured fish. Now, the industry is looking for alternative products that are derived from nature (organic) and have nutritional and health benefits. Marine nutraceuticals are naturally available organic substances that have greater health-promoting factors and are derived from seaweeds, marine microalgae, marine lipids, etc. So, directly or indirectly, marine nutraceuticals help in the near future to gratify everyone's (fish farmers and consumers) necessity of the aquaculture industry and overall enhance the aquaculture production.

References

1. Mouritsen, O.G., *Seaweeds: Edible, Available, and Sustainable*, University of Chicago Press, Chicago, IL, 2013, ISBN: 978-0-226-04436-1.
2. Pereira, L., Macroalgae. *Encyclopedia.*, 1, 1, 177–188, 2021. https://doi.org/10.3390/encyclopedia1010017.
3. Mehariya, S., Iovine, A., Casella, P., Musmarra, D., Chianese, S., Marino, T., Figoli, A., Sharma, N., Molino, A., Bio-based and agriculture resources for production of bioproducts, in: *Current Trends and Future Developments on (Bio-) Membranes: Membranes in Environmental Applications.* A. Figoli, Y. Li, A. Basile, (Eds.), pp. 263–282, Elsevier, Amsterdam, Netherlands, 2020. https://doi.org/10.1016/B978-0-12-816778-6.00012-6
4. de Souza, M.P., Hoeltz, M., Gressler, P.D., Benitez, L.B., Schneider, R.C.S., Potential of microalgal bioproducts: general perspectives and main challenges. *Waste Biomass Valoriz.*, 10, 2139e2156, 2019. https://doi.org/10.1007/s12649-018-0253-6.
5. Global Hunger Index. *GHI*, 2019. Bonn (Germany): Welthungerhilfe; Dublin (Ireland): Concern Worldwide. Available from: https://www.globalhungerindex.org/results.html (accessed 24/09/2020).
6. Bhatia, S.K., Mehariya, S., Bhatia, R.K., Kumar, M., Pugazhendhi, A., Awasthi, M.K., Atabani, A.E., Kumar, G., Kim, W., Seo, S.O., Yang, Y.-H., Wastewater based microalgal biorefinery for bioenergy production: progress and challenges. *Sci. Total Environ.*, 751, 141599, 2020. https://doi.org/10.1016/j.scitotenv.2020.141599.
7. Goswami, R.K., Agrawal, K., Mehariya, S., Molino, A., Musmarra, D., Verma, P., Microalgae-based biorefinery for utilization of carbon dioxide for production of valuable bioproducts, in: *Chemo-Biological Systems for CO_2 Utilization.* A. Kumar and S. Sharma (Eds.), pp. 203–228, CRC Press, Boca Raton, FL, USA, 2020.
8. Asha, K.K. and Susheela, M., *Marine Nutraceuticals.* ICAR-Central Institute of Fisheries Technology, Cochin, India, 2016.

9. Chaturvedi, V., Goswami, R.K., Verma, P., Genetic engineering for enhancement of biofuel production in microalgae, in: *Biorefineries: A Step towards Renewable and Clean Energy. Clean Energy Production Technologies*, P. Verma, (Ed.), p. 539e559, Springer, Singapore, 2020, https://doi.org/10.1007/978-981-15-9593-6_21.
10. Richmond, A., Biological principles of mass cultivation, in: *Handbook of Microalgal Culture*, A. Richmond, (Ed.), p. 125e177, Biotechnol. Appl. Phycol. Blackwell Sci., London, 2004, https://doi.org/10.1002/9780470995280.ch8.
11. Sathasivam, R., Radhakrishnan, R., Hashem, A., Abd_Allah, E.F., Microalgae metabolites: a rich source for food and medicine. *Saudi J. Biol. Sci.*, 26, 709e722, 2019. https://doi.org/10.1016/j.sjbs.2017.11.003.
12. Pereira, L., *Therapeutic and Nutritional Uses of Algae*, 1st, CRC Press/Taylor & Francis Group, Boca Raton, FL, 2018.
13. Nations, U., World population projected to reach 9.8 billion in 2050, and 11.2 billion in 2100, in: *Department of Economic and Social Affairs*, 2017.
14. Kiran, B.R. and Venkata, M.S., Microalgal cell biofactory therapeutic, nutraceutical and functional food applications. *Plants*, 10, 5, 836, 2021. https://doi.org/10.3390/plants10050836.
15. FAO, IFAD, UNICEF, WFP and WHO, *The state of food security and nutrition in the world 2020 transforming food systems for affordable healthy diets*, FAO, Rome, 2020.
16. Srimongkol, P., Sangtanoo, P., Songserm, P., Watsuntorn, W., Karnchanatat, A., Microalgae-based wastewater treatment for developing economic and environmental sustainability: current status and future prospects. *Front. Bioeng. Biotechnol.*, 7, 10, 904046, 2022. https://doi.org/10.3389/fbioe.2022.904046.
17. Sathasivam, R., Radhakrishnan, R., Hashem, A., Abd Allah, E.F., Microalgae metabolites: a rich source for food and medicine Saudi. *J. Biolog. Sci.*, 26, 4, 709–722, 2019. https://doi.org/10.1016/j.sjbs.2017.11.003.
18. Coulombier, N., Jauffrais, T., Lebouvier, N., Antioxidant compounds from microalgae: a review. *Mar. Drugs*, 19, 10, 549, 2021. https://doi.org/10.3390/md19100549.
19. Pai, S., Hebbar, A., Selvaraj, S., A critical look at challenges and future scopes of bioactive compounds and their incorporations in the food, energy, and pharmaceutical sector. *Environ. Sci. Poll Res.*, 29, 35518–35541, 2022. https://doi.org/10.1007/s11356-022-19423-4.
20. Martínez-Ruiz, M., Martínez-González, C.A., Kim, D.H., Santiesteban-Romero, B., Reyes-Pardo, H., Villaseñor-Zepeda, K.R., Parra-Saldivar, R., Microalgae bioactive compounds to topical applications products—a review. *MOLEFW*, 27, 11, 3512, 2022. https://doi.org/10.3390/molecules27113512.
21. Varfolomeev, S.D. and Wasserman, L.A., Microalgae as source of biofuel, food, fodder and medicine. *Appl. Biochem. Microbiol.*, 47, 789–807, 2011.
22. Pulz, O. and Gross, W., Valuable products from biotechnology of microalgae. *Appl. Microbiol. Biotechnol.*, 65, 6, 635–648, 2004. doi:10.1007/s00253-004-1647-x.

23. Wijffels, R.H. and Barbosa, M.J., Microalgae for the production of bulk chemicals and biofuels. *Biofuels Bioprod. Bioref.*, 4, 3, 287–295, 2010. doi:10.1002/bbb.215.
24. Boussiba, S., Carotenogenesis in the green alga Haematococcus pluvialis: cellular physiology and stress response. *Physiol. Plant.*, 108, 2, 111–117, 2000. doi:10.1034/j.1399-3054.2000.108002111.x. ISBN: 9781498755382.
25. Borowitzka, M.A., High-value products from microalgae—their development and commercialisation. *J. Appl. Phycol.*, 25, 3, 743–756, 2013. doi:10.1007/s10811-013-9983-9.
26. Boussiba, S. and Aflalo, C., Haematococcus pluvialis: a promising source of natural astaxanthin for the food industry. *Innov. Food Technol.*, 28, 37–39, 2005.
27. Priyadarshani, I. and Rath, B., Commercial and industrial applications of micro algae - A review. *J. Algal Biomass Util.*, 3, 4, 89–100, 2012.
28. Mimouni, V., Ulmann, L., Pasquet, V., Chénais, B., Morant-Manceau, A., Tremblin, G., Cadoret, J.-P., Antioxidant and anti-inflammatory activities of microalgae. *Curr. Pharmaceut. Biotechnol.*, 13, 14, 2733–2750, 2012. https://doi.org/10.2174/138920112804724828.
29. Shannon, E. and Abu-Ghannam, N., Seaweeds as nutraceuticals for health and nutrition. *Phycologia*, 58, 5, 563–577, 2019.
30. Katiyar, R. and Arora, A., Health promoting functional lipids from microalgae pool: a review. *Algal Res.*, 46, January, 101800, 2020. https://doi.org/10.1016/j.algal.2020.101800.
31. Quintana, N., Van der Kooy, F., Van de Rhee, M.D., Voshol, G.P., Verpoorte, R., Renewable energy from cyanobacteria: energy production optimization by metabolic pathway engineering. *Appl. Microbiol. Biotechnol.*, 91, 3, 471–490, 2011.
32. Rawat, I., Ranjith Kumar, R., Mutanda, T., Bux, F., Dual role of microalgae: phycoremediation of domestic wastewater and biomass production for sustainable biofuels production. *Appl. Energy*, 88, 10, 3411–3424, 2011.
33. Gouveia, L. and Empis, J., Relative stabilities of microalgal carotenoids in microalgal extracts, biomass and fish feed: effect of storage conditions. *Innov. Food Sci. Emerging Technol.*, 4, 227e233, 2003.
34. Christaki, E., Bonos, E., Giannenas, I., Florou-Paneri, P., Functional properties of carotenoids originating from algae. *J. Sci. Food Agric.*, 93, 5e11, 2013.
35. Freitas, A.C., Rodrigues, D., Rocha-Santos, T.A.P., Gomes, A.M.P., Duarte, A.C., Marine biotechnology advances towards applications in new functional foods. *Biotechnol. Adv.*, 30, 1506e1515, 2012.
36. Batista, A.P., Gouveia, L., Bandarra, N.M., Franco, J.M., Raymundo, A., Comparison of microalgal biomass profiles as novel functional ingredient for food products. *Algal Res.*, 2, 164e173, 2013.
37. Plaza, M., Herrero, M., Cifuentes, A., Ibanez, E., Innovative natural functional ingredients from microalgae. *J. Agric. Food. Chem.*, 57, 7159e7170, 2009.

38. Kadam, S.U., Tiwari, B.K., O'Donnell, C.P., Application of novel extraction technologies for bioactives from marine algae. *J. Agric. Food. Chem.*, 61, 4667e4675, 2013.
39. Reyes, F.A., Mendiola, J.A., Ibanez, E., del Valle, J.M., Astaxanthin extraction from Haematococcus pluvialis using CO2-expanded ethanol. *J. Supercrit. Fluid*, 92, 75e83, 2014.
40. Vanthoor-Koopmans, M. and Wijffels, R.H., Biorefinery of microalgae for food and fuel. *Biores. Technol.*, 135, 142e149, 2013.
41. Wijffels, R.H., Kruse, O., Klaas, J.H., Potential of industrial biotechnology with cyanobacteria and eukaryotic microalgae. *Curr. Opin. Biotechnol.*, 24, 405e413, 2013.
42. Bhuvana, P., Sangeetha, V., Anuradha, M., Ali, M.S., Spectral characterization of bioactive compounds from microalgae: N Oculata and C Vulgaris. *Biocatal. Agril. Biotechnol.*, 19, 101094, 2019. https://doi.org/10.1016/j.bcab.2019.101094.
43. Soto-Sierra, A., Stoykova, P., Nikolov, Z.L., Extraction and fractionation of microalgae-based protein products. *Algal Res.*, 36, 175–192, 2018. https://doi.org/10.1016/j.algal.2018.10.023.
44. Janssen, M., Wijffels, R.H., Barbosa, M.J., Microalgae based production of single-cell protein. *Curr. Opin. Biotechnol.*, 75, 102705, 2022. https://doi.org/10.1016/j.copbio.2022.102705.
45. Koyande, A.K., Chew, K.W., Rambabu, K., Tao, Y., Chu, D.T., Show, P.L., Microalgae: a potential alternative to health supplementation for humans. *Food Sci. Hum. Wellness*, 8, 1, 16–24, 2019. https://doi.org/10.1016/J.FSHW.2019.03.001.
46. Bishop, W.M. and Zubeck, H.M., Evaluation of microalgae for use as nutraceuticals and nutritional supplements. *J. Nutr. Food Sci.*, 2, 5e10, 2012.
47. Tandeau-de-Marsac, H.J., Adaptation of cyanobacteria to environmental stimuli: new steps towards molecular mechanisms. *FEMS Microbiol. Rev.*, 104, 119e90, 1993.
48. Blunt, J.W., Copp, B.R., Hu, W.-P., Munro, M.H.G., Northcote, P.T., Prinsep, M.R., Marine natural products. *Natural Product Rep.*, 25, 1, 35–94, Royal Society of Chemistry, Cambridge, United Kingdom, 2008.
49. Mata, T.M., Martins, A.A., Caetano, N.S., Microalgae for biodiesel production and other applications: a review. *Renew. Sustain. Energy Rev.*, 14, 1, 217e32, 2010. https://doi.org/10.1016/j.rser.2009.07.020.
50. Jain, R., Raghukumar, S., Tharanathan, R., Bhosle, N.B., Extracellular polysaccharide production by thraustochytrid protists. *Mar. Biotechnol.*, 7, 3, 184e92, 2005. https://doi.org/10.1007/s10126-004- 4025-x.
51. Kim, S.-K., Ravichandran, Y.D., Khan, S.B., Kim, Y.T., Prospective of the cosmeceuticals derived from marine organisms. *Biotechnol. Bioproc. Eng.*, 13, 5, 511e23, 2008. https://doi.org/10.1007/s12257-008- 0113-5.

52. Paliwal, C., Mitra, M., Bhayani, K., Bharadwaj, S.V., Ghosh, T., Dubey, S., Mishra, S., Abiotic stresses as tools for metabolites in microalgae. *Bioresour. Technol.*, 244, 1216–1226, 2017.
53. Mutanda, T., Naidoo, D., Bwapwa, J.K., Anandraj, A., Biotechnological applications of microalgal oleaginous compounds: current trends on microalgal bioprocessing of products. *Front. Energy Res.*, 8, 598803, 2020.
54. Shimizu, Y., Microalgal metabolites. *Chem. Rev.*, 93, 5, 1685–1698, 1993. https://doi.org/10.1021/cr00021a002.
55. Gouveia, L., Batista, A.P., Sousa, I., Raymundo, A., Bandarra, N.M., Microalgae in novel food products, in: *Food Chemistry Research Developments*, K.N. Papadopoulos, (Ed.), p. 75e112, Nova Science Publishers Inc., Hauppauge, NY, 2008.
56. de Jesus Raposo, M.F., de Morais, R.M.S.C., de Morais, A.M.M.B., Health applications of bioactive compounds from marine microalgae. *Life Sci.*, 93, 479e486, 2013.
57. Markou, G. and Nerantzis, E., Microalgae for high-value compounds and biofuels production: a review with focus on cultivation under stress conditions. *Biotechnol. Adv.*, 31, 1532e1542, 2013.
58. Barra, L., Chandrasekaran, R., Corato, F., Brunet, C., The challenge of ecophysiological biodiversity for biotechnological applications of marine microalgae. *Mar. Drugs*, 12, 1641e1675, 2014.
59. Eonseon, J., Polle, J.E.W., Lee, H.K., Hyund, S.M., Chang, M., Xanthophylls in microalgae: from biosynthesis to biotechnological mass production and application. *J. Microbiol. Biotechnol.*, 13, 165e174, 2003.
60. Guedes, A.C., Amaro, H.M., Malcata, F.X., Microalgae as sources of carotenoids. *Mar. Drugs*, 9, 625e644, 2011.
61. Lamers, P.P., Janssen, M., de Vos, R.C.H., Bino, R. J., Wijffels, R. H., Exploring and exploiting carotenoid accumulation in Dunaliella salina for cell-factory applications. *Trends Biotechnol.*, 26, 11, 631–638, 2008. https://doi.org/10.1016/j.tibtech.2008.07.002
62. Mulders, K.J.M., Lamers, P.P., Martens, D.E., Wijffels, R.H., Phototropic pigment production with microalgae: biological constraints and opportunities. *J. Phycol.*, 50, 229e242, 2014.
63. Chu, W.L., Biotechnological applications of microalgae. *Int. E-J. Sci. Med. Educ.*, 6, S24eS37, 2012.
64. Hosikian, A., Lim, S., Halim, R., Danquah, M.K., Chlorophyll extraction from microalgae: a review on the process engineering aspects. *Int. J. Chem. Eng.*, 2010, 11, 2010.
65. Stengel, D.B., Connan, S., Popper, Z.A., Algal chemodiversity and bioactivity: sources of natural variability and implications for commercial application. *Biotechnol. Adv.*, 29, 483e501, 2011.
66. Cuaresma, M., Janssen, M., Vilchez, C., Wijffels, R.H., Horizontal or vertical photobioreactors? How to improve microalgae photosynthetic efficiency. *Biores. Technol.*, 102, 5129e5137, 2011.

67. Miazek, K. and Ledakowicz, S., Chlorophyll extraction from leaves, needles and microalgae: a kinetic approach. *Int. J. Agric. Biol. Eng.*, 6, 107e115, 2013.
68. Eriksen, N.T., Production of phycocyaninda pigment with applications in biology, biotechnology, foods and medicine. *Appl. Microbiol. Biotechnol.*, 80, 1e14, 2008.
69. Hifney, A.F., Issa, A.A., Fawzy, M.A., Abiotic stress induced production of b-carotene, allophycocyanin and total lipids in *Spirulina* sp. *J. Biol. Earth Sci.*, 3, B54eB64, 2013.
70. Bule, M.H., Ahmed, I., Maqbool, F., Bilal, M., Iqbal, H.M.N., Microalgae as a source of high-value bioactive compounds. *Front. Biosci. - Scholar*, 10, 2, 197e216, 2018. https://doi.org/10.2741/s509.
71. Kim, S.-K. and Kang, K.-H., Medicinal effects of peptides from marine microalgae, in: *Advances in Food and Nutrition Research.* S.-K. Kim (Ed.), vol. 64, pp. 313–323, Academic Press, 2011. https://doi.org/10.1016/B978-0-12-387669-0.00025-9.
72. Talero, E., García-Maurino, S., Avila-Román, J., Rodríguez-Luna, A., Alcaide, A., Motilva, V., Bioactive compounds isolated from microalgae in chronic inflammation and cancer. *Mar. Drugs*, 13, 10, 6152e209, 2015. https://doi.org/10.3390/md13106152.
73. Zheng, L.-H., Wang, Y.-J., Sheng, J., Wang, F., Zheng, Y., Lin, X.-K., *et al.*, Antitumor peptides from marine organisms. *Mar. Drugs*, 9, 10, 1840e59, 2011. https://doi.org/10.3390/md9101840.
74. Manirafasha, E., Ndikubwimana, T., Zeng, X., Lu, Y., Jing, K., Phycobiliprotein: potential microalgae derived pharmaceutical and biological reagent. *Biochem. Eng. J.*, 109, 282e96, 2016. https://doi.org/10.1016/j.bej.2016.01.025.
75. Ryu, J., Park, S.-J., Kim, I.-H., Choi, Y., Nam, T.-J., Protective effect of porphyra-334 on UVA induced photoaging in human skin fibroblasts. *Int. J. Mol. Med.*, 34, 3, 796e803, 2014. https:// doi.org/10.3892/ijmm.2014.1815.
76. Galasso, C., Gentile, A., Orefice, I., Ianora, A., Bruno, A., Noonan, D.M., Sansone, C., Albini, A., Brunet, C., Microalgal derivatives as potential nutraceutical and food supplements for human health: A focus on cancer prevention and interception. *Nutrients*, 11, 6, 1226, 2019. https://doi.org/10.3390/nu11061226.
77. Hamed, I., The evolution and versatility of microalgal biotechnology: a review. *Compr. Rev. Food Sci. Food Saf.*, 15, 1104e1123, 2016. https://doi.org/10.1111/1541- 4337.12227.
78. Levasseur, W., Perre, P., Pozzobon, V., A review of high value-added molecules production by microalgae in light of the classification. *Biotechnol. Adv.*, 45, 107545, 2020. https://doi.org/10.1016/j.biotechadv.2020.107545.
79. Chen, C.-Y., Zhao, X.-Q., Yen, H.-W., Ho, S.-H., Cheng, C.-L., Lee, D.-J., *et al.*, Microalgae-based carbohydrates for biofuel production. *Biochem. Eng. J.*, 78, 1e10, 2013. https://doi.org/10.1016/j.bej. 2013.03.006.
80. Giordano, D., Costantini, M., Coppola, D., Lauritano, C., Núñez Pons, L., Ruocco, N., di Prisco, G., Ianora, A., Verde, C., Biotechnological

applications of bioactive peptides from marine sources, in: *Advances in Microbial Physiology*. R.K. Poole (Ed.), vol. 73, pp. 171–220, Academic Press, 2018. https://doi.org/10.1016/bs.ampbs.2018.05.002.

81. Mishra, A., Kavita, K., Jha, B., Characterization of extracellular polymeric substances produced by microalgae Dunaliella salina. *Carbohydr. Polym.*, 83, 2, 852e7, 2011. https://doi.org/10.1016/j.carbpol. 2010.08.067.
82. Raposo MF de, J., de Morais, A.M.M.B., de Morais, R.M.S.C., Influence of sulphate on the composition and antibacterial and antiviral properties of the exopolysaccharide from Porphyridium cruentum. *Life Sci.*, 101, 1e2, 56e63, 2014. https://doi.org/10.1016/j.lfs.2014.02.013.
83. Pignolet, O., Jubeau, S., Vaca-Garcia, C., Michaud, P., Highly valuable microalgae: biochemical and topological aspects. *J. Ind. Microbiol. Biotechnol.*, 40, 8, 781e96, 2013. https://doi.org/10.1007/s10295- 013-1281-7.
84. Wu, J., Gu, X., Yang, D., Xu, S., Wang, S., Chen, X., Wang, Z., Bioactive substances and potentiality of marine microalgae. *Food Sci. Nutr.*, 9, 9, 5279–5292, 2021. https://doi.org/10.1002/fsn3.2471.
85. Bisen, P.S., Nutritional therapy as a potent alternate to chemotherapy against cancer. *J. Cancer Sci. Ther.*, 8, 6, 168, 2016.
86. Spolaore, P., Joannis-Cassan, C., Duran, E., Isambert, A., Commercial applications of microalgae. *J. Biosci. Bioeng.*, 101, 87e96, 2006. https://doi.org/10.1263/ jbb.101.87.
87. Lin, W.N., Kapupara, K., Wen, Y.T., Chen, Y.H., Pan, I., Tsai, R.K., Haematococcus pluvialis-derived astaxanthin is a potential neuroprotective agent against optic nerve ischemia. *Mar. Drugs*, 18, 1e12, 2020. https://doi.org/10.3390/md18020085.
88. Riccio, G. and Lauritano, C., Microalgae with immunomodulatory activities. *Mar. Drugs*, 18, 1e18, 2020. https://doi.org/10.3390/md18010002.
89. Ariede, M.B., Candido, T.M., Jacome, A.L.M., Velasco, M.V.R., de Carvalho, J.C.M., Baby, A.R., Cosmetic attributes of algae- a review. *Algal Res.*, 25, 483e487, 2017. https://doi.org/10.1016/j.algal.2017.05.019.
90. Marino, T., Iovine, A., Casella, P., Martino, M., Chianese, S., Larocca, V., Musmarra, D., Molino, A., From Haematococcus pluvialis microalgae a powerful antioxidant for cosmetic applications. *Chem. Eng. Trans.*, 79, 271e276, 2020. https://doi.org/ 10.3303/CET2079046.
91. Priyadarshani, I. and Rath, B., Commercial and industrial applications of micro algae e a review. *J. Algal Biomass Util.*, 3, 89e100, 2012. https://www.semanticscholar.org/paper/Commercial-and-industrial-applications-of-micro-A-PriyadarshaniRath/9078cac2b6d51a9c.
92. Wang, H.M.D., Chen, C.C., Huynh, P., Chang, J.S., Exploring the potential of using algae in cosmetics. *Bioresour. Technol.*, 184, 355e362, 2015. https://doi.org/.
93. Georgakopoulou, M., The use of microalgae in biotechnological applications, in: *Case Studies*, p. 1e64, 2018.

94. Kim, S.K., *Marine Nutraceuticals: Prospects and Perspectives*, CRC Press, Boca Raton, FL, 20132013.
95. Wildman, R.E.C. and Taylor, C., *Wallace Handbook of Nutraceuticals and Functional Foods*, Second, CRC Press, Boca Raton, FL, 2006.
96. Robinson, C.B., Samocha, T.M., Fox, J.M., Gandy, R.L., McKee, D.A., The use of inert artificial commercial food sources as replacements of traditional live food items in the culture of larval shrimp, Farfantepenaeus aztecus. *Aquaculture*, 245, 135e147, 2005.
97. Molina Grima, E., Belarbi, E.H., Acien Fernandez, F.G., Robles Medina, A., Chisti, Y., Recovery of microalgal biomass and metabolites: process options and economics. *Biotechnol. Adv.*, 20, 491e515, 2003.
98. Linan-Cabello, M.A., Paniagua-Michel, J., Hopkins, P.M., Bioactive roles of carotenoids and retinoids in crustaceans. *Aquacult. Nutr.*, 8, 299e309, 2002.
99. Harrison, K.E., The role of nutrition in maturation, reproduction and embryonic development of decapod crustaceans: a review. *J. Shellfish Res.*, 9, 1e28, 1990.
100. Gouveia, L., Veloso, V., Reis, A., Fernandes, H., Novais, J., Empis, J., Evolution of pigment composition in *Chlorella* vulgaris. *Bioresour. Technol.*, 57, 157e159, 1996.
101. Baker, R. and Gunther, C., The role of carotenoids in consumer choice and the likely benefits from their inclusion into products for human consumption. *Trends Food Sci. Technol.*, 15, 484e488, 2004.
102. Kumari, A., Garima, Bharadvaja, N., A comprehensive review on algal nutraceuticals as prospective therapeutic agent for different diseases. *3 Biotech*, 13, 2, 44, 2023. Springer Nature, Heidelberg, Germany. https://doi.org/10.1007/s13205-022-03454-2.
103. Gouveia, L., Marques, A.E., Sousa, J.M., Moura, P., Bandarra, N.M., Microalgae–source of natural bioactive molecules as functional ingredients. *Food Sci. Technol. Bull. Funct. Foods*, *7*, 2, 21, 2010.
104. Norton, T.A., Melkonian, M., Andersen, R.A., Algal biodiversity. *Phycologia.*, 35, 308–326, 1996.
105. Singh, J. and Gu, S., Commercialization potential of microalgae for production of biofuels. *Renewable Sustainable Energy Rev.*, 14, 2596–2610, 2010.
106. Mobin, S. and Alam, F., Biofuel production from algae utilizing wastewater, in: *19th Australasian Fluid Mechanics Conference. Melbourne, Australia*, 2014, December.
107. Paniagua-Michel, J., Dujardin, E., Sironval, C., Le Tecuitlatl, concentre de Spirulines des Aztecs. *Bullacad. Roy Sci. Belgium*, 10, 253e263, 1992.
108. Gershwin, M.E. and Belay, A., *Spirulina in Human Nutrition and Health*, CRC Press, BocaRaton, FL, USA, 2008.
109. Asha, K.K. and Mathew, S., *Marine nutraceuticals*, ICAR-Central Institute of Fisheries Technology, Cochin, 2018.

110. Ambati, R.R., Phang, S.M., Ravi, S., Aswathanarayana, R.G., Astaxanthin: sources, extraction, stability, biological Activities and its commercial applications. *Mar. Drugs*, 12, 128e152, 2014.
111. Guerin, M., Huntley, M.E., Olaizola, M., Haematococcus astaxanthin: applications for human health and Nutrition. *Trends Biotechnol.*, 21, 210e216, 2003.
112. Lorenz, T.R. and Cysewski, G.R., Commercial potential for Haematococcus microalgae as a natural source of Astaxanthin. *Trends Biotechnol.*, 18, 160e167, 2000.
113. Sigamani, S., Ramamurthy, D., Natarajan, H., A review on potential biotechnological applications of microalgae. *J. Appl. Pharm. Sci.*, 6, 8, 179–184, 2016.
114. García-Sánchez, A., Miranda-Díaz, A. G., Cardona-Muñoz, E. G., The Role of Oxidative Stress in Physiopathology and Pharmacological Treatment with Pro- and Antioxidant Properties in Chronic Diseases. *Oxid. Med. Cell Longev.*, 2020, Hindawi Publishing Corporation, New York, United States 2082145. https://doi.org/10.1155/2020/2082145.
115. Montero-Lobato, Z., Vázquez, M., Navarro, F., JL, F., Bermejo, E., Garbayo, I., Cuaresma, M., Chemically induced production of anti-inflammatory Molecules in microalgae. *Mar. Drugs*, 16, 12, 478, 2018.
116. Pratt, R., Daniels, T.C., Eiler, J.J., Gunnison, J.B., Kumler, W.D., Oneto, J.F., Strait, L.A., Spoehr, H.A., Hardin, G.J., Milner, H.W., Smith, J.H.C., Chlorellin, an antibacterial substance from *Chlorella. Sci.*, 99, 2574, 351–352, 1944. https://doi.org/10.1126/science.99.2574.35.
117. Sung, H., Ferlay, J., Siegel, R.L., Laversanne, M., Soerjomataram, I., Jemal, A., Bray, F., Global cancer statistics 2020:GLOBOCAN estimates of incidence and mortality Worldwide for 36 cancers in 185 countries CA: a Cancer. *J. For. Clin.*, 71, 3, 209–249, 2021. https://doi.org/10.3322/caac.21660.
118. Fayyad, R.J., Ali, A.N., Dwaish, A.S., Abboodi, A.K., Anticancer activity of *Spirulina* platensis methanolic extracts against L20B and MCF7 human cancer cell lines. *Plant Arch.*, 19, 1, 1419–1426, 2019.
119. Bratchkova, A. and Kroumov, A.D., Microalgae as producers of biologically active compounds with antibacterial, antiviral, antifungal, antialgal, antiprotozoal, antiparasitic and anticancer activity. *Acta Microbiol. Bulg.*, 36, 3, 79–89, 2020.
120. Senousy, H.H., Abd Ellatif, S., Ali, S., Assessment of the antioxidant and anticancer potential of different isolated strains of cyanobacteria and microalgae from soil and agriculture drain water. *Environ. Sci. Pollut. Res.*, 27, 18463–18474, 2020.
121. Pereira, L., *Therapeutic and Nutritional Uses of Algae*, 1st, CRC Press/Taylor & Francis Group, Boca Raton, FL, 2018, ISBN: 9781498755382.
122. Pereira, L., Therapeutic and nutritional uses of marine algae: a pharmacy in the ocean. *Tradit. Med Res.*, 7, 3, 30, 2022. Doi: 10.53388/TMR20220310268.

123. Vo, T.S. and Kim, S.K., Potential anti-HIV agents from marine resources: an overview. *Mar. Drugs*, 8, 12, 2871–2892, 2010. MDPI, Basel, Switzerland. https://doi.org/10.3390/md8122871.
124. Ghannoum, M.A. and Rice, L.B., Antifungal agents: mode of action, mechanisms of resistance, and correlation of these mechanisms with bacterial resistance. *Clin. Microbiol. Rev.*, 12, 4, 501e17, 1999. https://doi.org/10.1128/CMR.12.4.501.
125. Mendiola, J.A., Torres, C.F., Tor e, A., Martín-Alvarez, P.J., Santoyo, S., Arredondo, B.O., *et al.*, Use of super- critical CO2 to obtain extracts with antimicrobial activity from Chaetoceros muelleri microalga. A correlation with their lipidic content. *Eur. Food Res. Tech.*, 224, 4, 505e10, 2007. https://doi.org/10.1007/s00217-006-0353-6.
126. Damonte, E.B., Pujol, C.A., Coto, C.E., Prospects for the therapy and prevention of dengue virus infections. *Adv. Virus Res.*, 63, 239–285, 2004.
127. Buck, C.B., Thompson, C.D., Roberts, J.N., Muller, M., Lowy, D.R., Schiller, J.T., Carrageenan is a potent inhibitor of Papilloma virus infection. *PLoS Pathog.*, 2, e69, 2006.
128. Talarico, L.B. and Damonte, E.B., Interference in dengue virus adsorption and uncoating by carrageenans. *Virology*, 363, 473–485, 2007.
129. Amaro, H., Guedes, A., Malcata, F., Antimicrobial activities of microalgae: an invited review, in: *Science against microbial pathogens: communicating current research and technological advances*, p. 1272e80, 2011.
130. Beyatli, Y., Aslim, B., Yuksekdag, Z., Screening for antimicrobial agent production of some microalgae in freshwater. *Internet J. Microbiol.*, 2, 2, 1–9, 2006. https://doi.org/10.5580/17b8.
131. Cherrington, C.A., Hinton, M., Mead, G.C., Chopra, I., Organic acids: chemistry, antibacterial activity and practical applications. *Adv. Microb. Physiol.*, 32, 87–108, 1991. doi:10.1016/S0065-2911(08)60006-5.
132. Smulders, F.J.M., Barendsen, P., Van Logtestijn, J.G., Mossel, D.A.A., Van der Marel, G.M., Review: lactic acid: considerations in favour of its acceptance as a meat decontamininant. *Int. J. Food Sci. Technol.*, 21, 4, 419e36, 2007. https://doi.org/10.1111/j.1365-2621.1986.tb00420.x.
133. Metwaly, F., Antibacterial and antifungal from *Spirulina*. *Glob. J. Biotechnol. Biochem.*, 3, 1, 22e31, 2008.
134. Ferdous, U.T. and Yusof, Z.N.B., Medicinal prospects of antioxidants from algal sources in cancer therapy. *Front. Pharmacol.*, 12, 157, 2021.
135. Lee, N.Y., Kim, Y., Kim, Y.S., Shin, J.H., Rubin, L.P., Kim, Y., Carotene exerts anti-colon cancer effects by regulating M2 macrophages and activated fibroblasts. *J. Nutr. Biochem.*, 82, 108402, 2020. https://doi.org/10.1016/jjnutbio2020108402.
136. Medina, R.A., Goeger, D.E., Hills, P., Mooberry, S.L., Huang, N., Romero, L.I., Ortega-Barría, E., Gerwick, W.H., McPhail, K.L., Coibamide A, a potent antiproliferative cyclic depsipeptide from the panamanian marine

cyanobacterium Leptolyngbya sp. *J. Amer. Chem. Soc.*, 130, 20, 6324–6325, 2008. https://doi.org/10.1021/ja801383.
137. Haq, S.H., Al-Ruwaished, G., Al-Mutlaq, M.A., Naji, S.A., Al-Mogren, M., Al-Rashed, S., Ain, Q.T., Al-Amro, A.A., Antioxidant, anticancer activity and phytochemical analysis of green algae, Chaetomorpha collected from the Arabian Gulf. *Sci. Rep.*, 9, 1, 17, 2019.
138. Singh, A. and Krishna, S., Immunomodulatory and therapeuticpotential of marine flora products in the treatment of cancer, in: *Bioactive natural products for the management of cancer: from bench to bedside*, pp. 139–166, Springer, Singapore, 2019.
139. Satyalakshmi, S., Determination of biological activities of three marine algae collected from Visakhapatnam coast. *Asian J. Pharm. Clin. Res.*, 10, 12, 274–279, 2012.
140. Saeedi, P., Petersohn, I., Salpea, P., Malanda, B., Karuranga, S., Unwin, N., Colagiuri, S., Guariguata, L., AA, M., Ogurtsova, K., Shaw, J.E., Global and regional diabetes prevalence estimates for 2019 and projections for 2030 and 2045: results from the international diabetes federation diabetes atlas. *Diabetes Res. Clin. Prac.*, 1, 157, 2019. https://doi.org/10.1016/j.diabres.2019.107843.
141. Lee, S.H. and Jeon, Y.J., Anti-diabetic effects of brown algae derived phlorotannins, marine polyphenols through diverse mechanisms. *Fitoterapia*, 86, 129–136, 2013. Elsevier Ltd; Amsterdam, Netherlands. doi:10.1016/j.fitote.2013.02.013.
142. Chou, Y.C., Prakash, E., Huang, C.F., Lien, T.W., Chen, X., Su, I.J., Chao, Y.S., Hsieh, H.P., Hsu, J.T., Bioassay-guided purification and identification of PPARα/γ agonists from Chlorella sorokiniana. *Phytother. Res.*, 22, 5, 605–13, 2008 May. doi:10.1002/ptr.2280.
143. Senthilkumar, T. and Ashokkumar, N., Impact of *Chlorella* pyrenoidosa on the attenuation of hyperglycemia-mediated oxidative stress and protection of kidney tissue in streptozotocin-cadmium induced diabetic nephropathic rats. *Biomed Preventative Nutr.*, 2, 125–131, 2012. https://doi.org/10.1016/J.BIONUT.2012.01.006.
144. Noguchi, N., Konishi, F., Kumamoto, S., Maruyama, I., Ando, Y., Yanagita, T., Beneficial effects of *Chlorella* on glucose and lipid metabolism in obese rodents on a high-fat diet. *ORCP*, 7, 2, e95–e105, 2013.
145. Jong-Yuh, C. and Mei-Fen, S., Potential hypoglycemic effects of *Chlorella* in streptozotocin-induced diabetic mice. *LifeSci*, 77, 9, 980–990, 2005.
146. Layam, A. and Reddy, C.L.K., Antidiabetic property of *Spirulina*. *Diabetol. Croat.*, 35, 29, 2006.
147. Nacer, W., Ahmed, F.Z.B., Merzouk, H., Benyaoub, W., Bouanane, S., Mebarek, K., Metabolic and antioxidant effects of micro algae in diabetic rats: implications for the prevention of human pathologies. *Arch. Cardiovasc. Dis.*, Suppl 11, 3, 360, 2019.

148. Guldas, M., Ziyanok-Demirtas, S., Sahan, Y., Yildiz, E., Gurbuz, O., Antioxidant and anti-diabetic properties of Spirulina platensis produced in Turkey. *Food Sci. Technol.*, 41, 615–25, 2020.
149. Fuertes, A., Alonso, D., Perez, D.I., Martinez, A., Luque, F.J., Medina, M., Evidence for a new binding mode to GSK-3: allosteric regulation by the marine compound palinurin. *Eur. J. Med. Chem.*, 60, 479–489, 2013.
150. Thariath, D.V., Divakaran, D., Chenicherry, S., Influence of salinity on the dimethylsulphoniopropionate production from Prymnesium simplex. *Sust. Environ. Res.*, 29, 1, 1, 2019.
151. Figueroa, F.L., Mycosporine-like amino acids from marine resource. *Mar. Drugs*, 19, 1, 18, 2021.
152. Raposo, M.F.D.J., De Morais, R.M.S.C., Bernardo de Morais, A.M.M., Bioactivity and applications of sulphated polysaccharides from marine microalgae. *Mar. Drugs*, 11, 1, 233–252, 2013. https://doi.org/10.3390/md11010233.
153. Cervantes-Llanos, M., Lagumersindez-Denis, N., Marín-Prida, J., Pavón-Fuentes, N., Falcon-Cama, V., Piniella-Matamoros, B., Camacho-Rodríguez, H., Fernández-Massó, J.R., Valenzuela Silva, C., Raíces-Cruz, I., Pentón-Arias, E., Beneficial effects of oral administration of C-phycocyanin and phycocyanobilin in rodent models of experimental autoimmune encephalomyelitis. *Life Sci.*, 194, 130–138, 2018. https://doi.org/10.1016/j.lfs.2017.12.032.
154. Pinto, L.F.R., Ferreira, G.F., Beatriz, F.P., Cabral, F.A., Maciel Filho R Lipid and phycocyanin extractions from *Spirulina* and economic assessment. *J. Supercrit. Fluids*, 184, 105567, 2022.
155. Speranza, L., Pesce, M., Patruno, A., Franceschelli, S., De Lutiis, M.A., Grilli, A., Felaco, M., Astaxanthin treatment reduced oxidative induced pro-inflammatory cytokines secretion in U937: SHP-1 as a novel biological target. *Mar. Drugs*, 10, 4, 890–899, 2012.
156. Guzmán, S., Gato, A., Lamela, M., Freire-Garabal, M., Calleja, J.M., Anti-inflammatory and immunomodulatory activities of polysaccharide from *Chlorella* stigmatophora and Phaeodactylum tricornutum. *Phytother. Res.*, 17, 6, 665–670, 2003.
157. Tabarzad, M., Atabaki, V., Hosseinabadi, T., Anti-inflammatory activity of bioactive compounds from microalgae and cyanobacteria by focusing on the mechanisms of action. *Mol. Biol. Rep.*, 47, 8, 6193–6205, 2020.
158. Mtaki, K., Kyewalyanga, M.S., Mtolera, M.S., Assessment of antioxidant contents and free radical-scavenging capacity of *Chlorella* vulgaris cultivated in lowcost media. *Appl. Sci.*, 10, 23, 8611, 2020.
159. Fujitani, N., Sakaki, S., Yamaguchi, Y., Takenaka, H., Inhibitory effects of microalgae on the activation of hyaluronidase. *J. Appl. Phycol.*, 13, 6, 489–492, 2001. https://doi.org/10.1023/A:1012592620347.
160. Ciccone, M.M., Cortese, F., Gesualdo, M., Carbonara, S., Zito, A., De Pascalis, F., Scicchitano, P., Riccioni, G., Dietary intake of carotenoids and their

antioxidant and anti-inflammatory effects in cardiovascular care. *Mediat. Inflamm.*, 2013, 782137, 2013. doi:10.1155/2013/782137.

161. Sánchez, J.F., Fernández, J.M., Acién, F.G., Rueda, A., Pérez-Parra, J., Molina, E., Influence of culture conditions on the productivity and lutein content of the new strain Scenedesmus almeriensis. *Process Biochem.*, 43, 4, 398–405, 2008.
162. Huheihel, M., Ishanu, V., Tal, J., Arad, S.M., Antiviral effect of red microalgal polysaccharides on Herpes simplex and Varicella zoster viruses. *J. Appl. Phycol.*, 13, 127–134, 2001.
163. Goiris, K., Muylaert, K., Fraeye, I., Foubert, I., De Brabanter, J., De Cooman, L., Antioxidant potential of microalgae in relation to their phenolic and carotenoid content. *J. Appl. Phycol.*, 24, 1477–1486, 2012.
164. Samarakoon, K.W., O-Nam, K., Ko, J.Y., Lee, J.H., Kang, M.C., Kim, D., Lee, J.B., Lee, J.S., Jeon, Y.J., Purification and identification of novel angiotensin-I converting enzyme (ACE) inhibitory peptides from cultured marine microalgae (Nannochloropsis oculata) protein hydrolysate. *J. Appl. Phycol.*, 25, 1595–1606, 2013.
165. El-Baky, H.H.A., El-Baz, F.K., El-Baroty, G.S., Production of phenolic compounds from *Spirulina* maxima microalgae and its protective effects. *Afr. J. Biotechnol.*, 8, 24, 7059–7067, 2009.
166. Abd El-Baky, H.H., El-Baz, F.K., El-Baroty, G.S., Enhancing antioxidant availability in wheat grains from plants grown under seawater stress in response to microalgae extract treatments. *J. Sci. Food Agric.*, 90, 2, 299–303, 2010. https://doi.org/10.1002/jsfa.3815.
167. Meireles, L.A., Guedes, A.C., Malcata, F.X., Lipid class composition of the microalga Pavlova lutheri: eicosapentaenoic and docosahexaenoic acids. *J. Agril. Food Chem.*, 51, 8, 2237–2241, 2003. https://doi.org/10.1021/jf025952y.
168. Hossain, Z., Kurihara, H., Hosokawa, M., Takahashi, K., Growth inhibition and induction of differentiation and apoptosis mediated by sodium butyrate in Caco-2 cells with algal glycolipids. *In Vitro Cell Dev. Biol. Anim.*, 41, 5, 154–159, 2005.
169. Maeda, N., Kokai, Y., Ohtani, S., Hada, T., Yoshida, H., Mizushina, Y., Inhibitory effects of preventive and curative orally administered spinach glycoglycerolipid fraction on the tumor growth of sarcoma and colon in mouse graft models. *Food Chem.*, 112, 1, 205–210, 2009.
170. Ahn, C.B., Jeon, Y.J., Kang, D.S., Shin, T.S., Jung, B.M., Free radical scavenging activity of enzymatic extracts from a brown seaweed Scytosiphon lomentaria by electron spin resonance spectrometry. *Food Res. Int.*, 37, 253–258, 2004. https://doi.org/10.1016/j.foodres.2003.12.002.
171. Hao, S., Yan, Y., Huang, W., Gai, F., Wang, J., Liu, L., Wang, C., C-phycocyanin reduces inflammation by inhibiting NF-jB activity through down regulating PDCD5 in lipopolysaccharide-induced RAW 2647 macrophages. *J. Funct. Foods*, 42, 21–29, 2018. https://doi.org/10.1016/j.jff.2018.01.008.

172. Yoshimoto, S., Okada, K., Hayashi, O., Immuno-regulatory and anti-infammatory actions of phycocyanin on Caco-2/U937 cells co-culture as a model of the intestinal barrier. *Funct. Foods Health Dis.*, 9, 7, 466–483, 2019.
173. Morris, H.J., Carrillo, O., Almarales, A., Bermudez, R.C., Lebeque, Y., Fontaine, R., Llaurado, G., Beltran, Y., Immunostimulant activity of an enzymatic protein hydrolysate from green microalga *Chlorella* vulgaris on undernourished mice Enzyme. *Microb. Technol.*, 40, 3, 456–460, 2007. https://doi.org/10.1016/j.enzmictec.2006.07.021.
174. Lopes, G., Sousa, C., Valentao, P., Andrade, P.B., Sterols in algae and health, in: *Bioactive compounds from marine foods: plant and animal sources*, B. Hernández-Ledesma and M. Herrero (Eds.), pp. 173–187, Wiley, Chichester, 2013.
175. Tarento, T.D., McClure, D.D., Vasiljevski, E., Schindeler, A., Dehghani, F., Kavanagh, J.M., Microalgae as a source of vitamin K1. *Algal Res.*, 36, 77–87, 2018. https://doi.org/10.1016/J.ALGAL.2018.10.008.
176. Day, A.G., Brinkmann, D., Franklin, S., Espina, K., Rudenko, G., Roberts, A., Howse, K.S., Safety evaluation of a highlipid algal biomass from *Chlorella* protothecoides. *Regul. Toxicol. Pharmacol.*, 55, 166–180, 2009.
177. Charles, C.N., Msagati, T., Swai, H., Chacha, M., Microalgae: an alternative natural source of bioavailable omega-3 DHA for promotion of mental health in East Africa. *Sci. Afr.*, 6, e00187, 2019. https://doi.org/10.1016/j.sciaf.2019.e00187.
178. Borowitzka, M., High-value products from microalgae –their development and commercialization. *J. Appl. Phycol.*, 25, 3, 743–756, 2013.
179. Gong, Y., Hu, H., Gao, Y., Xu, X., Gao, H., Microalgae as platforms for production of recombinant proteins and valuable compounds: progress and prospects. *J. Indus. Microb. Biotechnol.*, 38, 12, 1879–1890, 2011.
180. Nauroth, J.M., Liu, Y.C., Van Elswyk, M., Bell, R., Hall, E.B., Chung, G., Arterburn, L.M., Docosahexaenoic acid (DHA) and docosapentaenoic acid (DPAn-6) algal oils reduce inflammatory mediators in human peripheral mononuclear cells *in vitro* and paw edema *in vivo*. *Lipids*, 45, 5, 375–384, 2010.
181. Matsui, M.S., Muizzuddin, N., Arad, S., Marenus, K., Sulfated polysaccharides from red microalgae have anti-inflammatory properties *in vitro* and *in vivo*. *Appl. Biochem. Biotechnol.*, 104, 1, 13–22, 2003. https://doi.org/10.1385/abab:104:1:13.
182. Naviner, M., Berge, J.P., Durand, P., Le Bris, H., Antibacterial activity of the marine diatom Skeletonema costatum against aquacultural pathogens. *Aquac*, 174, 1, 15–24, 1999.
183. Das, B.K., Pradhan, J., Pattnaik, P., Samantaray, B.R., Samal, S.K., Production of antibacterials from the freshwater alga Euglena viridis (Ehren). *World J. Microbiol. Biotechnol.*, 21, 45–5, 2005.

184. Bhadury, P. and Wright, P.C., Exploitation of marine algae:biogenic compounds for potential antifouling applications. *Planta*, 219, 561–578, 2004. https://doi.org/10.1007/s00425-004-1307-5.
185. Ghasemi, Y., Moradian, A., Mohagheghzadeh, A., Antifungal and antibacterial activity of the microalgae collected from paddy fields of Iran: characterization of antimicrobial activity of Chroococcus disperses. *J. Biolog. Sci.*, 7, 904–910, 2007. https://doi.org/10.3923/jbs.2007.904.910.
186. Sibi, G., Inhibition of lipase and inflammatory mediators by *Chlorella* lipid extracts for antiacne treatment. *J. Adv. Pharm. Technol. Res.*, 6, 1, 7–12, 2015. https://doi.org/10.4103/2231-4040.150364.
187. Bhagavathy, S., Sumathi, P., Bell, I.J.S., Green algae Chlorococcum humicola – a new source of bioactive compounds with antimicrobial activity. *Asian Pacif. J. Trop. Med.*, 1, 1, 1–7, 2011.
188. Catarina, G.A., Barbosa, C.R., Amaro, H.M., Pereira, C.I., Xavier, M.F., Microalgal and cyanobacterial cell extracts for use as natural antibacterial additives against food pathogens. *Int. J. Food Sci. Technol.*, 46, 4, 862–870, 2011.
189. Najdenski, H.M., Gigova, L.G., Iliev, I.I., Pilarski, P.S., Lukavsky, J., Tsvetkova, I.V., Ninova, M.S., Kussovski, V.K., Antibacterial and antifungal activities of selected microalgae and cyanobacteria. *Int. J. Food Sci. Technol.*, 48, 7, 1533–1540, 2013. https://doi.org/10.1111/IJFS.12122.
190. Lee, J.B., Hayashi, K., Hirata, M., Kuroda, E., Suzuki, E., Kubo, Y., Hayashi, T., Antiviral sulfated polysaccharide from Navicula directa, a diatom collected from deep-sea water in Toyama Bay. *Biolog. Pharm. Bull.*, 29, 10, 2135–2139, 2006.
191. Fabregas, J., Garcıa, D., Fernandez-Alonso, M., Rocha, A.I., Gómez Puertas, P., Escribano, J.M., Otero, A., Coll, J.M., *In vitro* inhibition of the replication of haemorrhagic septicaemia virus (VHSV) and African swine fever virus (ASFV) by extracts from marine microalgae. *Antiviral Res.*, 44, 1, 67–73, 1999.
192. Hasui, M., Matsuda, M., Okutani, K., Shigeta, S., *In vitro* antiviral activities of sulfated polysaccharides from a marine microalga (Cochlodinium polykrikoides) against human immunodeficiency virus and other enveloped viruses. *Int. J. Biolog. Macromol.*, 17, 5, 293–297, 1995.
193. Huang, J., Chen, B., You, W., Studies on separation of extracellular polysaccharide from Porphyridium cruentum and its anti-HBV activity *in vitro*. *Chin. J. Mar. Drug.*, 24, 5, 18–20, 2001.
194. Ahmadi, A., Moghadamtousi, S.Z., Abubakar, S., Zandi, K., Antiviral potential of algae polysaccharides isolated from marine sources: a review. *Biomed. Res. Int.*, 2015, 825203, 2015. https://doi.org/10.1155/2015/825203.
195. Yim, J.H., Kim, S.J., Ahn, S.H., Lee, C.K., Rhie, K.T., Lee, H.K., Antiviral effects of sulfated exopolysacchride from the marine microalga Gyrodinium impudicum strain KG03. *Mar. Biotechnol.*, 6, 17–25, 2004.

196. Washida, K., Koyama, T., Yamada, K., Kitab, M., Uemura, D., Karatungiols A and B, two novel antimicrobial polyol compounds, from the symbiotic marine dinoflagellate Amphidinium sp. *Tetrahedron Lett.*, 47, 2521–2525, 2006. https://doi.org/10.1016/J.TETLET.2006.02.045.
197. Santoyo, S., Rodríguez-Meizoso, I., Cifuentes, A., Jaime, L., García-Blairsy Reina, G., Señorans, F.J., Ibáñez, E., Green processes based on the extraction with pressurized fluids to obtain potent antimicrobials from Haematococcus pluvialis microalgae. *LWT Food Sci. Technol.*, 42, 1213–1218, 2009.
198. Inan, B., Cakir Koc, R., Ozcimen, D., Comparison of the anticancer effect of microalgal oils and microalgal oilloaded electrosprayed nanoparticles against PC-3, SHSY5Y and AGS cell lines. *Artif. Cells Nanomed. Biotechnol.*, 49, 1, 381–389, 2021.
199. Wang, X. and Zhang, X., Separation, antitumor activities, and encapsulation of polypeptide from *Chlorella* pyrenoidosa. *Biotechnol. Prog.*, 29, 3, 681–687, 2013. https://doi.org/10.1002/btpr.1725.
200. Mendes, R.L., Fernandes, H.L., Coelho, J.P., Reis, E.C., Cabral, J.M.S., Novais, J.M., Palabra, A.F., Supercritical CO_2 extraction of carotenoids and other lipids from *Chlorella vulgaris*. *Food Chem.*, 53, 1, 99–103, 1995. doi:10.1016/0308-8146(95)95794-7.
201. Sheih, I.C., Fang, T.J., Wu, T.K., Lin, P.H., Anticancer and antioxidant activities of the peptide fraction from algae protein in waste. *J. Agril Food Chem*, 58, 1202–1207, 2010.
202. Luo, X., Su, P., Zhang, W., Advances in microalgae-derived phytosterols for functional food and pharmaceutical applications. *Mar. Drugs*, 13, 7, 4231–4254, 2015.
203. Ahmed, F., Zhou, W., Schenk, P.M., Pavlova lutheri is a high level producer of phytosterols. *Algal Res.*, 10, 210–217, 2015. https://doi.org/10.1016/j.algal.2015.05.013.
204. Giammanco, M., Di Majo, D., La Guardia, M., Aiello, S., Crescimannno, M., Flandina, C., Tumminello, F.M., Leto, G., Vitamin D in cancer chemoprevention. *Pharm. Biol.*, 53, 10, 1399–1434, 2015. https://doi.org/10.3109/13880209.2014.988274.
205. Kim, S.K. and Kang, K.H., Medicinal effects of peptides from marine microalgae. *Adv. Food Nutr. Res.*, 64, 313–323, 2011. https://doi.org/10.1016/B978-0-12-387669-0.00025-9.
206. de Morais, M.G., Vaz, B.D., de Morais, E.G., Costa, J.A., Biologically active metabolites synthesized by microalgae. *Biomed Res Int.*, 2015, 835761, 2015. doi:10.1155/2015/835761.
207. Mishra, V.K., Bacheti, R.K., Husen, A., Medicinal uses of chlorophyll: a critical overview, in: *Chlorophyll: Chlorophyll Structure Funct. Medicinal Uses*. H. Le, E. Salcedo, (Eds.), Nova Science Publishers, Hauppauge (NY), pp. 177–196, 2011.
208. Gürlek, C., Yarkent, C., Köse, A., Tuğcu, B., Gebeloğlu, I.K., Öncel, S.S., Elibol, M., Screening of antioxidant and cytotoxic activities of several

microalgal extracts with pharmaceutical potential. *Health Technol.*, 10, 1, 111–117, 2019. https://doi.org/10.1007/s12553-019-00388-3.

209. Zhang, J., Liu, L., Ren, Y., Chen, F., Characterization of exopolysaccharides produced by microalgae with antitumor activity on human colon cancer cells. *Int. J. Biolog. Macromol.*, 128, 761–767, 2019.
210. Abida, H., Ruchaud, S., Rios, L., Humeau, A., Probert, I., Vargas C., D., Bach, S., Bowler, C., Bioprospecting marine plankton. *Mar. Drugs*, 11, 11, 4594–4611, 2013.
211. Sadovskaya, I., Souissi, A., Souissi, S., Grard, T., Lencel, P., Greene, C.M., Duin, S., Dmitrenok, P.S., Chizhov, A.O., Shashkov, A.S., Usov, A.I., Chemical structure and biological activity of a highly branched (1? 3, 1? 6)-b-D-glucan from Isochrysis galbana. *Carbohydr. Polym.*, 111, 139–148, 2014.
212. Challouf, R., Trabelsi, L., Ben Dhieb, R., El Abed, O., Yahia, A., Ghozzi, K., Ben Ammar, J., Omran, H., Ben Ouada, H., Evaluation of cytotoxicity and biological activities in extracellular polysaccharides released by cyanobacterium Arthrospira platensis. *Braz. Arch. Biol. Technol.*, 54, 831–838, 2011.
213. Sheih, I.C., Fang, T.J., Wu, T.K., Isolation and characterization of a novel angiotensin-I converting enzyme (ACE) inhibitory peptide from the algae protein waste. *Food Chem.*, 115, 279–284, 2009.
214. Becker, E., Microalgae for human and animal nutrition, in: *Handbook of microalgae culture: applied phycology and biotechnology*, vol. 312, A. Richmond (Ed.), pp. 461–503, Blackwell Science, London, 2004.
215. Caporgno, M.P. and Mathys, A., Trends in microalgae incorporation into innovative food products with potential health benefits. *Front. Nutr.*, 5, 58, 2018. https://doi.org/10.3389/fnut.2018.00058.
216. Kleinegris, D.M.M., Janssen, M., Brandenburg, W.A., Wijffels, R.H., Continuous production of carotenoids from Dunaliella salina. *Enzyme Microb. Technol.*, 48, 3, 253–259, 2011. doi:10.1016/j.enzmictec.2010.11.005.
217. Long, S.F., Kang, S., Wang, Q.Q., Xu, Y.T., Pan, L., Hu, J.X., Li, M., Piao, X.S., Dietary supplementation with DHA-rich microalgae improves performance, serum composition, carcass trait, antioxidant status, and fatty acid profile of broilers. *Poult. Sci.*, 97, 6, 1881–1890, 2018.
218. Koller, M., Muhr, A., Braunegg, G., Microalgae as versatile cellular factories for valued products. *Algal Res.*, 6, 52–63, 2014. https://doi.org/10.1016/j.algal.2014.09.002.
219. Chiranjeevi, P. and Venkata, M.S., Diverse acidogenic effluents as feedstock for microalgae cultivation: dual phase metabolic transition on biomass growth and lipid synthesis. *Bioresour. Technol.*, 242, 191–196, 2017.
220. Alzahrani, M., Proteins and their enzymatic hydrolysates from the marine diatom Nitzschia laevis and screening for their *in vitro* antioxidant, anti-hypertension, anti-inflammatory and antimicrobial activities [PhD diss.]. Auckland (New Zealand): Research Space Auckland; 2018.

221. Priatni, S., Ratnaningrum, D., Kosasih, W., The screening of antidiabetic activity and the cultivation study of local marine microalgae, in: *IOP Conf. Ser. Mater. Sci. Eng*, vol. 1011, p. 012066, 2021.
222. Rohit, M.V. and Venkata, M.S., Quantum yield and fatty acid profile variations with nutritional mode during microalgae cultivation. *Front. Bioengg. Biotechnol.*, 6, 111, 2018.
223. Gong, M. and Bassi, A., Carotenoids from microalgae: a review of recent developments. *Biotechnol. Adv.*, 34, 8, 1396–1412, 2016.
224. Rasmussen, H.M. and Johnson, E.J., Nutrients for the aging eye. *Clin. Interven. Aging*, 8, 741–748, 2013.
225. Liu, T., Liu, W.H., Zhao, J.S., Meng, F.Z., Wang, H., Lutein protects against β-amyloid peptide-induced oxidative stress in cerebrovascular endothelial cells through modulation of Nrf-2 and NF-κb. *Cell Biol. Toxicol.*, *33*, 57–67, 2017.
226. Capelli, B., Bagchi, D., Cysewski, G.R., Synthetic astaxanthin is significantly inferior to algal-based astaxanthin as an antioxidant and may not be suitable as a human nutraceutical supplement. *Nutrafoods*, 12, 145–152, 2013.
227. Seo, C., Sohn, J.H., Oh, H., Kim, B.Y., Ahn, J.S., Isolation of the protein tyrosine phosphatase 1B inhibitory metabolite from the marine-derived fungus Cosmospora sp SF-5060. *Bioorg. Med. Chem. Lett.*, 19, 6095–6097, 2009.
228. Peng, J., Yuan, J.P., Wu, C.F., Wang, J.H., Fucoxanthin, a marine carotenoid present in brown seaweeds and diatoms: metabolism and bioactivities relevant to human health. *Mar. Drugs*, 9, 10, 1806–1828, 2011. https://doi.org/10.3390/md9101806.
229. Simoons, F.J., *Food in China: A Cultural and Historical Inquiry*, CRC Press, Boca Raton, FL, USA, 1991.
230. Kirk, E.A. and Behrens, P.W., Commercial developments in microalgal biotechnology. *J. Phycol.*, 35, 215e226, 1999.
231. Arterburn, L.M., Hall, E.B., Oken, H., Distribution, interconversion, and dose response of n-3 fatty acids in humans. *Am. J. Clin. Nutr.*, 83, 1467e1476, 2006.
232. Robles Medina, A., Molina Grima, E., Gimenez Gimenez, A., Ibanez Gonzalez, M.J., Downstream processing of algal polyunsaturated fatty acids. *Biotechnol. Adv.*, 16, 517e580, 1998.
233. Cardozo, K.H.M., Guaratini, T., Barros, M., Falcao, V., Tonon, A.P., Lopes, N.P., Campos, S., Torres, M.A., Souza, A.O., Pinto, E., Metabolites from algae with economical impact. *Comp. Biochem. Physiol. Part C*, 146, 60e78, 2007.
234. Sayanova, O.V. and Napier, J.A., Eicosapentaenoic acid: biosynthetic routes and the potential for synthesis in transgenic plants. *Phytochemistry*, 65, 147e158, 2004.
235. Qiu, X., Biosynthesis of docohexanoic acid (DHA): two distinct pathways. *Prostaglandisn leukotrienes essential fatty acids*, 68, 181e186, 2003.
236. Funk, C.D., Prostaglandins and leukotrienes: advances in eicosanoids biology. *Science*, 294, 1871e1875, 2001.

237. Petrie, J.R., Shrestha, P., Mansour, M.P., Nichols, P.D., Liu, Q., Singh, S.P., Metabolic engineering of omega-3 long-chain polyunsaturated fatty acids in plants using an acyl-CoA [delta] 6-desaturase with 33 3- preference from the marine microalga micromonas pusilla. *Metab. Eng.*, 12, 3, 233e240, 2010.
238. Gellenbeck, K.W., Utilization of algal materials for nutraceutical and cosmeceutical aplicationsdwhat do manufacturers need to know? *J. Appl. Phycol.*, 24, 309e313, 2012.
239. Desmorieux, H. and Decaen, N., Convective drying of *Spirulina* in thin layer. *J. Food Eng.*, 66, 497e503, 2005.
240. Muller-Feuga, A., The role of microalgae in aquaculture: situation and trends. *J. Appl. Phycol.*, 12, 527e534, 2000.
241. Borowitzka, M.A., Microalgae for aquaculture: opportunities and constraints. *J. Appl. Phycol.*, 9, 393e401, 1991.
242. Becker, W., Microalgae for aquaculture. The nutritional value of microalgae for aquaculture, in: *Handbook of Microalgal Culture*, A. Richmond, (Ed.), p. 380e391, Blackwell, Oxford, 2004.
243. Aragao, C., Conceiçao, L.E.C., Dinis, M.T., Fyhn, H.-J., Amino acid pools of rotifers and Artemia under different conditions: nutritional implications for fish larvae. *Aquaculture*, 234, 429e445, 2004.
244. Chuntapa, D., Powtongsook, S., Menasveta, P., Water quality control using *Spirulina* platensis in shrimp culture tanks. *Aquaculture*, 220, 355e366, 2003.
245. Lio-Po, G.D., Leano, E.M., Penaranda, M.M.D., Villa-Franco, A.U., Sombito, C.D., Guanzon, N.G., Antiluminous Vibrio factors associated with the 'green water' growout culture of the tiger shrimp Penaeus monodon. *Aquaculture*, 250, 1e7, 2005.
246. Rodolfi, L., Chini Zittelli, G., Barsanti, L., Rosati, G., Tredici, M.R., Growth medium recycling in Nannochloropsis sp. mass cultivation. *Biomol. Eng*, 20, 243e248, 2003.
247. Waldenstedt, L., Inborr, J., Hansson, I., Elwinger, K., Effects of astaxanthin-rich algal meal (*Haematococcus pluvalis*) on growth performance, caecal campylobacter and clostridial counts and tissue astaxanthin concentration of broiler chickens. *Anim. Feed Sci. Technol.*, 108, 119e132, 2003.
248. Zittelli, C.G., Lavista, F., Bastianini, A., Rodolfi, L., Vincenzini, M., Tredici, M.R., Production of eicosapentaenoic acid by nannochloropsis sp. cultures in outdoor tubular photobioreactors. *J. Biotechnol.*, 70, 299e312, 1999.
249. Yuan, J.P., Peng, J., Yin, K., Wang, J.H., Potential health promoting effects of astaxanthin: a high-value carotenoid mostly from microalgae. *Mol. Nutr. Food Res.*, 55, 150e165, 2011.
250. Kim, S.H., Jean, D.I., Lim, Y.P., An, G., Weight gain limitations and liver protection by long term feeding of astaxanthin in murines. *J. Korean Soc. Appl. Biol. Chem.*, 52, 180e185, 2009.
251. Gradelet, S., Le Bon, A.M., Berges, R., Suchelet, M., Astorg, P., Dietary carotenoids inhibit aflotoxin B1-induced liver preneoplastic foci and DNA

damage on the rats: role of the modulation of aflatoxin B1 metabolism. *Carcinogenesis*, 19, 403e411, 1998.
252. Jiang, Y., Chen, F., Liang, S.-Z., Production potential of docosahexaenoic acid by the heterotrophic marine dinoflagellate Crypthecodinium cohnii. *Process Biochem.*, 34, 633e637, 1999.
253. Wen, Z.Y. and Chen, F., Heterotrophic production of eicosapentaenoic acid by microalgae. *Biotechnol. Adv.*, 21, 273e294, 2003.
254. Kroes, R., Schaefer, E.J., Squire, R.A., Williams, G.M., A review of the safety of DHA45-oil. *Food Chem. Toxicol.*, 41, 1433e1446, 2003.
255. Gershwin, M.E. and Belay, A., *Spirulina in Human Nutrition and Health*, CRC Press, Boca Raton, FL, 2008.
256. Iwata, K., Inayama, T., Kato, T., Effects of *Spirulina* platensis on plasma lipoprotein lipase activity in fructose-induced hyperlipidemic rats. *J. Nutr. Sci. Vitaminol (Tokyo)*, 36, 65e171, 1990.
257. Mishima, T., Murata, J., Toyoshima, M., Fujii, H., Nakajima, M., Inhibition of tumor invasion and metastasis by calcium spirulan (Ca-SP), a novel sulfated polysaccharide derived from a blue-green alga, *Spirulina* platensis. *Clin. Exp. Metastasis*, 16, 541e550, 1998.
258. de Caire, G.Z., de Cano, M.S., de Mule, C.Z., Steyerthal, N., Piantanida, M., Effect of *Spirulina* platensis on glucose, uric acid and cholesterol levels in the blood of rodents. *Intern. J. Exp. Bot.*, 57, 93e96, 1995.
259. Romay, C., Armesto, J., Remirez, D., Gonzalez, R., Ledon, N., *et al.*, Antioxidant and anti-inflammatory properties of C-phycocyanin from blue-green algae. *Inflamm. Res.*, 47, 36e41, 1998.
260. Archer, D.L. and Glinsmann, W.H., Intestinal infection and malnutrition initiate acquired immune deficiency syndrome (AIDS). *Nutr. Res.*, 5, 9e19, 1985.
261. Morita, K., Matsueda, T., Iida, T., Hasegawa, T., *Chlorella* accelerates dioxin excretion in rats. *J. Nutr.*, 129, 1731e1736, 1999.
262. Shim, J.Y., Shin, H.S., Han, J.G., Park, H.S., Lim, B.L., Protective effects of *Chlorella* vulgaris on liver toxicity in cadmium-administered rats. *J. Med. Food*, 11, 479e485, 2008.
263. Shibata, S., Natori, Y., Nishihara, T., Tomisaka, K., Matsumoto, K., Antioxidant and anti-cataract effects of *Chlorella* on rats with streptozotocin induced diabetes. *J. Nutr. Sci. Vitaminol (Tokyo)*, 49, 334e339, 2003.
264. Gouveia, L., Batista, A.P., Miranda, A., Empis, J., Raymundo, A., *Chlorella* vulgaris biomass used as colouring source in traditional butter cookies. *Innovative Food Sci. Emerging Technol.*, 8, 433e436, 2007.
265. Wang, J., Qin, S., Lin, J., Wang, Q., Li, W., Gao, Y., Phycobiliproteins from microalgae: research progress in sustainable production and extraction processes. *Biotechnol. Biofuels Bioprod.*, 16, 170, 2023.
266. Jacob-Lopes, E., Maroneze, M.M., Deprá, M.C., Sartori, R.B., Dias, R.R., Zepka, L.Q., Bioactive food compounds from microalgae: An innovative framework on industrial biorefineries. *Curr. Opin. Food Sci.*, 25, 1–7, 2019.

267. de Jesus Raposo, M.F., De Morais, A.M.M.B., De Morais, R.M.S.C., Bioactivity and applications of polysaccharides from marine microalgae, in: *Polysaccharides Bioactivity Biotechnol.* K.G. Ramawat, J.-M. Mérillon (Eds.), pp. 1683–1727, Cham, Switzerland, Springer Nature Switzerland AG, 2015. doi:10.1007/978-3-319-16298-0_47.
268. Bernal, P. and Llamas, M.A., Promising biotechnological applications of antibiofilm exopolysaccharides. *Microb. Biotechnol.*, 5, 6, 670, 2012.
269. Moreira, J.B., Vaz, B.S., Barcelos Cardias, B., Gonzales Cruz, C., Araujo de Almeida, A.C., Vieira Costa, J.A., Greque de Morais, M., Microalgae polysaccharides: an alternative source for food production and sustainable agriculture. *Polysaccharides*, 3, 2, 441–457, 2022.
270. de Jesus Raposo, M.F., Miranda Bernardo De Morais, A.M., Santos Costa De Morais, R.M., Emergent sources of prebiotics: Seaweeds and microalgae. *Mar. Drugs*, 14, 2, 27, 2016.
271. Moreno-Garcia, L., Adjallé, K., Barnabé, S., Raghavan, G.S.V., Microalgae biomass production for a biorefinery system: recent advances and the way towards sustainability. *Renewable Sustainable Energy Rev.*, 76, 493–506, 2017.
272. Pierre, G., Delattre, C., Dubessay, P., Jubeau, S., Vialleix, C., Cadoret, J.-P., Probert, I., Michaud, P., What is in store for EPS microalgae in the next decade? *Molecules*, 24, 23, 4296, 2019.
273. Barbarino, E. and Lourenço, S.O., An evaluation of methods for extraction and quantification of protein from marine macro-and micro-algae. *J. Appl. Phycol.*, 17, 447–460, 2005.
274. Gellenbeck, K.W., Utilization of algal materials for nutraceutical and cosmeceutical application what do manufacturers need to know? *J. Appl. Phycol.*, 24, 309e313, 2012.
275. Chen, Z., Wang, L., Qiu, S., Ge, S., Determination of Microalgal Lipid Content and Fatty Acid for Biofuel Production. *BioMed Res. Int.*, 2018, Article ID 1503126, https://doi.org/10.1155/2018/1503126.
276. Schmidt, E.B. and Skou, H.A., Christensen JH, Dyerberg. N-3 fatty acids from fish and coronary artery disease: implications for public health. *J. Public Health Nutr.*, 3, 1, 91e98, 2000.
277. Patil, V., Kallqvist, T., Olsen, E., Vogt, G., Gislerød, H.R., Fatty acid composition of 12 micro-algae for possible use in aquaculture feed. *Aquacul. Int.*, 15, 1e9, 2007.
278. Kaur, M., Bhatia, S., Gupta, U., Decker, E., Tak, Y., Bali, M., Gupta, V.K., Dar, R.A., Bala, S., Microalgal bioactive metabolites as promising implements in nutraceuticals and pharmaceuticals: inspiring therapy for health benefits. *Phytochem Rev.*, 1–31, 2023 Jan 14. doi:10.1007/s11101-022-09848-7.

4

Extraction of Drugs and Food Ingredients from Marine Macroalgae

P. Ruban[1]*, Vinitha Ebziba C.[2], K. Ramalashmi[3], S. Arul Diana Christie[4], Mohammed Rafiqkhan K.[5] and Kavitha M.B.[6]

[1]*Department of Biotechnology, Nehru Arts and Science College, Coimbatore, India*
[2]*Department of Biotechnology and Research, Shri Nehru Maha Vidyalaya College of Arts and Science, Coimbatore, India*
[3]*Department of Biotechnology, Bharathiar University, Coimbatore, India*
[4]*Department of Microbiology, Sri Ramakrishna College of Arts and Science for Women, Coimbatore, India*
[5]*Department of Biology, School of Life Sciences, St Joseph's University, Bengaluru, Karnataka, India*
[6]*Department of Microbiology, Shri Nehru Maha Vidyalaya College of Arts and Science, Coimbatore, India*

Abstract

Marine macroalgae have gained significant attention lately due to their bioactive compounds, which are used in pharmaceutical and food applications. Increasing demand for sustainable and natural alternatives in several industries has led to the recognition of marine macroalgae as a promising solution. This is attributed to their abundance, diversity, and unique chemical composition.

In this chapter review, marine macroalgae, extraction of drugs from marine macroalgae, and extraction of food ingredients from marine macroalgae are discussed.

Keywords: Marine macroalgae, drugs from marine algae, food ingredients from marine macroalgae, green algae, brown algae

**Corresponding author*: rubanbiotechz@gmail.com

Shaju S. S., Vipin. P. M. and Visakh. P.M. (eds.) Marine Biotechnology: Studies and Applications, (111–138) © 2025 Scrivener Publishing LLC

4.1 Introduction

Macroalgae, otherwise known as seaweeds, are large, plant-like organisms thriving in marine environments. These organisms are integral parts of the ecosystems that exist on coastlines and in oceans all over the world. These underwater forests have remarkable diversity in terms of size, shape, and color ranging from simple filamentous forms to very complex structures. Three broad categories are identified as follows: green (Chlorophyta), brown (Phaeophyceae), and red (Rhodophyta) algae depending on their pigmentation and biochemical properties.

Macroalgae provide habitats and serve as food for marine animals but, more importantly, participate in nutrient circulation and shore protection. Some marine organisms, from tiny invertebrates and small crustaceans up to great fish species, rely upon the macroalgae shelters and feeding grounds on account of the strong, more robust structure. In sum, seaweeds strengthen shorelines, thus promoting balance within an ecosystem with lessened coastal erosion brought about by nutrient absorption coupled with carbon dioxide elimination through water [1].

Ecological Role of Marine Macroalgae

Marine macroalgae are also termed ecological keystones within the marine environment. Marine macroalgae contribute to biodiversity and stability to the ecosystem in addition to general health. The most significant roles provided by marine macroalgae include the following:

Habitat Provision: Seaweed beds, especially if kelp species form it, are very complex and, as such, serve as nurseries to young fish and crustaceans.
Food Source: Macroalgae are a direct and indirect food source for many marine organisms ranging from the smallest herbivores to large fish and mammals.
Nutrient Cycling and Water Quality: Seaweeds play a significant role in nutrient cycling absorbing excess nitrogen and phosphorus from the water. This can reduce eutrophication, which is the process leading to oxygen depletion and harmful algal blooms.
Carbon Sequestration and Ocean Health: Macroalgae utilize much of the carbon dioxide in the atmosphere through photosynthesis, which lowers ocean acidity and sequesters carbon for the good of marine ecosystems and reduction of atmospheric CO_2.

Biochemical Composition of Marine Macroalgae

The biochemical composition of marine macroalgae is diverse and complex allowing them to thrive in all kinds of settings. These diverse biochemical compositions offer an abundance of compounds of significant importance.

Polysaccharides: Among the most studied constituents, seaweed polysaccharides include alginate, carrageenan, and agar. These are hybrid carbs of great utility in numerous industrial uses.
Alginate: Alginate is commonly utilized in food preparation as a gelling and suspending agent, as well as in pharmacy for drug delivery systems, primarily within brown algae.
Carrageenan: Carrageenan is a gelling agent extracted from red algae; food preparations often employ carrageenan to add texture to products like dairy and meat substitutes.
Agar: The gelling agent obtained from red algae, agar is used in food preparation and also plays a vital role in microbiological culture media. It promotes the growth of cultures.
Marine macroalgal proteins are a rich source of essential amino acids and are very nutritious, thereby gaining much attention as an alternative source of conventional protein sources. Recently, this quality has contributed to interest from the food industry where such demand is becoming popular as a source of plant-based, sustainable protein.
Polyphenols: Marine macroalgae, specifically brown algae, contain a wealth of polyphenolic compounds, phlorotannins, renowned for their antioxidant, anti-inflammatory, and antimicrobial properties and, thus, great promise in disease prevention and functional food products and nutraceuticals.
Pigments: There are pigments, including chlorophyll, carotenoids, and phycobiliproteins, that exist in macroalgae and display distinct individual properties. Such pigments provide colors and contribute to photosynthetic efficiency and have possible antioxidant effects that are applied in health supplements and cosmetics.
Lipids: The seaweed lipids consist of omega-3 and omega-6 types of fatty acids that are of essential benefits to humans. The compounds serve as an alternative source replacing oils extracted from fish. In the pharmaceuticals and functional food sectors, the products are promising.

Marine Macroalgae: Possible Applications

Biochemical contents in marine macroalgae have made them applicable in many sectors from food to pharmaceutical, cosmetic, and biotechnology industries.

Food Industry: Application of seaweeds in the food industry: it is a thickener, stabilizer, and gelling agent in processed foods. The protein and nutrient content also make it useful for dietary supplements and health foods, especially in the rapidly growing market for plant-based diets.
Pharmaceuticals: Bioactive compounds of seaweeds, like polysaccharides and polyphenols, have drawn much attention due to possible application in drug development. Among the applications, these bioactive compounds have been explored to work as antioxidants, anti-inflammatory agents, and antimicrobial agents.
Cosmetics: There are applications in cosmetics for moisturizing, anti-aging, and soothing of the skin. Polysaccharides create textures in skincare formulations, and polyphenols and pigments are added for antioxidant purposes.
Agriculture and Animal Feed: This compositional quality renders seaweed extracts as natural fertilizers. It also forms animal feed to improve both growth and immunity among livestock.
Alginate is used as a bioprinting material and for cell encapsulation, while the property of seaweed to absorb carbon and nutrients makes it ideal for environmental remediation in reducing eutrophication and ocean acidification [2, 3].

4.2 Marine Macroalgae

Marine macroalgae have gained significant attention lately due to their bioactive compounds used in pharmaceutical and food applications. Increasing demand for sustainable and natural alternatives in several industries has led to the recognition of marine macroalgae as a promising solution. This is attributed to their abundance, diversity, and unique chemical composition. Extensive research has been done to identify the bioactive compounds of marine macroalgae to fully exploit their therapeutic and nutritional potential through efficient extraction techniques. These molecules include polysaccharides, proteins, polyphenols, pigments, and lipids, which exhibit different pharmacological and nutritional properties [4].

Marine macroalgae extraction of bioactive compounds is a multidisciplinary activity that involves chemistry, biology, and engineering principles. Advanced techniques, including supercritical fluid extraction, ultrasound-assisted extraction, and enzyme-assisted extraction, have evolved from the traditional methods that include Soxhlet extraction and maceration. The advanced techniques are more efficient and selective in the extraction of target compounds. The bioactive compounds that are extracted from marine macroalgae are diversified through their pharmacological activities as follows: antioxidant, anti-inflammatory, antimicrobial, antiviral, and anticancer. The compounds also support nutritional aspects, including dietary fiber, proteins, essential amino acids, omega-3 fatty acids, and vitamins. Consequently, marine macroalgae have immense potential as a source of functional food ingredients and dietary supplements [5].

4.3 Classification of Marine Macroalgae

The three main taxonomic groups of seaweeds are green algae (Chlorophyta), brown algae (Phaeophyceae), and red algae (Rhodophyta). Within each group, there are many species that vary in their characteristics, habitats, and biochemical composition [6–8].

Green Algae (Chlorophyta)
Green algae include several genera, such as *Ulva*, *Enteromorpha*, and *Codium*, classified under the class Chlorophyta.

- *Ulva* is known as sea lettuce and often occurs in the intertidal areas. With fragile, thin, almost translucent sheets, it takes on a bright green hue. In addition, it is quite flexible to the conditions in its environment.
- It is a filamentous, green alga known by the name Enteromorpha. It grows in large patches of dense mats in shallow coastal waters and forms more or less abundant populations, usually in estuaries, tide pools, or high nutrient areas.
- Codium, also known as dead man's fingers is one species of conspicuous green algae. Codium has branching finger-like growth and typically occurs on rock-associated intertidal sites. Codium is robust because it tolerates low tide where it dries exposed to air.

Brown Algae (Phaeophyceae)

- Fucus: Fucus is one of the widespread brown algae found within intertidal rocky areas on all continents. The flattened, strap-like fronds of Fucus are accompanied by air bladders enabling float and wave impact resistivity.
- Sargassum (Sargasso weed): Sargassum species are large brown algae that form floating mats known as sargassum rafts. These rafts are home to many types of marine life and are regularly observed in tropical and subtropical waters.
- The algae thrives in cold temperate marine environment, and their usual hue is brown because of the pigment fucoxanthin.
- Laminaria (Kelp): Laminaria species, or kelps, are large brown algae that form dense underwater forests in the temperate and polar zones. They have long ribbons attached to a holdfast anchored to the seabed.

All these different species of marine macroalgae in the oceans and seas of the world demonstrate varied ecological roles, adaptations, and economic importance. They play an important role in enhancing the biodiversity and functionality of marine ecosystems.

Geographical Distribution of Seaweed Diversity

This has the consequence that the spatial structure of seaweed species varies in accordance with the influence of several factors, which range from water temperature, salinity, nutrient availability, substrate composition, to wave action. Thus, specific species have been adapted for unique environmental conditions and thus are found in other areas of the world [9–12].

Temperate Regions

Brown algae, also known as Phaeophyta, are often found in temperate regions worldwide. These are mostly marine algae and are identified by their characteristic brown color, which is caused by the pigment fucoxanthin. They are usually seen in coastal regions, especially in cold, nutrient-rich waters. The temperate regions offer ideal conditions for the growth of brown algae due to their moderate temperatures and richness in nutrients. Brown algae are commonly found in intertidal regions where they attach to rocks or other substrata with holdfasts. Some species can even form large underwater forests called kelp forests, which offer habitats and food to various marine organisms. Apart from their ecological significance, brown algae are also of economic importance. They are harvested for a variety of purposes including food (e.g., seaweed snacks, sushi wraps), cosmetics, and pharmaceuticals. Additionally, some species of brown algae are under research for their use in biofuel production due to the fact that they contain high amounts of carbohydrates. Brown algae, such as kelp (Laminaria), are common in temperate coastal areas with cold water temperatures and rocky substrates. These are the coastlines of North America, Europe, and parts of Asia.

Tropical Regions

Red algae are also referred to as Rhodophyta and inhabit tropical oceans throughout the world. They generally inhabit marine surroundings and, because of the existence of pigment phycoerythrin, are typically red. They inhabit predominantly warm, shallow waters, usually attached to a rock or some other substance in a coral reef ecosystem. These conditions are ideal for red algae growth because the warmth, as well as large exposure to sunlight, help foster photosynthesis. It thrives in the abundance found within coral reef communities where it plays important roles. Red algae offer a dwelling and food source that supports different marine species while stabilizing the structure of these complex ecosystems.

Besides their ecological significance, red algae also hold economic value. They are utilized to produce a wide array of products, which find extensive use in food products in the form of food additives, thickeners, and gelling agents. Algae extracts are used in various cosmetics, pharmaceutical, and biotechnology products. Summarily, red algae are an important part of tropical marine ecosystems that provide essential ecological diversity as well as economic value. The species of red algae, for example, *Gracilaria*, are often seen in tropical and subtropical areas with warm water temperature and coral reef ecosystems. Some of the areas include the Caribbean, Southeast Asia, and the Pacific Islands.

Polar Regions

Only a few species of brown algae have been found in the polar region, which includes the Arctic and Antarctic. Brown algae are generally associated with temperate regions, but some species have adapted to the extreme conditions of polar environments. The brown algae commonly grow in intertidal zones where they attach to rocks or other surfaces. They have developed special adaptations to survive the cold temperatures, low light conditions, and seasonal changes that characterize polar regions. For instance, they could have mechanisms that allow them to tolerate freezing temperatures as well as strategies to effectively utilize the scarce nutrients present. Brown algae are important ecologically in polar ecosystems as they are habitats and sources of food for many marine organisms. These actually constitute a part of the whole biodiversity and functioning of marine ecosystems that exist within polar regions. Nevertheless, the brown algae are often not as diverse as those found within the temperate or tropical. However, since there are some Phaeophyta members that survived under these harsh conditions, it, too, indicates some tenacity as well as variability. Notably, such cold waters, rich in nutrients at the polar regions, such as the Arctic and Antarctic, support Fucus species.

Intertidal Zones

Intertidal regions are dynamically located regions along the coast with periods of being covered in water during high tide and exposed to air through low tide. Seaweeds, which encompass algae of different species, thrive in intertidal regions all over the world. These seaweeds adapt to the severe conditions characterizing this habitat, namely, exposure to air, changed temperatures, and wave turbulence. Some of the noticeable seaweeds that might be seen in intertidal regions include the following:

- Green algae (Chlorophyta): These types of green algae are commonly seen in intertidal zones, attaching to rocks, sand, or other surfaces. They often appear as green mats or clumps that form lush vegetation providing important functions to the ecosystem in intertidal environments.
- Red algae (Rhodophyta): Red algae are also typical in the intertidal zone, but most of them prefer warmer or subtropical locations. They may appear in light pink, deep red, or any other shades and might present themselves as crusts, filaments, or even leafy growths along the coastline.
- Brown algae (Phaeophyta): Several species of brown algae tolerate intertidal habitats and are usually fixed to rocks or other substrata; they may form brownish, mat-like masses or grow upright.

Intertidal seaweeds are important suppliers of shelter and food for a wide range of intertidal animals, from snails and sea urchins to crustaceans. They also play a crucial role in stabilizing sediments, preventing erosion, and promoting nutrient exchange along the coast. In general, these seaweeds face desiccation due to periodic exposure to air when the tide is low. To counter this problem, many species have developed mechanisms, such as mucilage production or specialized tissues, to retain moisture and survive dehydration. Several species of seaweed, including *Ulva* and

Corallina, have adapted to the intertidal zones where they are subjected to alternating periods of air and water due to tidal fluctuations.

4.4 Extraction of Drugs from Marine Macroalgae

The extraction method is crucial in getting bioactive substances from their natural sources, especially marine macroalgae. The technique is essential for effective separation and concentration of desired compounds from complex mixtures like plant matter, algae, and many biological specimens. In a more literal sense, it would also be very beneficial if there were ways to understand extraction techniques regarding bioactive compounds found within marine macroalgae using different perspectives such as those found in refs. [13–17].

1. Efficiency: The efficiency of the recovery of bioactive compounds from marine macroalgae depends on extraction techniques. It can be determined by how an extraction method, solvent, temperature, and duration may affect yield and purity of the obtained compounds. It is critical to optimize extraction conditions so that maximum recovery of the target compound can be ensured with minimum degradation of the sensitive molecules.
2. Selectivity: Different extraction techniques provide various selectivities to the desired classes of bioactive compounds. For instance, polar solvents are suitable for the extraction of water- or ethanol-soluble compounds, such as polysaccharides and polyphenols, while hexane or chloroform and other non-polar solvents are more suitable for lipid extraction. Such selective extraction techniques help eliminate interfering substances that result in a purer product or better-quality extracted product.
3. Safety: The extraction techniques must be safe taking into account the compounds extracted and the environment. Green extraction techniques include SFE and UAE, where the solvents are non-toxic, and energy consumption is relatively minimal, which minimizes the environmental footprint of the extraction process. Proper handling and proper disposal of solvents and wastes are vital in the conservation of well-being among personnel in a laboratory setting, as well as prevention against environmental contamination.

4. Scalability: Extraction methods should be scalable to handle massive industrial production demands. Certain methods, like Soxhlet extraction and maceration, work relatively well in small-scale lab experiments, but others, such as SFE and continuous extraction systems, are quite adaptable for large-scale processing. Scalable extraction methods have the potential to facilitate commercial-level production of bioactive compounds from marine macroalgae for pharmaceutical, food, and other industrial applications.
5. Economic Feasibility: The extraction efficiency of methods is very important, especially for business. Other factors that must be considered include the cost of solvents, initial equipment cost, labor cost, and energy consumption. Optimizing extraction variables and using low-cost extraction methods can help reduce manufacturing costs and increase the competitiveness of products obtained from marine macroalgae in the market.

Extraction Techniques for Bioactive Compounds

Various techniques are applied to the extraction of bioactive compounds from natural sources, such as seaweeds, aiming to isolate and purify the compounds of interest, with reduced risk of degradation and contamination [18–20].

1. Solvent Extraction: Organic solvents, like ethanol, methanol, or acetone, can be used to dissolve and extract bioactive compounds from seaweed biomass. The choice depends on how polar the compounds of interest are. The crude extract, thus formed after solvent evaporation, could be further purified.
2. Supercritical Fluid Extraction (SFE): This method employs a supercritical fluid as the solvent. A supercritical fluid is a liquid that exists at critical pressure and temperature, and this can be either carbon dioxide or water. SFE has several advantages as follows: it is selective, the solvent is not toxic, and the extraction parameters may be controlled to extract a specific compound.
3. UAE (Ultrasound-Assisted Extraction): Ultrasound waves in a solvent, applied to seaweed biomass, lead to cultivation and mechanical effects that cause the release of bioactive compounds. They increase the efficiency of extraction and decrease the extraction time compared to traditional techniques.

4. MAE: MAE is the process of producing microwave energy in the biomass seaweed, thus making it accelerate extraction faster than other extraction techniques. This increases yield as the solvent required is less.
5. Enzyme-Assisted Extraction: Utilization of enzymes, such as cellulases or proteases, breaks down cell walls so that the bioactive compounds in seaweed biomass can be released. The extraction efficiency is improved, yet harsh chemical solvents need not be used.
6. Pressurized Liquid Extraction (PLE): Also called accelerated solvent extraction (ASE), PLE uses elevated temperature and pressure to enable penetration of the solvent into the matrix and improve the efficiency of extraction. The method is amenable to automation saving extraction time and reducing solvent consumption.
7. Solid-Phase Extraction (SPE): SPE is a technique in which target compounds are adsorbed onto a solid adsorbent material and then eluted with a solvent to recover the desired compounds. SPE can be used for sample cleanup and concentration before analysis or further use. Extraction methods can then be tailored depending on the properties of the target compounds and the characteristics of the source biomasses, as well as on the desired purity and amount of the final product. Optimization of extraction parameters, such as choice of solvent, temperature, pressure, and time, ensures maximum efficiency of the procedure. Moreover, purification can be achieved through chromatographic or filtration techniques to selectively separate specific bioactive agents within complex seaweed extracts.

Chromatography

Column Chromatography

Column chromatography is a process of separation where compounds are separated based on differential affinities toward stationary and mobile phases. In column chromatography, the stationary phase consists of an adsorbent solid, which interacts with the compounds differently, and the mobile phase is usually a liquid solvent or mixtures of solvents passing through the column [21–23].

Steps Involved in Column Chromatography

1. Preparation of the Sample

Extraction
Extraction is the initial extraction of bioactive compounds from marine macroalgae through the use of appropriate solvents such as ethanol, methanol, or water, with techniques such as maceration, Soxhlet extraction, or ultrasonic-assisted extraction.

Concentration
Concentration is done through the concentration of the crude extract by removal of the solvent using a rotary evaporator or similar device.

2. Column Preparation
    - The column is saturated with an adsorbent material, such as silica gel, alumina, or a polymer resin. It depends on the characteristics that need to be isolated regarding the choice of the type of adsorbent to be used.
    - The column saturated with the mobile phase helps in ensuring consistent filling while preventing the formation of any air pockets.

3. Loading Sample
    - The concentrated extract is dissolved in a small volume of the mobile phase and applied on top of the column.
    - The sample is allowed to adsorb onto the stationary phase.

4. Elution
    - The column is constantly being traversed by the mobile phase causing compounds present in the extract to travel down the column depending on their interactions with both the stationary and mobile phases.
    - Gradient elution allows changing the polarity of the mobile phase in a sequence for separating compounds that have different affinities.

5. Collection and Detection
    - The eluate is collected in fractions at regular time intervals.
    - Each fraction is analyzed by a technique that can include TLC, HPLC, or even spectrophotometry to search for the presence of bioactive compounds.

6. Purification
    - The fractions containing the desired compounds are pooled together, and if necessary, further purified. This might require more chromatography or even some crystallization steps.

a. Thin-Layer Chromatography (TLC):
TLC refers to the process in which the crude extract is applied to a thin layer of adsorbent material usually made up of silica gel on a glass plate. The plate is then placed in a solvent chamber. The solvent moves up the plate, and the compounds in the extract segregate according to their attraction to the adsorbent and the solvent.
b. Techniques, such as UV light and chemical staining, help recognize the separable compounds. Bands of the desired compound are scraped off the plate for further purification.
c. HPLC refers to a highly advanced form of liquid chromatography that also provides higher resolution and increased sensitivity. A crude extract is pumped under high pressure through a column packed by a stationary phase. Compound separation occurs as a function of their interactions with a stationary phase and solvent used, and the detection is preferably done using UV-Vis or other detectors. HPLC has the advantage of being able to control separation parameters very accurately, and a significant amount of bioactive compounds are being purified through this method.

1. Filtration

a. Gravity Filtration: This is the process wherein crude extract is passed through a filter medium by the action of gravity. This is good enough for removing solid impurities or particulate matter from the extract but is insufficient for separating complex mixtures of compounds.
b. Vacuum Filtration: Vacuum filtration is similar to gravity filtration except that the vacuum pump is used to increase the rate of filtration. It is ideal for fast filtration of large volumes of liquid and, depending on the nature of the sample, can be used with a variety of filter media, including filter paper or glass fiber filters.
c. Membrane Filtration: In membrane filtration, crude extracts are passed through specific membrane filters. A particular particle or molecule is split with the help of separation, which can remove macromolecules or large impurities. Ultrafiltration as well as microfiltration are techniques in which membranes with different pore sizes are used for even further separation.

These purification methods can be used independently or in combination to reach a level of purity and concentration for the bioactive compound isolated from seaweeds or obtained from any natural source. It is equally important in purification to optimize parameters in purification, which may be the right selection of a stationary phase or of a solvent system in the case of chromatography.

Bioactive Compounds in Marine Macroalgae

Marine macroalgae, including seaweeds, are very rich in bioactive compounds that provide a range of health-promoting effects. Bioactive compounds in seaweeds are a key factor in the ecological and economic importance of these algae, and their utilization in pharmaceuticals, nutraceuticals, cosmeceuticals, and functional foods is gaining increased interest [24–26].

1. Polyphenols: Marine macroalgae are rich sources of polyphenolic compounds, such as flavonoids, phenolic acids, and tannins; all of these have antioxidant, anti-inflammatory, and anticancer properties. They may perhaps protect against cardiovascular diseases and neurodegenerative disorders.
2. Phycobiliproteins: These are the water-soluble pigments of red and blue-green algae, that possess very potent antioxidant activity. These proteins are taken under consideration for therapeutic usage in anticancer therapy, photodynamic therapy, as well as food and cosmetic coloring.
3. Fucoidans: Brown algae contain sulfated polysaccharides known as fucoidans primarily located in cell walls. These compounds possess several biological activities, including anticoagulant, antiviral, anti-inflammatory, and anticancer effects. There is also promise of wound healing and skin health by using fucoidans.
4. Phlorotannins: Phlorotannins are exclusive compounds of brown algae. These are potent polyphenolic compounds with antioxidant, anti-inflammatory, antiviral, and antibacterial activities. They may find their applications in the prevention and treatment of diseases such as cancer, diabetes, and cardiovascular disorders.
5. Laminarin: Laminarin is a β-glucan polysaccharide extracted from brown algae, which has immunomodulatory activities

that enhance the immunity system. It can be useful for the enhancement of immunity and prevention of infection.

6. Chlorophylls and Carotenoids: Chlorophylls and carotenoids, respectively, are responsible for the green and orange-red colors in algae. Being antioxidants, these compounds contribute to photosynthesis and therefore are essential for algae cells. Chlorophyll may also provide health benefits related to detoxification and healing of wounds.
7. Essential Fatty Acids: Some species of red algae contain essential omega-3 fatty acids such as EPA (eicosapentaenoic acid) and DHA (docosahexaenoic acid). These are the good health-related fatty acids.
8. Vitamins and Minerals: Seaweeds, within the larger category of marine macroalgae, constitute an exceptional source of several vitamins for human well-being. These comprise vitamin C and vitamin K, as well as some B vitamins. Furthermore, they contain various minerals such as iodine, calcium, magnesium, and iron. All of these micronutrients play key roles in immune functions, bone health, and even energy metabolism within the organism.

Generally, marine macroalgae are very rich and eco-friendly bioactive compounds providing broad health benefits. Current ongoing research into the bioactive compounds of these algae is finding new therapeutic applications and uses in various fields.

Pharmacological Properties and Health Benefits

Marine macroalgae, which are seaweeds, are known to possess various pharmacological properties and health-promoting properties due to their high content of bioactive compounds [27, 28].

1. Antioxidant Activity: Marine macroalgae contain various bioactive compounds, such as polyphenols, phycobiliproteins, and phlorotannins, which show excellent antioxidant activity. Antioxidants neutralize free radicals within the body and minimize the effects of oxidative stress and inflammation, possibly preventing various chronic diseases like cardiovascular disease, cancer, and neurodegenerative disorders.

2. Anti-Inflammation Activity: Anti-inflammatory activity is found in very few of the extract-derived compounds from marine macroalgae such as fucoidans, phlorotannins, and essential fatty acids. Through pathways and tissues/organ systems, these modulate inflammation and, thereby, may be useful in alleviating inflammatory diseases like arthritis, inflammatory bowel disease, and asthma.
3. Anticancer Activity: Some of the bioactive compounds from marine macroalgae showed promising anticancer activity in preclinical studies. Some of these compounds could inhibit the growth of tumor cells, induce apoptosis, and suppress angiogenesis. In this way, they could prevent or delay various types of cancers.
4. Antiviral and Antibacterial Activity: It was found that compounds of a few marine macroalgae, including fucoidans and phlorotannins, can perform a role in exhibiting antiviral and antibacterial properties. They may limit virus and bacteria multiplication that ensures the prevention of infectious diseases, such as viral diseases related to influenza and herpes infections, and bacterial diseases concerning *Staphylococcus aureus* and *Escherichia coli* infections.
5. Immunomodulatory Activities: Some of the polysaccharides of seaweeds, including fucoidans and laminarin, have been proven to exhibit immunomodulatory properties. These polymers are found to modify immune cell activity and can enhance immunity against infections or disease-producing pathogens to overall health and immunity. As a result, infection or autoimmune condition may be easily controlled.
6. Cardioprotective Effects: Marine macroalgae bioactive compounds contain cardioprotective compounds, including omega-3 fatty acids and phlorotannins. The compounds have effects that prevent the development of heart conditions by lowering blood pressure and cholesterol levels, preventing clotting in blood vessels, and improving the function of the blood vessels.
7. Skin Health and Wound Healing. Fucoidans and phlorotannins, compounds existing in marine macroalgae, have manifested significant benefits such as stimulating human skin health and promoting wound healing. These are capable of

inducing stimulation of collagen synthesis, stimulation of wound closure, inhibition of the inflammation process, and counteracting UV-induced damages to the skin, thus promoting human skin health and its renewal.
8. Bone Health: Marine macroalgae are rich in minerals, like calcium, magnesium, and iodine, which are known to be essential for maintaining healthy bones. Regular intake of seaweeds can help in strengthening the bones, preventing osteoporosis, and maintaining overall skeletal health.

This emphasizes the potential of marine macroalgae as sources of valuable natural bioactive compounds for the promotion of human health and prevention of diverse diseases. Further research is needed into their bioactive constituents and mechanisms of action to unlock their full therapeutic potential.

4.5 Extraction of Food Ingredients from Marine Macroalgae

In food industries, marine macroalgae play diverse roles offering unique flavors, textures, and nutritional benefits. The following are ways in which seaweeds are utilized in the food industry [29–33]:

1. Direct Consumption: Several countries in the world directly consume seaweeds as food. They are used in multiple culinary purposes, such as in salads, soups, stews, sushi, and side dishes. According to the different species of the seaweeds, taste varies, from mild and salty to savory and umami, and gives depth and density to the dishes.
2. Ingredient in Food Products: Seaweeds are used as ingredients in a large number of food products because of their nutritional importance, flavoring ability, and functional characteristics. They are incorporated in processed foods such as snacks, noodles, bread, sauces, dressings, crackers, and condiments. The extracts of seaweeds also find application as natural flavors, colorants, and thickeners in formulations of food products.

3. Nutritional Enhancement: Seaweeds are nutrient-rich foodstuffs with vitamins, minerals, proteins, fiber, and bioactive compounds. This addition of seaweeds into food products enhances the nutritional value with such valuable nutrients such as iodine, calcium, iron, vitamin C, and omega-3 fatty acids. The associated health benefits for the consumer of seaweed-based foodstuffs include increased immunity, better bone health, and improved cardiovascular health.
4. Natural Food Additives: Some compounds extracted from sea plants, like carrageenan, alginate, and agar, are used as natural additives and stabilizing agents in the food industry. They act as gelling agents, emulsifiers, and thickeners and alter the texture of various products as well as enhance its sensorial quality and stable period. It is used as part of desserts and beverage, dairy, and processed meat products.
5. Seafood and Plant-Based Meat Alternatives: Seaweed ingredients are blended into formulations for seafood as well as plant-based meat alternatives. Incorporation of seaweed extracts into food conveys meaty, fishy flavors and textures plus the nutritionality of meats and fish; hence, such products lend themselves well to vegan-, flexitarian-, and vegetarian-type diets. Plant-based protein sources have an umami flavor profile enhanced by seaweed-based umami flavor boosters.
6. Functional Foods and Health Supplements: Seaweed is utilized in food manufacturing for the production of dietary supplements, functional foods, and nutraceuticals with specific health benefits. Health supplements and functional foods are prepared from preparations of seaweed extracts in the form of bioactive compounds, including phycocyanins, fucoidans, and polyphenols, related to weight management, detoxification, immune support, and anti-aging conditions.
7. Seaweed Farming and Aquaculture: Seaweed culturing through aquaculture is an industry that contributes to sustainable environmental conservation for food production. Large scales of seaweeds used in feed, biofuels, food, and bioremediation production are produced in seaweed farms. It offers various economic opportunities for the people by the coast while helping address the environmental problems associated with agriculture and aquaculture.

Marine macroalgae contain a rich source of nutrients and are increasingly gaining attention for their role as functional foods [34–36].

1. Rich in Vitamins and Minerals: Seaweed is a good source of Vitamins A, C, E, and K and minerals such as calcium, magnesium, iodine, and iron. The functions of these nutrients include energy metabolism, immune support, and bone health.
2. Fiber Rich Seaweeds: They contain a lot of dietary fibers: the contents are a rich mix of soluble and insoluble fiber from agar, carrageenan, and alginate. They also enhance the digestion process through the regulation of blood sugar and enhancement of satisfaction, which elevates the cholesterol levels as well.
3. Omega-3 Fatty Acids: Species of marine algae, such as *Saccharina japonica*, *Porphyra*, and *Ulva*, have been reported to contain rich sources of omega-3 fatty acids, especially EPA (eicosapentaenoic acid) and DHA (docosahexaenoic acid). These are essential fatty acids that are useful for brain activity, heart health, and in the reduction of body inflammation.
4. Antioxidants: Seaweeds are highly antioxidant in nature as they contain carotenoids, phycocyanin, and flavonoids. All these compounds may neutralize the harmful free radicals of our body by reducing the risk of developing chronic diseases, such as cardiovascular diseases, cancer, and even aging conditions.
5. Iodine: Seaweeds (*Porphyra* spp., *Undaria pinnatifida*, *Laminaria* spp.) contain a group of the richest available sources of natural iodine; however, this mineral assumes critical importance in human thyroid function as well as in thyroid hormone formation. Adequate nutrition with iodine can significantly ensure healthy metabolism without occurrence of iodine deficiency disorders.
6. Prebiotics: Various species of seaweed have prebiotic fibers from *Enteromorpha* spp., *Gelidium* spp., and *Gracilaria* spp., which feed good gut bacteria. Seaweeds will nourish your overall immune system and intestines since they promote growth through good probiotic bacteria.

7. Anti-Inflammatory Activities: Certain types of compounds of the *Gracilaria* spp., *Porphyra* spp., and *Chondrus crispus* have been proven to possess anti-inflammatory properties. Intake of seaweeds may reduce the inflow of inflammatory responses that lead to the risk of developing inflammatory diseases, especially those of inflammatory bowel disease and arthritis.
8. Weight Loss: Due to the presence of plenty of fiber and low saturated fat and cholesterol content, the overall energy present in seaweed can contribute to weight loss as it can make an individual feel full without contributing a higher calorie content.
9. Blood Glucose Regulation: Studies show that some chemicals in *Ascophyllum nodosum*, *Laminaria japonica*, and *Gracilaria* spp. can regulate blood glucose levels and increase the body's sensitivity to insulin making them useful for people who have diabetes or are likely to develop it.
10. Functional Applications in Food Products: Mixing of seaweed with various nutritional and functional properties into various food products and dietary supplements is a very popular trend today. It has become an excellent ingredient to add to soups, sushi, snacks, condiments, salads, and drinks.

Generally, seaweeds have a unique combination of nutrients and bioactive compounds that makes them functional foods with numerous health benefits. However, they should be taken in moderation, with caution about contamination with heavy metals, among other contaminants, especially in seaweeds.

Extraction Methods [37–41]

1. Mechanical Techniques
    - Milling and Grinding: It is the process of reducing seaweed into a fine powder that enhances the surface area for extraction.
    - Pressing: Mechanical pressing can be applied for extracting liquid and juice fractions from the large biomass.

2. Solvent Extraction
    - Solvent-based extraction with organic solvents, such as methanol, acetone, hexane, and ethanol, is common for extracting lipophilic compounds.
    - Aqueous Extraction: Water is used to extract hydrophilic compounds such as proteins and polysaccharides.

3. Enzyme-Assisted Extraction (EAE)

a. Specific cell wall-breaking enzymes lead to the release of bioactive compounds. For example, cellulases and proteases can be applied to polysaccharides and proteins, respectively.

4. Ultrasonic-Assisted Extraction (UAE):

a. Ultrasonication

This is based on the disruption of cell walls through sound waves, hence releasing more bioactive compounds. The process is efficient in cutting down extraction time and using solvents.

5. Microwave-Assisted Extraction (MAE)

Microwaves produce heat and raise the penetration of a solvent as well as the mass transfer rate. This can enhance the efficiency of extraction of thermally stable compounds.

6. Supercritical Fluid Extraction (SFE)

a. Supercritical CO_2 is used for the extraction of non-polar compounds, thus being environmentally friendly and does not use toxic organic solvents.

7. Pressurized Liquid Extraction (PLE)

a. Solvents at high temperature and pressure enhance the extraction of polar and nonpolar compounds.

8. Membrane Filtration and Purification

a. Ultrafiltration and reverse osmosis membrane technologies can be implemented after extraction for the purification and concentration of the compounds extracted.

Applications in the Food Industry

1. Thickeners and Gelling Agents: Polysaccharides, such as agar, alginate, and carrageenan, are used as thickeners, stabilizers, and gelling agents in many food products.
2. Natural Colorants: Pigments, like chlorophyll, carotenoids, and phycobiliproteins, are used as natural food colorants.
3. Nutritional Supplements: Extracted vitamins, minerals, proteins, and omega-3 fatty acids are formulated into dietary supplements.
4. Functional Foods: Bioactive compounds with antioxidant, anti-inflammatory, and antimicrobial properties are incorporated into functional foods for increased benefits.
5. Flavor Enhancers: Some extracts can be applied to flavor food products.

Marine Macroalgae Challenges and Future Perspectives

Despite their tremendous potential in food, pharmaceutical, and environmental industries, marine macroalgae are still facing many challenges [42–46].

Challenges

1. Environmental Issues: Seaweed farming and harvesting might result in habitat alteration, nutrient runoff, and competition with native species. These concerns with respect to the environment may be minimized through sustainable practices and careful management to maintain the health of marine ecosystems for a very long period.
2. Regulatory Barriers: The regimes of regulations governing seaweed farming, harvesting, and marketing vary significantly between countries and regions. Simplification of regulations and clarification of rules guiding the activities associated with seaweeds can enhance the development and innovation of the industry while promoting safety and sustainability.
3. Resource Constraint: There are limited good coastal areas, water resources, and infrastructure to expand the seaweed farming operations. Technological advancements, innovative farming methods, and strategic planning will help overcome these resource constraints for sustainable increases in seaweed production.

4. Biological Challenges: Seaweed farming is very susceptible to pests, diseases, and invasive species. Improvements in breeding programs, disease management, and genetic engineering will strengthen and improve the productivity of seaweed crops.
5. Market Development: Despite the increasing popularity of seaweed-based food products among consumers, relative to other agricultural commodities, market development and consumer awareness still lag behind. With effective marketing strategies, diversification of products, and culinary innovation, the reach of seaweed-derived foods and products will be expanded further.
6. Technological Limitations: The seaweed industry faces the challenges of technological cultivation, harvesting, processing technologies, and product development. Huge investment in research and development to overcome these limitations and adoption of the latest technology will drive innovation and competitiveness in the sector.

Future Perspectives

1. Seaweed Culture Expansion: Since the demand for other alternative bioresources and sustainably sourced food continues to increase, this is substantial growth potential for seaweed culture worldwide. Expansion of operations for seaweed farming creates more food security, economic opportunities, and alleviates many of the environmental pressures previously associated with traditional agriculture and aquaculture.
2. Product Diversification: Behind feed applications and old-school food lie the future aspects of seaweed. This includes diversified bioplastic sources, biofuel sources, functional ingredient sources, and pharmaceutical sources represented by bioactive compounds present in seaweed.
3. Integrated Aquaculture: The integrated production of aquaculture may benefit from the added advantage of improved ecosystem services, nutrient cycling, and overall productivity in aquaculture operations. Multi-trophic systems are synergistic benefits particularly when seaweed culture combines fish farming with shellfish culture.

4. Biorefinery Approaches: Biorefinery approaches use resource utilization and value extraction from the seaweed biomass to process seaweed into useful products. The biochemical, biophysical, or bio-chemical/biophysical processes vaporize carbohydrate, lipid, protein, and bioactive compounds for multiple high-value products and significantly less waste and environmental effects.
5. Collaborative Research and Innovation: It is, therefore, critical that not-for-profit organizations, the government, academia, and industry collaborate to advance research and develop the technological and market take-up process in the seaweed industry. Private–public partnerships, knowledge funding initiatives, and the sharing networks accelerate the step process with common challenges being addressed as per the industry's needs.

4.6 Conclusion

Finally, the extraction of drugs and food ingredients from marine macroalgae promises different industries, with food industries, pharmaceuticals, and biotechnology on the list. Careful techniques in extraction and processing result in a variety of compounds from the seaweeds, such as nutritional benefits and therapeutical functions.

Marine macroalgae are rich sources of bioactive compounds that show different pharmacological activities such as anticancer, anti-inflammatory, antioxidant, antimicrobial, and immunomodulatory effects. These bioactive compounds may offer a chance for the development of therapeutic agents and novel drugs to address the unmet medical needs as well as combat various diseases.

Seaweeds are becoming important sources of flavor and nutritional enhancement with functional properties contributing to the production of new foods and dietary supplements in the food industry. While ingredients sourced from seaweeds are used in foods as natural stabilizers, additives, and thickeners, only a few of the bioactive compounds, like polyphenols, fucoidans, and phycocyanins, provide health benefits, thus forming part of a rapidly growing functional food and nutraceutical market.

However, the extraction of drugs and nutrients from seaweeds has many challenges. Among the major barriers are environmental constraints, regulatory bottlenecks, resource availability, biological barriers, and technological impediments to sustainable and scalable manufacturing of seaweed-based products. These bottlenecks can be overcome by collaboration, investment in R&D activities, application of state-of-the-art technologies, and proper implementation of procedures to ensure that the economics, ecology, and societal benefits are well realized from the industry.

Though obstacles exist, there is a reason to believe a bright future exists for the extraction of drugs and food ingredients from marine macroalgae. Continued advancement in scientific understanding, technology, and industry collaboration will help unlock all the potential in using seaweeds as resources for human health, nutrition, and sustainable development. With effective stewardship and responsible management, marine macroalgae can go a long way in answering global issues and meeting the needs of the present and future generations.

References

1. Sharma, S., Gogoi, D., Gogoi, B.K., Emerging extraction technologies for bioactive compounds from marine algae: A review, in: *Separation and Purification Reviews*, pp. 1–19, 2023.
2. Sun, Y. and Zhou, C., Emerging extraction technologies for bioactive compounds from marine algae: A review, in: *Separation Science and Technology*, pp. 1–17, 2023.
3. Wang, Z., Wang, Z., Zhu, L., Zhou, H., Wang, S., Emerging extraction technologies for bioactive compounds from marine algae: A review. *Sep. Purif. Technol.*, 286, 120824, 2023.
4. Li, Z., Zhao, D., Wang, Y., Wang, J., Emerging extraction technologies for bioactive compounds from marine algae: A review. *Food Chem.*, 381, 132388, 2023.
5. Ma, H., Li, D., Li, S., Sun, Y., Emerging extraction technologies for bioactive compounds from marine algae: A review. *Food Bioprod. Process.*, 130, 144–155, 2023.
6. Dawes, C., Macroalgae systematics, in: *Seaweed in health and disease prevention*, pp. 107–148, Academic Press, San Diego, CA, 2016.
7. Lagourgue, L. and Payri, C.E., Large-scale diversity reassessment, evolutionary history, and taxonomic revision of the green macroalgae family Udoteaceae (Bryopsidales, Chlorophyta). *J. Syst. Evol.*, *60*, 1, 101–127, 2022.
8. Bolton, J.J., The biogeography of kelps (Laminariales, Phaeophyceae): a global analysis with new insights from recent advances in molecular phylogenetics. *Helgol. Mar. Res.*, *64*, 263–279, 2010. (Arita and Rodríguez 2004; Sihvonen and Siljander 2005; De Oliveira and Daly 1999; Gixhari *et al.*, 2012).

9. Arita, H.T. and Rodríguez, P., Local-regional relationships and the geographical distribution of species. *Global Ecol. Biogeogr.*, *13*, 1, 15–21, 2004.
10. Sihvonen, P. and Siljander, M., Species diversity and geographical distribution of Scopulini moths (Lepidoptera: Geometridae, Sterrhinae) on a worldwide scale. *Biodivers. Conserv.*, *14*, 703–721, 2005.
11. De Oliveira, A.A. and Daly, D.C., Geographic distribution of tree species occurring in the region of Manaus, Brazil: implications for regional diversity and conservation. *Biodivers. Conserv.*, 8, 1245–1259, 1999.
12. Gixhari, B., Ismaili, H., Vrapi, H., Dias, S., Sulovari, H., Elezi, F., Geographic distribution and diversity of fruit tree species in Albania. *Int. J. Ecosyst. Ecol. Sciences*, 2, 4, 355–360, 2012.
13. Wu, H., Zhang, J., Li, Y., Zhou, H., Emerging extraction technologies for bioactive compounds from marine algae: A review. *Chem. Eng. J.*, 439, 135953, 2023.
14. Zhang, H., Liu, X., Yang, Y., Liu, X., Emerging extraction technologies for bioactive compounds from marine algae: A review. *Ultrason. Sonochem.*, 89, 105554, 2023.
15. Li, W., Sun, Y., Wang, H., Li, Y., Emerging extraction technologies for bioactive compounds from marine algae: A review. *J. Cleaner Prod.*, 343, 130364, 2023.
16. Chen, X., Wang, S., Wang, L., Zhang, J., Emerging extraction technologies for bioactive compounds from marine algae: A review. *Front. Chem.*, 11, 798164, 2023.
17. Jiang, Y., Li, H., Li, T., Dong, Y., Emerging extraction technologies for bioactive compounds from marine algae: A review. *J. Food Eng.*, 342, 110450, 2023.
18. Azmir, J., Zaidul, I.S.M., Rahman, M.M., Sharif, K.M., Mohamed, A., Sahena, F., Omar, A.K.M., Techniques for extraction of bioactive compounds from plant materials: A review. *J. Food Eng.*, *117*, 4, 426–436, 2013.
19. Alam, M.A., Sarker, M.Z., II, Ghafoor, K., Happy, R.A., Ferdosh, S., Bioactive compounds and extraction techniques. *Recovering Bioactive Compounds Agric. Wastes*, 28, 33–53, 2017.
20. Martins, R., Barbosa, A., Advinha, B., Sales, H., Pontes, R., Nunes, J., Green extraction techniques of bioactive compounds: a state-of-the-art review. *Processes*, *11*, 8, 2255, 2023.
21. He, L., Yin, Q., Zhang, L., Jin, J., Wang, Y., Purification and characterization of polysaccharides from marine macroalgae by column chromatography. *Int. J. Biol. Macromol.*, *148*, 1234–1241, 2020.
22. Kim, S.K. and Wijesekara, I., Column chromatography for bioactive compound purification from marine algae. *Mar. Drugs*, *17*, 3, 120, 2019.
23. Holdt, S.L. and Kraan, S., Bioactive compounds in seaweed: Functional food applications and extraction techniques. *J. Appl. Phycol.*, *23*, 4, 543–597, 2011.

24. Vijayaraj, R., Altaff, K., Kumaran, N.S., Volatile bioactive compounds from marine macro-algae and their pharmacological properties. *Indian J. Geo-Mar. Sci. (IJMS)*, *51*, 10, 832–842, 2022.
25. Sudhakar, M.P., Kumar, B.R., Mathimani, T., Arunkumar, K., A review on bioenergy and bioactive compounds from microalgae and macroalgae-sustainable energy perspective. *J. Cleaner Prod.*, *228*, 1320–1333, 2019.
26. Falkenberg, M., Nakano, E., Zambotti-Villela, L., Zatelli, G.A., Philippus, A.C., Imamura, K.B., Graminha, M.A., Bioactive compounds against neglected diseases isolated from macroalgae: a review. *J. Appl. Phycol.*, *31*, 797–823, 2019.
27. Nisar, M.F., Khadim, M., Rafiq, M., Chen, J., Yang, Y., Wan, C.C., Pharmacological properties and health benefits of eugenol: A comprehensive review. *Oxid. Med. Cell. Longevity*, *2021*, 2497354, 2021.
28. Priyanka, K.R., Rajaram, R., Sivakumar, S.R., A critical review on pharmacological properties of marine macroalgae. *Biomass Convers. Biorefin.*, 10, 1–25, 2022.
29. Balina, K., Romagnoli, F., Blumberga, D., Seaweed biorefinery concept for sustainable use of marine resources. *Energy Procedia*, 128, 504–511, 2017.
30. Cherry, P., O'Hara, C., Magee, P.J., McSorley, E.M., Allsopp, P.J., Risks and benefits of consuming edible seaweeds. *Nutr. Rev.*, 77, 5, 307–329, 2019.
31. Gupta, S. and Abu-Ghannam, N., Recent developments in the application of seaweeds or seaweed extracts as a means for enhancing the safety and quality attributes of foods. *Innovative Food Sci. Emerg. Technol.*, 12, 4, 600–609, 2011.
32. Lourenço, S.O., Barbarino, E., De-Paula, J.C., Pereira, L.O.S., Marquez, U.M.L., Amino acid composition, protein content and calculation of nitrogen-to-protein conversion factors for 19 tropical seaweeds. *Phycol. Res.*, 50, 3, 233–241, 2002.
33. Rajapakse, N. and Kim, S.K., Nutritional and digestive health benefits of seaweed. *Adv. Food Nutr. Res.*, 64, 17–28, 2011.
34. Aumeerun, S., Soulange-Govinden, J., Driver, M.F., Ranga, R.A., Ravishankar, G.A., Hudaa, N., Macroalgae and microalgae: novel sources of functional food and feed, in: *Handbook of algal technologies and phytochemicals*, pp. 207–219, CRC Press, Boca Raton, FL, 2019.
35. Siddiki, S.Y.A., Mofijur, M., Kumar, P.S., Ahmed, S.F., Inayat, A., Kusumo, F., Mahlia, T.M., II, Microalgae biomass as a sustainable source for biofuel, biochemical and biobased value-added products: An integrated biorefinery concept. *Fuel*, *307*, 121782, 2022.
36. Torres-Tiji, Y., Fields, F.J., Mayfield, S.P., Microalgae as a future food source. *Biotechnol. Adv.*, *41*, 107536, 2020.
37. Browne, R., Hayes, M., Lonergan, B., Extraction and utilization of bioactive compounds from marine macroalgae: A review of recent developments. *J. Food Sci. Technol.*, *58*, 9, 3453–3467, 2021.

38. Cian, R.E., Drago, S.R., De Medina, F.S., Martínez-Augustin, O., Bioactive properties of peptides obtained by enzymatic hydrolysis from protein byproducts of Porphyra columbina. *Food Res. Int.*, *76*, 549–556, 2015.
39. Harnedy, P.A. and FitzGerald, R.J., Extraction of protein from the macroalga Palmaria palmata. *LWT Food Sci. Technol.*, *51*, 2, 375–382, 2013.
40. Murphy, C. and Bohrer, B.M., The effect of dietary inclusion of marine macroalgae on the composition and quality of muscle foods: A review. *Crit. Rev. Food Sci. Nutr.*, *61*, 22, 3673–3685, 2021.
41. Parjikolaei, B.R., Bruhn, A., Eybye, K.L., Larsen, M.M., Protein extraction from Danish seaweeds: Comparison of three extraction methods. *J. Appl. Phycol.*, *28*, 6, 3527–3536, 2016.
42. Barbier, M., Charrier, B., Araujo, R., Holdt, S.L., Jacquemin, B., Rebours, C., Marine macroalgae: A sustainable biomass feedstock for biorefineries. *Biotechnol. Adv.*, 37, 3, 6–12, 2019.
43. Buschmann, A.H., Camus, C., Infante, J., Neori, A., Israel, A., Hernández-González, M.C., Critchley, A.T., Seaweed production: Overview of the global state of exploitation, farming and emerging research activity. *Eur. J. Phycol.*, 52, 4, 391–406, 2017.
44. Cai, J., He, Y., Yu, X., Tan, J., Marine macroalgae: An untapped resource for the discovery of new drugs. *Mar. Drugs*, 19, 10, 507, 2021.
45. Mac Monagail, M., Cornish, L., Morrison, L., Araújo, R., Critchley, A.T., Sustainable harvesting of wild seaweed resources. *Eur. J. Phycol.*, 52, 4, 371–390, 2017.
46. Pérez-Lloréns, J.L., Hernández, I., Vergara, J.J., Brun, F.G., Martínez-López, J., Perspectives on the use of seaweeds as a promising source of bioactive compounds for health and disease prevention. *Mar. Drugs*, 18, 2, 115, 2020.

5

Biotechnological Applications of Marine Bacteria: Marine Biotechnology

Nayomi John*, Anjala Shirin, Roby K. Mathew, Muhammed Faiz, Sneha P. Thomas, Fathima Safa and Lena Joseph

Department of Microbiology, Mar Athanasius College (Autonomous), Kothamangalam, Kerala, India

Abstract

Marine bacteria, a vast and diverse group of microorganisms, play crucial roles in marine ecosystems and significantly impact human and animal health. These bacteria are broadly classified into obligate and opportunistic categories. Marine biotechnology has opened up a vast array of applications for marine bacteria, including bioremediation, biofertilizers, biomedical applications, enzyme production, food and feed, bioenergy, bioproducts, marine pollution monitoring, biocatalysis, and cosmetics and personal care. Marine bacteria are crucial in the marine environment playing a vital role in the ecosystem and offering numerous benefits for marine biotechnology. They produce bioactive compounds, clean pollutants, and develop novel enzymes and biocatalysts. By focusing on the positive aspects of marine bacteria and unlocking their potential, we can discover a wealth of opportunities to improve human health, promote sustainability, and protect the environment ultimately leading to a brighter, more sustainable future.

Keywords: Marine bacteria, bioremediation, enzymes, drugs and pharmaceutics, marine bacterial EPS, bioactive pigment, biosurfactant

**Corresponding author*: nayomijohn@macollege.in

Shaju S. S., Vipin. P. M. and Visakh. P.M. (eds.) Marine Biotechnology: Studies and Applications, (139–168) © 2025 Scrivener Publishing LLC

5.1 Introduction

The sea is where all life on Earth began. Half of the world's population depends heavily on the oceans for their food, and they also influence the climate. The marine environment is a vast reservoir of the biodiversity that contains over 97% of all the water on earth and spans over 70% of the planet's surface. The ocean places a crucial role in human society through its ecosystem services [1]. Researchers have been studying the richness and functioning of these fascinating marine environment for decades because of the creatures that call it home. The phylogenetic diversity of sediments in marine environments is higher than that of soil in terrestrial contexts [2]. Numerous ecological services offered by marine habitat assist society [3]. These services are mostly associated with provisioning services (like food), cultural services (like tourism), and supporting services (like primary production and nutrient cycling). With significant environmental gradients in physical, chemical, and hydrological characteristics, like temperature, light intensity, salinity, and pressure, the marine ecosystem supports a varied range of life. To survive, adapt, and thrive the variety of these competing ecosystems, marine animals have developed a wide range of forms, functions, and methods in response to these varied conditions [4].

The diversity of species found in our oceans and seas, which makes up the beautiful tapestry of marine biodiversity, is essential for the health and adaptation of marine ecosystems. A remarkable diversity of creatures, each with distinct biological systems and traits, may be found in the marine environment, despite it being one of the most under-utilized biological resources. From the simplest microalgae and bacteria to the complex sponges and fish, marine organisms have evolved remarkable adaptations to thrive in environments with varying salinity, pressure, temperature, and light conditions making them a treasure trove of untapped potential. More than 650 novel marine compounds were identified in 2003 alone from marine microorganisms and phytoplankton, as well as from green, brown and red algae, sponges, coelenterates, bryozoans, mollusks, tunicates, and echinoderms [5]. Most marine organisms, primarily microorganisms are still unknown [6, 7]. Maintaining a delicate balance and resilience of marine ecosystems depends on marine biodiversity, which includes an extraordinarily wide variety of creatures from majestic whales to microscopic planktons. Moreover, it serves as a vital foundation for industries, like tourism and recreation, generating employment opportunities and economic growth for local communities while also providing countless benefits for human well-being and the health of our planet. A key element of the Earth's ecosystem, marine biodiversity is essential for controlling the planet's weather and climate. It also supplies vital resources, like fish and shell fish, which are the main sources of protein for millions of people around the world. The significance of protecting marine biodiversity for the welfare of the Earth and human populations is further demonstrated by the fact that coral reefs serve as natural storm and hurricane barriers protecting coastal towns from the destructive consequences of extreme weather events. In addition to coral reefs, mangroves and seagrass beds also provide crucial coastal protection by stabilizing shorelines and preventing erosions. Additionally, because they produce oxygen through photosynthesis and absorb carbon dioxide from the atmosphere, marine plants and phytoplankton are essential in reducing the effect of climate change. The depth and features of the bottom, together with changes in water temperature, salinity, nutrients levels, and currents, all affect patterns of marine biodiversity. Additionally, latitudinal and seasonal changes in sunlight also play a significant role. The biodiversity of the water column includes both surface and deep-sea ecosystems. From deep hydrothermal vents with temperatures ranging from −1.5°C to 350°C in Antarctic seas and pressure ranging from one to more than a thousand atmospheres, sea bed biodiversity spans a variety of habitats. Cold seeps, mid ocean ranges, very photic and non-photic zones, and oligotrophic to eutrophic nutrient conditions are all present.

Marine organisms encompass a vast array of life forms dwelling in oceans, seas, and saline environments. They are vital for sustaining marine ecosystems and contribute significantly to the planet's overall well-being. Ranging from microscopic plankton to colossal whales, these organisms exhibit diverse adaptations to various environmental conditions. Among them, phytoplankton stand out as a crucial group comprising tiny, drifting organisms essential for marine ecosystems. They absorb carbon dioxide, use light energy to produce carbohydrates, and exhale oxygen, just like terrestrial plants do. It is said to be the ocean's primary producers for this reason. There are two distinct types of organisms that make up phytoplankton. Protists are single-celled algae that fall under the broader group. These forms, which are most prevalent close to the coast, include diatoms. These organisms occasionally respond to shifting seasons and the availability of nutrients like nitrogen, iron, and phosphorus by forming blooms, which are sudden increases in population. Photosynthetic bacteria are the other kind of phytoplankton cells. Cyanobacteria are also a member of this category. Many marine organisms, such as fish, whales, and zooplanktons, depend on phytoplanktons as food supply. The entire sea environment would be destroyed without phytoplanktons [8]. Zooplanktons, another vital component of marine ecosystems, fulfill a crucial role by consuming phytoplanktons and serving as prey for larger predators like fish and whales. Maintaining the health of marine ecosystems depends on their role in moving energy up the marine food chain. In pelagic food web, zooplanktons comprise a varied collection of protistan and metazoan eaters that occupy several tropic levels. They influence the efficiency of export, transfer to higher tropic levels, nutrient cycling, and production–grazing coupling. The grazing impact of microzooplankton and the respiratory requirements of mesozooplankton have significant global impact on the flow of carbon. Nevertheless, this equilibrium is susceptible to trophic structure uncertainties. It is predicted that climate change would drastically disrupt how zooplankton cycle carbon, which will have an immediate effect on important species. In addition to recycling nutrients and particulate carbon into dissolved forms, zooplanktons play a crucial role in connecting primary producers with higher trophic levels. Understanding the various roles that zooplanktons play in carbon transformations and forecasting future changes are becoming more and more important as a result of climate change and other stressors that impact their abundance, distribution, size, and life cycles. Fish are crucial as both a food source for humans and other marine animals and as key players in marine ecosystems. They help regulate populations of other marine organisms, contribute to nutrient cycling, and support marine biodiversity. Marine mammals, including whales, dolphins, and seals, constitute another vital group within marine ecosystems. As apex predators, they regulate the populations of other marine organisms. Because they store large amounts of carbon in their bodies and release it into the environment when they die, whales, in particular, are essential to the carbon cycle. Pelagic dolphins, found offshore, generally face fewer human interactions due to their small size, limited commercial value, wide distribution, and distance from human activities. While many seal species breed on land and spend substantial time foraging on continental shelves, some, especially in the southern hemisphere, venture into areas beyond national jurisdiction. The crab eater seal, the world's most abundant marine mammal, inhabits pack ice and primarily feeds on krill. Marine reptiles trace their origins back to ancient land-dwelling ancestors that eventually transitioned to life in the sea. Only a small fraction, roughly 1%, of all reptile species are truly marine. These include sea turtles, marine iguanas, and sea snakes. While some other reptiles, such as certain turtles, lizards, snakes, and crocodiles, inhabit estuarine environments, they still rely on freshwater to some extent. The marine reptiles we observe today evolved from terrestrial ancestors that successfully adapted to aquatic life long ago, benefiting from physiological adaptations acquired during their time on land. The oceans, spanning from coastlines to the deepest depths, contain a remarkable richness of microbial life. These microbes are crucial for driving productivity, maintaining ecosystem functions, and shaping global biogeochemical cycles. However, our understanding of marine microbial diversity, including its patterns, drivers, genetic and metabolic capabilities, and ecological roles, remains limited. Oceanic waters are home to a vast and varied variety of bacteria, viruses, protists, and fungi. Oceans, estuaries, lagoons, mangroves, coral reefs, the deep sea, and the seafloor are all home to marine microorganisms, as are the lower and abyssal depths from coastal to offshore regions, as well as the general oceanic and specialized niches like blue waters of coral reefs to black smokers of hot thermal vents on the seafloor [9].

Marine microorganisms play a vital role in the ecosystem of the ocean by promoting important biogeochemical processes, such the cycling of carbon, nitrogen and phosphorus, and aiding in the creation of organic matter. They also interact with other marine organisms, such as pathogens, or in symbiotic interactions. Despite their great diversity, they can be divided into groups based on evolutionary links, metabolic capacities, and morphology. Ocean currents, upwelling, photosynthesis, and nutrient recycling are some of the physical and biological elements that influence the availability of nutrients in the ocean. Different temperatures are needed by bacteria for growth and activity; warmer waters support more bacterial biomass and higher growth rates, whereas colder waters support less bacterial biomass and slower growth rates. A diverse range of microorganisms, mainly bacteria and archaea, have been discovered thriving in extreme environments. These include thermophiles, halophiles, alkalophiles, psychrophiles, piezophiles, and polyextremophiles. In the marine environment, certain bacterial groups are particularly abundant, such as cyanobacteria, proteobacteria, and actinobacteria. These microorganisms play various roles, including decomposing organic matter (saprobes), forming symbiotic relationships, and even acting as parasites, highlighting their adaptability and importance in the ocean's ecosystem. They are marked by extreme conditions, like high pressure and temperature, creating a hostile environment for most forms of life. The base of the food chain is made up of bacteria, which give other creatures, like shrimp, crabs, and tube worms, their nourishment. Deep-sea vent microbial communities are dominated by sulfur and hydrogen-respiring epsilon proteobacteria, as well as methanogenic and sulfate-reducing archaea, notwithstanding their great variability [10].

In the oceans, marine fungi are common and occupy a variety of ecological niches. They form associations with organisms across all trophic levels playing various roles such as decomposers (saprobes), symbiotic partners, and parasites. This versatility enables marine fungi to thrive in various marine environments, from the seafloor to the surface waters, and from coral reefs to polar regions. A variety of substrates, including invertebrates, decomposing wood, seawater, silt, seaweeds, and mangrove debris, can yield marine fungal strains. A distinct protist group of eukaryotic microbes, the thraustochytrids produce bioactive substances such as antibacterial agents. Because they lack plastids and are not photosynthetic, these marine heterotrophic protists—a fungus-like clade of Stramenopiles, class Labyrinthula of the Chromista kingdom—are not microalgae, despite the fact that they are sometimes referred to as such [11]. Viruses are recognized as significant pathogens in numerous marine organisms, with the majority being bacteriophages. These viruses can exist in two main forms: as virioplankton, freely circulating in the water column, or in association with macro- or microorganisms. With its enormous genetic variety, this viral community acts as a gene bank for prokaryotic communities.

Marine biotechnology uses biotechnology, bioinformatics, and molecular and cell biology to create goods and processes from marine creatures. The marine microbial biosphere represents a vast, untapped resource with significant biotechnological potential. Marine biotechnology is a multidisciplinary field utilizing marine resources, like organisms, algae, and microbes, for diverse biotechnological applications. It spans various research areas, including marine drug discovery, bioremediation, biomaterial production, bioenergy generation, and aquaculture. This field also focuses on advancing aquaculture ensuring seafood safety, producing biofuels, and more. Marine biotechnology includes the development of new pharmaceuticals derived from marine organisms, such as antimicrobial compounds, antifungal agents, and anti-inflammatory drugs. Marine biotechnology is also used in the development of novel biomaterials for medical and industrial applications, such as biodegradable plastics, adhesives, and coatings. In addition, marine biotechnology plays a crucial role in environmental sustainability efforts, as it can be used for the development of bioremediation technologies to clean up polluted marine environments. Additionally, the development of renewable energy sources, like biofuels made from marine algae, depends on marine biotechnology. All things considered, the sustainable utilization of marine resources and a variety of businesses stand to benefit greatly from marine biotechnology. With significant environmental gradients in physical, chemical, and hydrological characteristics like temperature, light intensity, salinity, and pressure, marine ecosystems support a varied range of life. Marine biotechnology is still in its early stages, and although though this discipline has been around since the 1960s and 1970s, it is still seen as an emerging field [1, 12]. Marine bacteria can degrade various pollutants, including oil spills, heavy metals, and organic contaminants. Utilizing these bacteria to clean up contaminated water and soil helps mitigate environmental damage from human activities and restore ecosystems. Antimicrobial secondary metabolites and antibiotics are also produced in large quantities by marine bacteria, especially actinomycetes. Numerous marine microorganisms have strong antibacterial properties and are regarded as a promising source for the development of novel treatment classes, especially in the fields of infectious illnesses and oncology. These bacteria are special because they can catalyze processes at temperatures close to water's freezing point. For the study of marine bacteria and other microorganisms in a biotechnological setting, the surfaces of numerous marine macroorganisms, including sponges, algae, tropical stony corals, cold-water corals, hydroids, crabs, and fish, provide particularly intriguing niches. These surfaces host abundant and diverse microbial communities.

5.2 Marine Bacteria

A wide variety of micro- and macro-organisms may be found in the aquatic environment. Oceanic seas are home to marine microorganisms. It includes neutral and helpful bacterial strains as well as a wide and varied range of obligatory and opportunist bacterial pathogens. Marine bacteria are a broad category of microorganisms that are essential to the upkeep and proper operation of marine environments. Among the 10^9 bacterial taxa found on Earth, 10^6 are found in the oceans. Understanding the distribution pattern and community structure demands an understanding of marine bacterial diversity. Their existence in the marine environment is shaped by a diverse array of chemical and physical and hydrological factors, including pressure, intensity of light, ionic concentration, temperature, salinity, oxygen levels, nutrient availability, and water quality.

5.2.1 Marine Bacteria Adaptation to Environmental Shifts

Bacteria can thrive even in harsh environments due to their adaptive strategies that encompass a variety of metabolic processes in cells and are evident in their biochemical and physiological characteristics. An increasing number of alkalophilic, thermophilic, halophilic, psychrophilic, polyextremophilic, and piezophilic microorganisms, primarily consisting of bacteria and archaea, have been identified so far. The comparatively large amount of Na+ ions in seawater has impacted the necessity for these ions among the majority marine bacteria [13]. Even 10% (w/v) NaCl can provide ideal development conditions for extreme halophiles [14]. Additionally, marine bacteria can endure extended periods of starvation and thrive in environments with reduced nutrient availability [15]. Furthermore, the great majority of the marine environment is defined by high pressures (greater than 100 × 105 Pa) and low temperatures (less than 4°C), which lead to the development of barophilic and psychrophilic bacteria [16]. Some bacterial strains are regarded as poly-extremophiles because they have both high temperature and alkaline pH optimalities [14]. The majority of hyperthermophiles are archaea, whose ideal growth temperatures are over 80°C. Extremophiles are common among archaea. Extremozymes, which are enzymes produced from extremophiles, have excellent thermal stability and high specific activity at high temperatures. Numerous marine bacteria have developed the ability to thrive best at either low (oligotrophic) or high (copiotrophic) nutrient concentrations allowing various species to inhabit diverse oceanic trophic environments. Certain bacterial groups, such as cyanobacteria and actinobacteria, often known as actinomycetes, are notable for their ability to produce secondary metabolites.

Numerous bacterial groups found in mangrove ecosystems are fed by detritus and support the ecosystem through a variety of processes, including photosynthesis, methanogenesis, nitrogen fixation, agarolysis, antibiotic synthesis and the production of enzymes (L-glutaminase, arylsulfatase, L-asparaginase, cellulase, chitinase, phosphatase, and protease). One of the primary sources of nutrients in coral reef sediments are bacteria, particularly nitrogen and phosphorus, which are produced by bacterial activity. The mucus that coral produces, which is mostly composed of protein and polysaccharides, provides an excellent growth medium for bacteria. Deep-sea bacteria are mostly heterotrophic, meaning they may use the plentiful organic carbon or dissolved organic matter at an optimum concentration. They are able to break down polymeric materials into monomeric and oligomeric molecules, which other species can then use for metabolism. Therefore, the deep-sea bacteria have well evolved to severe oligotrophy, and the supply of substrates mostly limits their activity [17].

Group	Physiology	Function in the marine environment
Archaebacteria Sulfate-reducing bacteria	Chemoautotrophs, anaerobic, thermophilic, and mesophilic	Participate in the cycling of sulfur compounds in seawater and contribute more than 50% of the carbon turnover of sediments along the coast
Methanogenic bacteria	Chemoautotrophs, strict anaerobes, utilize a restricted range of simple carbon compounds for their carbon and energy sources in the process of methanogenesis	Produce methane as a byproduct of their energy-generating metabolism and use trimethylamine as a substrate in the marine environment
Halophilic bacteria	Even at saturation concentrations, they need at least 12%–15% NaCl to thrive	In environments with high salinity, including salt lakes and salterns, red colonies that are developed because of a high carotenoid content control osmotic pressure and fend off the denaturing effects of salt
Eubacteria Luminous bacteria	Light is produced by a basic protein-like substance called luciferin in contact with the oxygen molecule: Gram-negative and motile heterophilic rods	In deep water, creatures use bioluminescence as a protective mechanism to attract food, confuse and divert predators, and blend in with their surroundings. Luminous bacteria aid in the sea's nutrient cycle and provide food for marine life as gut microbiota

(*Continued*)

(*Continued*)

Group	Physiology	Function in the marine environment
Nitrifying bacteria	Oxidize either ammonia to nitrite (*Nitrosococcus*) or nitrite to nitrate (*Nitrococcus*) and convert nitrogen to a form readily available for other biological processes	A highly significant process because acidic sediment particles are bound by positively charged ammonium ions. They are more prevalent in nearshore seas than in offshore areas, where they become accessible for biological activities

5.2.2 Marine Bacteria Isolation

The procedure for separating marine bacteria emphasizes the value of the field of marine biotechnology and the methods used to reveal the mysteries of the ocean's microbial ecosystem. Pure cultures are required for species delineation for microbial taxonomy based on physiological and biochemical tests. Cultures can be investigated for the discovery of novel natural products, such as antibiotics, for the molecular design of biosynthetic elements, and for the potential application of the isolates to bioremediation and environmental protection. Bringing marine bacteria into culture offers a platform for experimental testing of their physiological and metabolic function. The steps involve are the following: sample collection and preparation, cultivation and incubation, and identification and characterization.

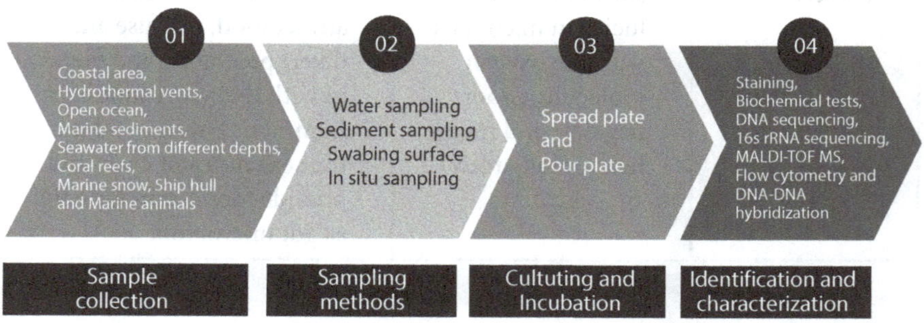

5.2.3 Sample Collection and Preparation

Sample collection of marine bacteria involves collecting seawater or sediment samples from various marine environments, such as coastal areas (beaches, estuaries, mangroves), hydrothermal vents, open ocean (surface waters, deep-sea sediments), marine sediments (sand, mud, silt), seawater from different depths (surface, mesopelagic, bathypelagic), coral reefs, marine snow (organic matter sinking from surface waters), ship hull, and marine animals. The different sampling methods consist of the following:

1. Water Sampling: Samples can be collected from coastal areas, open oceans, and coral reefs. Collect samples using a Niskin bottle, Go-Flo bottle, or pumping system, which are the types of water samplers that allow for the collection of water samples at different depths. They are lowered into the water column at a desired depth and then closed remotely to collect the water sample. These samples are stored in sterile containers or bags and kept cool (4°C) or frozen (−80°C).
2. Sediment Sampling: Samples are collected from coastal area, deep sea, etc. Collect the sediment samples using a box corer for surface sediments, gravity corer for deeper sediments, and dredge for larger sediment samples. Keep them in sterile containers or bags. Add preservatives, like formalin, to prevent degradation. Store samples in cool (4°C) or frozen (−80°C) conditions.
3. Swabbing Surface: Sampling locations include artificial reefs, ship hulls, and marine animals. Use a sterile swab to collect bacteria from the surface. Then, store swabs in sterile containers. Keep samples cool (4°C) or frozen (−80°C).
4. *In situ* sampling: Use a submersible or remotely operated vehicle (ROV) to collect water or sediment samples. Store samples in sterile containers or bags. Keep samples cool (4°C) or frozen (−80°C).

Sample Handling and Storage: Label and document samples with location, depth, and time. Store samples in a cool, dark place (4°C) or frozen (−80°C). Process samples as soon as possible to prevent degradation.

Remember to follow proper sampling and handling protocols to prevent contamination and ensure the integrity of the samples for further analysis and isolation of marine bacteria. Once collected, the samples are transported to the laboratory in sterile containers or bags taking care to minimize contamination and preserve the natural environment of the bacteria. This step involves filtering the samples to remove larger particles and adding nutrients to enrich bacterial growth.

5.2.4 Culturing and Incubation

The prepared samples are then spread on suitable agar plates or into liquid media providing a suitable environment for bacterial growth. To isolate the bacterium, four culture mediums are utilized, namely, starch casein agar (SCA), standard marine broth 2216 (MB), a medium of aqueous sediment extracts as the only source of nutrients, water-extracted matter (WEM), and a medium of methanolic sediment extracts as the only source of nutrients, methanol-extracted matter (MEM). MB is made up of 1.5% agar (BD) and 3.7% marine broth 2216 (BD) [18]. WEM is similar to water-based soil extract [19, 20]. After shaking 200 g of sediment with an equivalent amount of clean water at 160 rpm for the whole night at 30°C, the water-sediment slurry was centrifuged for 10 min at 3,000 rpm, and the supernatant was filtered through a filter paper (d = 15 cm) to remove the sediment. After adding 1.5% agar (BD) to the filtrate, it was sterilized. This medium had a pH of 6.5–7.0 and a salinity of 0.5%–0.8%. MEM was created after earlier research [20]. The slurry was cleaned by filter paper filtration as previously said. A total of 200 g of sediment was extracted with an equivalent volume of 80% methanol (AR) by shaking (160 rpm) overnight at 30°C. Then, the slurry was mixed with another 200 ml of 80% methanol, and the step of filtering was repeated. In a water bath set at 37°C, the filtrates were combined and allowed to evaporate. A 0.22-μm pore-size membrane filter (Millipore) was used to filter the residue after it was dissolved in 50 ml of clean water. The medium was made by combining 1.5% agar (BD), 2.5% methanol extracted matter, and half-strength artificial saltwater. It was then autoclave sterilized. Actinobacteria from sea samples were often treated with SCA, which was dissolved in coastal seawater/pure water (1:1, by volume) and comprised 10 g/L of starch, 0.3 g/L of casein, and 1.5% agar (BD) [21, 22]. Screening of marine isolates for their antibacterial activity was done by point inoculation method and inoculation method using Zobell Marine Broth/agar (ZMB/ZMA).

For the enumeration of bacteria in marine waters, extensive use is made of the pour plate and spread plate. A 0.1-ml/1-ml (pour plate) dilution of the 0.2-g samples was applied to each of the four culture medium plates after they had been serially diluted in 0.9 ml of sterile natural seawater. For 30 days, the plates were incubated at 28°C to 30°C. After that, colonies were selected and separated according to their growth patterns, color, and shape. To produce pure isolates and recover sufficient cell biomass for identification and preservation, the colonies were streaked twice onto MB agar. The cells were maintained in 20% glycerol (v/v) at −80°C after the isolates were inoculated in 30 ml of MB using a 100-ml glass flask at 30°C and shaken for 3–7 days to get cell biomass.

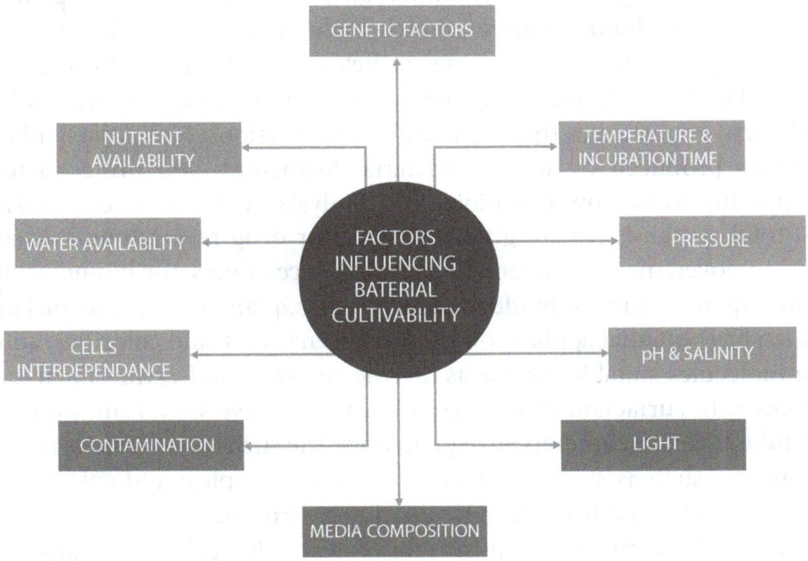

5.2.5 Identification and Characterization

The selected colonies are re-cultured and re-isolated to ensure purity. They can be analyzed using a variety of techniques, such as staining, biochemical test, DNA sequencing, DNA–DNA hybridization, MALDI-TOF mass spectrometry, 16S rRNA gene sequencing, and flow cytometry, which are used to classify bacteria and understand their properties. Alignment of sequences can be compared with those available at the GenBank database of the NCBI by using the advanced BLAST search tool.

5.2.6 Marine Bacteria: A Key to Biotechnology

In both pelagic and benthic ecosystems, bacteria are typically the most abundant and diverse members of the microbial community, and they are essential to the richness and functionality of the marine ecosystem. Around the world, a large variety of bacterial species may be found in marine environments. The marine bacteria *Candidatus pelagibacter* is acknowledged as one of the most prevalent species on the planet [23, 24]. Majority of the bioactive substances that have been extracted from marine sources are obtained from bacteria, and marine microbes have been widely used for a variety of biotechnological applications. Notably, marine bacteria, particularly actinomycetes, generate a wide range of secondary metabolites. Some have the potential to be developed into therapeutic medicines, many of which show biological activity. Additionally, many marine microorganisms have strong antibiotic qualities, which make them a valuable source for developing new therapeutic classes in fields like infectious disorders and cancer. Furthermore, marine cyanobacteria develop cytotoxic and biologically active substances that are useful as medications. Novel biosynthetic enzymes produced by marine bacteria, firmicutes, and proteobacteria are proving to be powerful biological catalysts in therapeutic chemistry and total synthesis, creating new avenues for drug research and discovery. Moreover, marine bacteria are a rich source of versatile biomolecules, including biopolymers, biodegradable plastics, pigments such as melanin (which has potential applications in sunscreens, dyes, and coloring), adhesive molecules suitable for use as marine cements, and extracellular substances with surfactant properties, all of which have significant potential for industrial and commercial exploitation. The surfaces of various marine organisms, such as sponges, algae, corals (both tropical and cold water), hydroids, crabs, and fish offer a fascinating environment to explore marine bacteria and other microorganisms for biotechnological applications.

In actuality, flourishing and varied colonies of microbes may be found on the surfaces of almost all aquatic species. As an example, a complex and diverse microbial community that includes bacteria, eukaryotes, and archaea usually coexists and interacts with its host organism in the tissues of stony corals. Archaea, bacteria, fungi, and microalgae are among the abundant and varied microbial populations that inhabit marine sponges. These microorganisms can make up as much as 40% of the sponge's volume and are crucial to the host's metabolic processes, which, in turn, greatly enhances the sponge's general health and functionality. Sponges' microbial partners are thought to perform various essential functions, including reinforcing the sponge's skeleton, acquiring nutrients, processing waste products, and producing secondary metabolites. Research has increasingly revealed that prokaryotes and higher eukaryotic organisms, such as sponges, form intimate associations, often mutually beneficial or symbiotic in nature. These connections between epibiotic microbes and their eukaryotic hosts are thought to be established mostly through chemical signals and interactions. As an example, soft-bodied sea creatures, lacking physical defense mechanisms, rely on chemical defense strategies, such as producing bioactive compounds, to survive and thrive in their environments. Bioactive metabolites are abundant in marine invertebrates, but a new study has shown that some of these substances, which were previously believed to be created by the host, are really made by their microbial symbionts. Numerous bioactive substances, such as antibiotics, antiviral agents, anticancer agents, and biopolymers, can be produced by epibiotic microbes. It is necessary to conduct further research on these epibiotic microbial populations since they may provide important discoveries. Similarly, investigating other understudied ecosystems, such as marine species' guts, is probably going to produce significant contributions to biotechnological developments.

Moreover, marine bacteria hold vast potential for applications in bioremediation offering promising solutions to mitigate pollution in marine environment. Marine bacteria have a wide range of potential uses, including the remediation of oil spills through the use of hydrocarbon-degrading bacteria, the biosynthesis of surfactants and emulsifiers for various industrial uses, environmental cleanup, and even pharmaceutical applications. Moreover, hydrocarbon pollution poses a significant environmental concern, particularly in coastal regions that are heavily impacted by human activities, making the development of effective bioremediation strategies a pressing need. Pollutants, like hydrocarbons, pesticides, toxic metals, and herbicides, enter the marine ecosystem through various human activities, including urban runoff, oil spills, and industrial and shipping processes. These hydrocarbons pose serious risks to human health and the environment because they frequently build up in sediments. Establishing ideal circumstances that boost the metabolic activity of bacteria that break down hydrocarbons is essential for effective petroleum bioremediation. Harnessing the power of native bacterial communities can facilitate the bioremediation of contaminated sediments by supplementing them with suitable substrates that stimulate their natural degradation processes.

5.3 Biotechnological Applications of Marine Bacteria

The marine environment contains large bacterial communities that have several applications in the biotechnological field such as waste management and bioremediation. In addition, these bacteria are used in pharmaceutical industries to make several drugs, and some are used as bio surfactants. Marine microorganisms have significant roles in marine biotechnology because of their distinct properties. They can withstand extreme temperature and pressure as well as high salt conditions. These characteristics provide the ability to breakdown complex products into simple forms, recycle nutrients, and support biodegradation. The metabolic routes and physiological modifications made by marine bacteria demonstrate their adaptability to change under biological, chemical, and physical circumstances [25, 26].

5.3.1 Application of Marine Bacteria in Waste Management

Waste management is a critical environmental issue that occurs due to human activities. To overcome this issue, the usage of marine organisms is an efficient alternative. Bacterial communities can break down, change, or produce a variety of substances that are produced either naturally or artificially. These support helps to maintain the equilibrium of the environment by managing the concentration of pollutants.

Degradation of Wastewater
Primary applications of marine bacteria in waste management are in the treatment of wastewater. Organic debris, nutrients, heavy metals, and pathogens are just a few of the toxins found in wastewater that, if left untreated, can seriously endanger the environment and public health. Marine bacteria have been found to be highly effective in degrading organic pollutants in wastewater through a process called biodegradation. These bacteria are able to break down complex organic compounds into simpler, less harmful substances, which can then be safely discharged into the environment.

Decomposition of Organic materials
Bacteria are essential for breaking down organic materials into their most basic constituents. While cellulolytic bacteria, like *Cytophaga* and *Sporocytophaga*, aid in the breakdown of cellulose, proteolytic bacteria, like *Eubacteria* and *Pseudomonas* sp., aid in the breakdown of proteins. Certain bacteria, such as *Bacillus*, *Pseudomonas*, and *Vibrio*, may degrade chitin, a structural polysaccharide, with their exoenzyme chitinase, despite the fact that it is resistant to degradation and can even be preserved in fossils. Additionally, *Clostridium pectinovorum* is responsible for decomposing pectin yielding pectic acid and methanol as end products.

5.3.2 Bioremediation

The persistent contamination of freshwater ecosystems by industrial, agricultural, and urban activity often finds its way into rivers and the ocean through runoff. Bioremediation is an environmentally benign method for cleaning up the environment that uses living organisms or their structures or products to immobilize, decrease harm, inactivate, or eliminate toxins.

The kind of contamination, including its structure, oxidation stage, and complexation form, determines the bioremediation strategy to be employed. In bioremediation processes, such as the biodegradation of plastics, petroleum, and diesel, as well as polyaromatic and halogenated hydrocarbon and heavy metal removal, bacteria are acknowledged as key players.

Elimination of Heavy Metals

One of the most urgent environmental problems is heavy metal contamination, which comes from both natural and human activity. The physical and chemical remediation techniques now in use have been shown to be expensive, inefficient, and frequently produce hazardous consequences. However, even at low concentrations, marine microorganisms effectively remove heavy metals without generating any hazardous byproducts making them a plausible alternative. Particularly noteworthy is the discovery that the naturally occurring marine bacteria *Vibrio harveyi* has a significant capacity for cadmium bioaccumulation reaching up to 23.3 mg Cd^{2+}/g of dry cells. In a bioreactor, a group of marine bacteria can effectively remove mercury via a process that is independent of disturbances [27]. Furthermore, marine bacteria possess a unique combination of genetic systems that enable them to degrade phenol and heavy metals. As demonstrated by the marine bacterium *Enterobacter cloaceae*, they can also chelate heavy metals eliminating them from contaminated environments by secreting exopolysaccharides. This bacterium can chelate significant amounts of heavy metals, including 65% of cadmium, 20% of copper, and 8% of cobalt at a metal concentration of 100 mg/L [27]. Similarly, it has been discovered that several purple nonsulfur marine bacteria, like *Rhodobium marinum* and *Rhodobacter sphaeroides*, have the capacity to biosorb or biotransform heavy metals, including copper, zinc, cadmium, and lead, from contaminated environments. Because of their ability to withstand and absorb heavy metals, marine bacteria have been found to be valuable markers for evaluating marine contamination.

Polyaromatic Hydrocarbon Degradation (PAHs)

Widespread in the environment, polyaromatic hydrocarbons (PAHs) are dangerous because of their toxicity, persistence, mutagenicity, and carcinogenicity [28]. Through bioremediation, numerous marine bacteria may, however, degrade PAHs and transform them into CO_2 and metabolic intermediates, which supply carbon and energy for cell growth.

Latha and Lalitha Kumari [29] successfully transferred a catabolic plasmid from *Pseudomonas putida*, which contains genes for hydrocarbon degradation, into a marine bacterium, increasing its efficiency in degrading PAHs. This study demonstrates that the bioremediation potential of these marine bacteria can be enhanced. To improve the biodegradation of PAHs in marine environments, a number of new marine bacterial species have been used, such as *Cycloclasticus spirillensus, Lutibacterium anuloederans*, and *Neptunomonas naphthovorans*. Furthermore, a wide variety of marine bacteria have been isolated from marine sources and demonstrated the ability to mineralize naphthalene, a major component of PAHs, including *Achromobacter denitrificans, Bacillus cereus, Corynebacterium renale, Cyclotrophicus* sp., *Moraxella* sp., *Mycobacterium* sp., *Burkholderia cepacia, Pseudomonas fluorescens, Pseudomonas paucimobilis, P. putida, Brevundimonas vesicularis, Comamonas testosteroni, Rhodococcus* sp., *Streptomyces* sp., and *Vibrio* sp. [30].

Biodegradation of Diesel and Petroleum

An estimated 1.7 to 8.8 million tons of petroleum hydrocarbons are discharged into marine and estuarine habitats annually making crude oil the most major organic contaminant in the marine environment. These organic pollutants may be broken down by oil-eating bacteria in the marine environment, which then use the carbon they create as fuel. *Acinetobacter, Marinococcus, Methylobacterium, Micrococcus, Nocardia, Planococcus*, and *Rhodococcus* are a few of the significant marine bacterial genera that may break down oils [31]. Agreia, *Marinobacter, Pseudoalteromonas, Pseudomonas, Psychrobacter*, and *Shewanella* are among the bacteria that are linked to the substantial breakdown of crude oil and its constituents [32]. A more powerful bacterium has been found in Arabian Sea sediments that can degrade oil by 39% in just 8 days in experimental circumstances [33].

Biofilm Formation

Biofilms are a type of surface-attached, matrix-enclosed community of microbes that are highly organized and dynamic, and they are found in a wide variety of ecosystems. These formations, which are frequently created by several bacterial species, are highly prevalent in marine habitats. Numerous subtidal and intertidal hard surfaces, including rocks, ships, and loops, are covered in biofilms in the maritime environment. Naturally occurring biofilm structures serve as a shielding habitat for bacteria that can withstand a range of environmental stressors seen in marine environments, such as desiccation, pH changes, UV radiation, and osmotic shock.

Petroleum oil can be removed from contaminated oceans or marine systems by using marine bacterial biofilms, which have been shown to be an effective method for remediating a range of organic and inorganic pollutants [34].

Degradation of Plastic

Fishing packing and other applications use a variety of plastic classes, including polyvinyl chloride, polystyrene, polyethylene terephthalate, polyethylene, and polypropylene, which eventually pollute the environment. Plastic may be broken down into non-toxic forms by microorganisms. In 30 days, *Rhodococcus ruber* breaks down 8% of the dry weight of plastic in an *in vitro* concentrated liquid culture [35]. It was also discovered that some bacterial species associated with mangroves, such as *Micrococcus*, *Moraxella*, *Pseudomonas*, *Streptococcus*, and *Staphylococcus*, could break down 20% of plastic in a month [36].

5.3.3 Enzymes

Enzymes, vital biocatalysts, facilitate numerous biochemical processes crucial for cellular function and sustaining life. Their exceptional catalytic efficiency, eco-friendliness, and cost effectiveness make them a preferred alternative to chemical catalysts in industrial applications. As a result, enzymes have seen widespread adoption across various sectors, including food, agriculture, textiles, pharmaceuticals, biofuel production, and bioremediation, leading to significant growth in their usage. The extremely diverse physiological condition in the marine environment results in the physiochemical and genetic manipulation of marine bacterial species. As a result, extremely stable enzyme systems are created that can operate in a variety of conditions, including high or extremely low temperatures, alkaline or acidic pH, and osmotic and hydrostatic pressures. A marine *Janibacter* sp. strain HTCC2649 has been found to produce a highly versatile mono and diacylglycerol lipase enzyme, which exhibits optimal activity at pH 7 and 30°C but retains 50% of its activity even at a low temperature of 5°C. Hydrolase enzymes, in particular lipase, are the most widely isolated and industrially utilized enzymes derived from marine microorganisms. They also show stability in the presence of detergents and various metal ions [37]. They have several advantages over traditional mesophilic enzymes making them ideal candidates for developing detergents suitable for both cold- and high-temperature washing conditions. Their stability makes them valuable assets in various industries, such as pharmaceuticals, food processing, and paper production [38]. Many marine isolates, including *Nocardiopsis* sp., *Bacillus* sp., *Wangia* sp., and *Pseudoalteromonas* sp., demonstrate cold-active amylase activity demonstrating the special characteristics of marine-derived amylases. Marine bacteria have also been found to produce amylases with remarkable pH tolerance, thermostability, and salt tolerance, further expanding the potential applications of these enzymes. Protease is an enzyme that is used in many different industries, such as the production of detergent and leather, as well as in applications related to food, health, and waste management. Protease isolation has been facilitated by the use of marine bacteria, including *Pseudoalteromonas* sp., *Bacillus* spp., and *Pseudomonas* sp. Furthermore, because these polysaccharides are prevalent in the marine environment and provide microorganisms with a carbon source, agarase and chitinase enzymes are also obtained from marine bacteria. Other enzymes that have been isolated from the marine environment include cellulase, ligninase, esterase, laccase, and gelatinase.

5.3.4 Application in Drugs and Pharmaceutics

It is believed that a wide variety of natural products and unique chemical structures with unique biological activities can be found in oceans. These can be exploited to find possible drugs that have high specificity and effectiveness in treating human diseases. Pharmaceuticals include new anti-inflammatory drugs (like manoalide, topsentins, and pseudopterosins), anticancer drugs (like bryostatins and sarcodictyin), and antibiotics (like marinone). The main way that probiotic bacteria, such as *lactobacilli* and *bifidobacteria*, can control harmful microbes is by generating antibacterial proteins like bacteriocin and anticancer agents [39]. Dietary supplements containing lactobacilli are reportedly lowering the induction of experimental colon cancer [40].

5.3.5 Possible Uses for Marine Bacterial EPS

Because of their strict physical and chemical requirements, bacterial extracellular polymeric substances (EPS) have found application in a variety of industries, including food, pharmaceutical, biomedical, and bioremediation. Bacterial exopolysaccharides' rheological, emulsifying, and solidifying qualities have proven crucial for their many uses.

Th1 cell-mediated immunity is strongly stimulated by EPS released by *B. licheniformis* and *Geobacillus thermodenitrificans*, which has medical and pharmacological uses. These may therefore be employed therapeutically as immunomodulatory agents. The EPS compounds generated by *A.*

infernus can be used to treat arteriosclerosis and lipemia after being sulfated and depolymerized [41, 42]. *Vibrio diabolicus's* HE800EPS was discovered to be a potent bone-healing substance free of hypersensitivity and inflammatory responses [43]. In an engineering connective tissue model, it was shown that HE800 EPS improved collagen structure and encouraged fibroblasts to settle in the extracellular matrix.

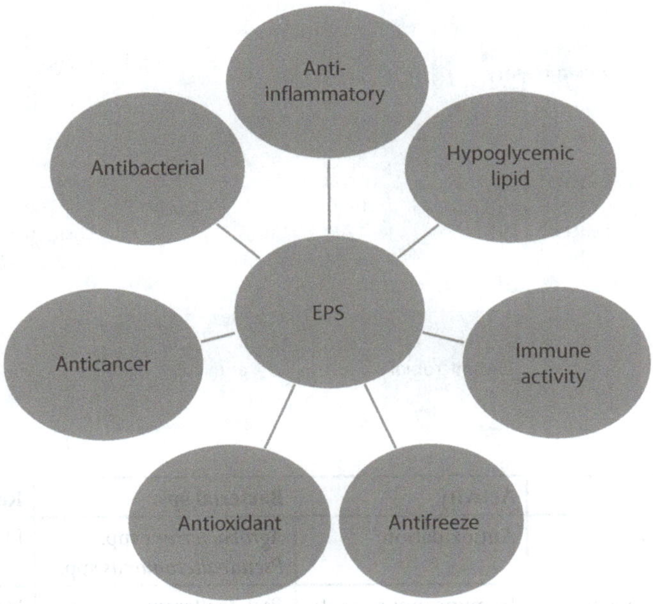

5.3.6 Bioactive Pigment

Numerous actions, such as antioxidant, anti-inflammatory, anticancer, antibacterial, and antiviral qualities, make the biologically active pigment that was separated from marine bacteria useful in food, cosmetics, and pharmaceutical products. A health benefit of the biologically active pigment isolated from marine bacteria is its antioxidant activity. Numerous chronic illnesses, such as cancer, heart disease, and neurological problems, are significantly influenced by oxidative stress. By reducing oxidative damage and neutralizing free radicals, the pigment's antioxidant qualities shield cells from harm and minimize the chance of contracting certain illnesses. The pigment has been shown to decrease tumor development, prevent cancer cells from proliferating, and cause apoptosis or programmed cell death. These results imply that the pigment might be useful as a therapeutic agent to treat different kinds of cancer. The cosmetics business uses biologically active pigments that have been separated from marine microbes. Because of their anti-inflammatory and antioxidant qualities, carotenoid pigments—like astaxanthin and fucoxanthin—have been used in skincare products to help shield the skin from UV rays and other environmental stressors.

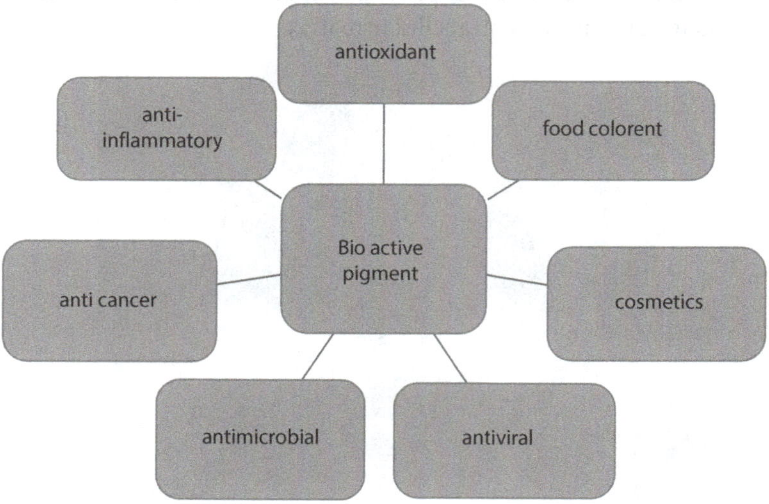

Pigments	Activity	Bacterial sps.	References
Astaxanthin	Antioxidation	*Agrobacterium* spp. *Pseudoalteromonas* spp.	[44]
Cycloprodigiosin	Immunosuppressant; anticancer; antimalarial	*Pseudoalteromonas* sps.	[45–47]
Melanins	Protection from UV irradiation	*Vibrio* spp., *Shewanella* spp., *Alteromonas* spp., *Cellulophaga* spp.,	[48–51]
Prodigiosin	Antibacterial; anticancer; algicidal	*Pseudoalteromonas* spp.	[52, 53]
Undecylprodigiosin	Anticancer	*Streptomyces* spp.	[54]
Violacein	Antibiotic; antiprotozoan; anticancer	*Pseudoalteromonas* spp; *Collimonas* CT	[55, 56]
Methyl saphenate	Antibiotic	*Pseudonocardia* sp.	[57]

(Continued)

(*Continued*)

Pigments	Activity	Bacterial sps.	References
Phenazine derivatives	Cytotoxic	*Bacillus* sp.	[58]
Pyocyanin and pyorubril	Antibacterial	*Pseudomonas* sp.	[59]
Heptyl prodigiosin	Antiplasmodial	*α-Proteobacteria*	[60]
Scytonemin	Protection from UV irradiation, anti-inflammatory, antiproliferative	Cyanobacteria	[61]
Tryptanthrin	Antibiotic	Cytophaga/Flexibacteria AM13,1 strain	[62]

5.3.7 Biosurfactant

Marine bacteria have emerged as a promising source of biosurfactants, which are surface-active compounds that have the ability to reduce surface tension and interfacial tension between liquids. This property makes them highly effective in applications such as emulsification, wetting, foaming, and dispersion, which are essential in various industries such as petroleum, food, pharmaceuticals, and cosmetics. One of the key benefits of using marine bacteria in biosurfactant production is their ability to thrive in extreme environments such as high salinity, temperature, and pressure conditions. This versatility allows for the discovery of unique biosurfactants that may be more effective than those produced by conventional bacteria. Additionally, marine bacteria are often less susceptible to contamination and more environmentally friendly compared to synthetic surfactants. Several marine bacteria have been identified as potent producers of biosurfactants, including strains of *Pseudomonas*, *Bacillus*, and *Vibrio*. These bacteria can produce a wide range of biosurfactants with varying properties such as glycolipids, lipopeptides, and phospholipids. These biosurfactants have shown promising results in applications such as enhanced oil recovery, bioremediation of oil spills, and antimicrobial activities. Furthermore, marine bacteria can be easily cultured in bioreactors using seawater as a medium making the production process more cost effective and sustainable. By utilizing marine bacteria in biosurfactant production, companies can reduce their reliance on synthetic surfactants derived from petroleum, which are non-biodegradable and can be harmful to the environment. Applications of biosurfactants produced from marine bacteria include increasing the efficiency of oil extraction from oil wells, acting as emulsifiers, foaming agents, and stabilizers in food products, being used in personal care products such as soaps, shampoos, and lotions, serving as drug delivery agents, and contributing to the production of pharmaceuticals.

5.4 Conclusion

Marine bacteria, a vast and diverse group of microorganisms, play crucial roles in marine ecosystems and significantly impact human and animal health. These bacteria are broadly classified into obligate and opportunistic categories. Obligate marine bacteria are adapted to specific ocean conditions such as high salinity and particular temperatures. In contrast, opportunistic marine bacteria are more adaptable, thriving in various environments and sometimes becoming pathogenic. Marine bacteria possess a dual nature, harboring both beneficial and harmful properties. While they have the potential to harm human health and the environment, their positive aspects can be leveraged and amplified through various applications to greatly benefit human well-being and meet human needs.

Advancements in genomic sequencing, bioinformatics, and biotechnology have enhanced our understanding and utilization of marine bacteria. These technologies have revealed the potential of marine bacteria to produce novel antibiotics, enzymes, and bioactive compounds, which have significant implications for medical and veterinary applications. Marine biotechnology offers a powerful tool for harnessing the positive aspects of marine bacteria amplifying their benefits to greatly enhance human well-being and address pressing human needs. Through various applications, marine biotechnology can unlock the full potential of marine bacteria leading to significant improvements in human health and quality of life, sustainable solutions for food security and environmental conservation, innovative products and technologies for a better future, enhanced human well-being through cosmetics and personal care, and new bioenergy sources and sustainable industrial processes. By leveraging marine biotechnology, we can maximize the positive impacts of marine bacteria and create a brighter, more sustainable future for all.

Marine biotechnology has opened up a vast array of applications for marine bacteria, including bioremediation, biofertilizers, biomedical applications, enzyme production, food and feed, bioenergy, bioproducts, marine pollution monitoring, biocatalysis, and cosmetics and personal care. By harnessing the beneficial qualities of marine bacteria, we can develop innovative solutions that produce life-saving medicines and bioactive compounds, enhance food security and sustainable agriculture, develop eco-friendly products and bioremediation techniques, support human health through cosmetics and personal care products, and explore new bioenergy sources and sustainable industrial processes.

Marine bacteria are also critical for environmental applications like bioremediation and plastic degradation. Certain species can degrade pollutants and plastics offering sustainable solutions for marine pollution management and contributing to ecosystem health. By leveraging the natural capabilities of these bacteria, innovative methods are being developed to clean up oil spills, detoxify hazardous substances, and reduce oceanic plastic waste. Marine bacteria are crucial in the marine environment, playing a vital role in the ecosystem and offering numerous benefits for marine biotechnology. They produce bioactive compounds, clean pollutants, and develop novel enzymes and biocatalysts. Additionally, they enhance food security through aquaculture, discover new medicines, and provide innovative products and technologies. Marine bacteria also contribute to environmental monitoring, sustainable energy, improved water quality, and novel materials. Their significance cannot be overstated, and their potential for innovation is vast and untapped making them a vital area of research and exploration.

In conclusion, marine bacteria are a dynamic and integral component of marine life offering substantial benefits for environmental sustainability and health. The division into obligate and opportunistic categories highlights their adaptability and specialized functions. Through advanced technological applications, marine bacteria are being harnessed to tackle critical challenges, such as pollution and disease, underscoring their invaluable role in both natural ecosystems and biotechnological innovations. By focusing on the positive aspects of marine bacteria and unlocking their potential, we can discover a wealth of opportunities to improve human health, promote sustainability, and protect the environment ultimately s future.

References

1. Rayner, R., Jolly, C., Gouldman, C., Ocean Observing and the Blue Economy. *Front. Mar. Sci.*, 6, 2019, doi: 10.3389/fmars.2019.00330.
2. Verma, A., Chettri, D., Verma, A., Bioprospecting of Marine Bacteria for their Diversity and Biotechnological Application. *Open Access J. Microbiol. Biotechnol.*, 7, 1–24, 2022. 10.23880/oajmb-16000231.

3. Townsed, M., Davies, K., Hanley, N., Hewitt, J.E., Lundquist, C.J., Lohrer, A.M., The challenge of implementing the marine ecosystem service concept. *Front. Mar. Sci.*, 5, 359, 2018. doi: 10.3389/fmars.2018.00359.
4. Rotter, A., Barbier, M., Bertoni, F., Bones, A.M., Cancela, M.L., Carlsson, J., Vasquez, M.I., The essentials of marine biotechnology. *Front. Mar. Sci.*, 8, 629629, 2021.
5. Blunt, J.W., Copp, B.R., Munro, M.H.G., Northcote, P.T., Prinsep, M.R., Marine natural products. *Nat. Prod. Rep.*, 22, 1, 15–61, 2005.
6. Colwell, R.R., Fulfilling the promise of biotechnology. *Biotechnol. Adv.*, 20, 3–4, 2002.
7. Rasmussen, R.S. and Morrissey, M.T., Marine biotechnology for production of food ingredients. *Adv. Food Nutr. Res.*, 52, 237–292, 2007.
8. Baharum, S.N., Beng, E.K., Mokhtar, M.A.A., *Marine Microorganisms: Potential Application and Challenges*, Selangor, Malaysia, 2010.
9. Qasim, S.Z., *The Indian Ocean: Images and Realities*, pp. 57–90, Oxford and IBH, New Delhi, 1999.
10. Huber, J.A., Welch, D.M., Morrison, H.G., Huse, S.M., Neal, P.R., Butterfield, D.A., *et al.*, Microbial population structures in the deep marine biosphere. *Science*, 318, 97–100, 2007.
11. Leyland, B., Leu, S., Boussiba, S., Are thraustochytrids algae? *Fungal Biol.*, 121, 835–840, 2017. doi: 10.1016/j.funbio.2017.07.006.
12. Rotter, A., Gaudencio, S., Dubnika, A., Vasquez, M.I., Marine Biotechnology, Revealing an Ocean of Opportunities. *Front. Mar. Sci.*, 9, 887630, 2022.
13. Macleod, R.A., Question of existence of specific marine bacteria. *Bacteriol. Rev.*, 29, 9–23, 1965.
14. Bowers, K., Mesbah, N., Wiegel, J., Biodiversity of poly-extremophilic Bacteria: Does combining the extremes of high salt, alkaline pH and elevated temperature approach a physicochemical boundary for life. *Saline Syst.*, 5, 9, 2009.
15. Wai, S.N., Mizunoe, Y., Yoshida, S., How Vibrio cholerae survive during starvation. *FEMS Microbiol. Lett.*, 180, 123–131, 1999.
16. Delong, E.F. and Yayanos, A.A., Properties of the glucose-transport system in some deep-sea bacteria. *Appl. Environ. Microbiol.*, 53, 527–532, 1987.
17. Deming., J.W. and Yager, P., Natural bacterial assemblages in deep-sea sediments: towards a global view, in: *Deep Sea Food Chains and the Global Carbon Cycles*, G.T. Rowe, and V. Pariente, (Eds.), pp. 11–28, Kluwer, Dordrecht, 1992.
18. Sanz-Sáez, I., Salazar, G., Sánchez, P., Lara, E., Royo-Llonch, M., Sà, E.L., *et al.*, Diversity and distribution of marine heterotrophic bacteria from a large culture collection. *BMC Microbiol.*, 20, 207, 2020. doi: 10.1186/s12866-020-01884-7.
19. Taylor, C.B., The nutritinal requirements of the predominant bacterial flora of the soil. *Proc. Soc. Appl. Bacteriol.*, 14, 101–111, 1951b.

20. Nguyen, T.M., Seo, C., Ji, M., Paik, M.J., Myung, S.W., Kim, J., Effective soil extraction method for cultivating previously uncultured soil bacteria. *Appl. Environ. Microbiol.*, 84, e01145-18, 2018. doi: 10.1128/AEM.01145-18.
21. Bukhari, M.A.A., Thomas, A.N., Wong, N.K., Culture conditions for optimal growth of Actinomycetes from marine sponges, in: *Developments in Sustainable Chemical and Bioprocess Technology*, R. Pogaku, A. Bono, C. Chu (Eds.), pp. 203–210, Springer, Boston, MA, 2013.
22. Salim, F.M., Sharmili, S.A., Anbumalarmathi, J., Umamaheswari, K., Isolation, molecular characterization and identification of antibiotic producing actinomycetes from soil samples. *J. Appl. Pharm. Sci.*, 7, 69–75, 2017.
23. Giovannoni, S.J., Tripp, H.J., Givan, S., Podar, M., Vergin, K.L., Baptista, D., Bibbs, L., Eads, J., Richardson, T.H., Noordewier, M., Rappe, M.S., Short, J.M., Carrington, J.C., Mathur, E.J., Genome streamlining in a cosmopolitan oceanic bacterium. *Science*, 309, 1242–1245, 2005.
24. Luna, G.M., *Biotechnological potential of marine microbes*, Springer handbook of marine biotechnology, 651–661, 2015.
25. Sivaperumal, P., Kamala, K., Rajaram, R., Bioremediation of industrial waste through enzyme producing marine microorganisms. *Adv. Food Nutr. Res.*,, 80, 165–179, 2017.
26. Urbanek, A.K., Rymowicz, W., Mirończuk, A.M., Degradation of plastics and plastic-degrading bacteria in cold marine habitats. *Appl. Microbiol. Biotechnol.*, 102, 18, 7669–7678, 2018. https://doi.org/10.1007/s00253-018-9195-y.
27. Canstein, H., Kelly, S., Li, Y., Wagner-Dobler, I., Species diversity improves the efficiency of mercury-reducing biofilms under changing environmental conditions. *Appl. Environ. Microbiol.*, 68, 2829–2837, 2002.
28. Cerniglia, C.E., Biodegradation of polycyclic aromatic hydrocarbons. *Biodegradation*, 3, 351–368, 1992.
29. Latha, K. and Lalithakumari, D., Transfer and expression of a hydrocarbon degrading plasmid pHCL from Pseudomonas putida to marine bacteria. *World J. Microbiol. Biotechnol.*, 17, 523–528, 2001.
30. Samanta, S.K., Singh, O.V., Jain, R.K., Polycyclic aromatic hydro- carbons: environmental pollution and bioremediation. *Trends Biotechnol.*, 20, 6, 243–248, 2002.
31. Sakalle, K. and Rajkumar, S., *Isolation of crude oil degrading marine bacteria and assessment for biosurfactant production*, 2009.
32. Deppe, U., Richnow, H.H., Michaelis, W., Antranikian, G., Degradation of crude oil by an arctic microbial consortium. *Extremophiles*, 9, 461–470, 2005.
33. Mukherji, S., Jagadevan, S., Mohapatra, G., Vijay, A., Biodegradation of diesel oil by an Arabian Sea sediment culture isolated from the vicinity of an oil field. *Bioresour. Technol.*, 95, 281–286, 2004.
34. Vu, B., Chen, M., Crawford, R.J., Ivanova, E.P., Bacterial extracellular polysaccharides involved in biofilm formation. *Molecules*, 14, 2535–2554, 2009.
35. Andrady, A.L., Microplastics in the marine environment. *Mar. Poll Bull.*, 62, 1596–1605, 2011.

36. Kathiresan, K., Polythene and Plastics-degrading microbes from the mangrove soil. *Rev. Biol. Trop*, 51, 629–634, 2003.
37. Yuan, D., Lan, D., Xin, R., Yang, B., Wang, Y., Biochemical properties of a new cold-active mono-and diacylglycerol lipase from marine member Janibacter sp. strain HTCC2649. *Int. J. Mol. Sci.*, 15, 6, 10554–10566, 2014.
38. Navvabi, A., Razzaghi, M., Fernandes, P., Karami, L., Homaei, A., Novel lipases discovery specifically from marine organisms for industrial production and practical applications. *Process Biochem.*, 70, 61–70, 2018.
39. DeVugst L and Vandamme, E.J., *Bacteriocins of lactic acid bacteria. Microbiol Genet Appl*, vol. 75, pp. 140174–140179, Blackie Acadamic & Profession, London, 1994.
40. Goldin, B.R. and Gorbach, S.L., Probiotics for humans, in: *Probiotics*, R. Fuller (Ed.), pp. 355–376, Chapman and Hall, London, UK, 1992.
41. Guezennec, J., Pignet, P., Lijour, Y., Gentric, E., Ratiskol, J., Colliec Jouault, S., Sulfation and depolymerization of a bacterial exopolysaccharide of hydrothermal origin. *Carbohydr. Polym.*, 37, 19–24, 1998.
42. Zhenming, C. and Yan, F., Exopolysaccharides from marine bacteria. *J. Ocean Univ. China*, 4, 1, 67–74, 2005.
43. Zanchetta, P., Lagarde, N., Guezennec, J., A new bone-healing material: a hyaluronic acid-like bacterial exopolysaccharide. *Calcif. Tissue Int.*, 72, 74–9, 2003.
44. Misawa, N., Satomi, Y., Kondo, K., *et al.*, Structure and functional analysis of a marine bacterial carotenoid biosynthesis gene cluster and astaxanthin biosynthetic pathway proposed at the gene level. *J. Bacteriol.*, 177, 22, 6575–6584, 1995.
45. Yamamoto, C., Takemoto, H., Kuno, K., *et al.*, Cycloprodigiosin hydrochloride, a new H+/Cl– symporter, induces apoptosis in human and rat hepatocellular cancer cell lines *in vitro* and inhibits the growth of hepatocellular carcinoma xenografts in nude mice. *Hepatology*, 30, 4, 894–902, 1999.
46. Kawauchi, K., Shibutani, K., Yagisawa, H., *et al.*, A possible immunosuppressant, cycloprodigiosin hydrochloride, obtained from Pseudoalteromonas denitrificans. *Biochem. Biophys. Res. Commun.*, 237, 3, 543–547, 1997.
47. Kim, H.S., Hayashi, M., Shibata, Y., *et al.*, Cycloprodigiosin hydrochloride obtained from Pseudoalteromonas denitrificansis a potent antimalarial agent. *Biol. Pharm. Bull.*, 22, 5, 532–534, 1999.
48. Kotob, S.I., Coon, S.L., Quintero, E.J., Weiner, R.M., Homogentisic acid is the primary precursor of melaninsynthesis in Vibrio cholerae, a Hyphomonas strain, and Shewanella colwelliana. *Appl. Environ. Microbiol.*, 61, 4, 1620–1622, 1995.
49. Ruzafa, C., Sanchez-Amat, A., Solano, F., Characterization of the melanogenic system in Vibrio cholerae, ATCC14035. *Pigm. Cell Res.*, 8, 3, 147–152, 1995.

50. Ivanova, E.P., Kiprianova, E.A., Mikhailov, V.V., *et al.*, Characterization and identification of marine Alteromonas nigrifaciens strains and emendation of the description. *Int. J. Syst. Bacteriol.*, 46, 1, 223–228, 1996.
51. Fuqua, W.C. and Weiner, R.M., The melA gene is essential for melanin biosynthesis in the marine bacterium Shewanella colwelliana. *J. Gen. Microbiol.*, 139, 5, 1105–1114, 1993.
52. Gerber, N.N. and Gauthier, M.J., New prodigiosin-like pig ment from Alteromonas rubra. *Appl. Environ. Microbiol.*, 37, 6, 1176–1179, 1979.
53. Kim, D., Lee, J.S., Park, Y.K., *et al.*, Biosynthesis of antibiotic prodiginines in the marine bacterium Hahella chejuensis KCTC 2396. *J. Appl. Microbiol.*, 102, 4, 937–944, 2007.
54. Gerber, N.N., Prodigiosin-like pigments. *CRC Crit. Rev. Microbiol.*, 3, 4, 469–485, 1975.
55. Yada, S., Wang, Y., Zou, Y., *et al.*, Isolation and characterization of two groups of novel marine bacteria producing violacein. *Mar. Biotechnol.*, 10, 2, 128–132, 2008.
56. Hakvag, S., Fjærvik, E., Klinkenberg, G., *et al.*, Violacein producing Collimonas sp. from the sea surface microlayer of costal waters in Trøndelag, Norway. *Mar. Drugs*, 7, 4, 576–588, 2009.
57. Maskey, R.P., Kock, I., Helmke, E., Laatsch, H., Isolation and structure determination of phenazostatin D, a new phenazine from a marine actinomycete isolate Pseudonocardia sp. B6273. *Z. fur Naturforschung*, 58b, 7, 692–694, 2003.
58. Li, D., Wang, F., Xiao, X., Zeng, X., Gu, Q.Q., Zhu, W., A new cytotoxic phenazine derivative from a deep sea bacterium Bacillus sp. *Arch. Pharmacal Res.*, 30, 5, 552–555, 2007.
59. Saha, S., Thavasi, R., Jayalakshmi, S., Phenazine pigments from Pseudomonas aeruginosa and their application as antibacterial agent and food colourants. *Res. J. Microbiol.*, 3, 3, 122–128, 2008.
60. Lazaro, J.E., Nitcheu, J., Predicala, R.Z., *et al.*, Heptyl prodigiosin, a bacterial metabolite, is antimalarial *in vivo* and nonmutagenic *in vitro*. *J. Nat. Toxins*, 11, 4, 367–377, 2002.
61. Stevenson, C.S., Capper, E.A., Roshak, A.K., *et al.*, Scytonemin—a marine natural product inhibitor of kinases key in hyperproliferative inflammatory diseases. *Inflammation Res.*, 51, 2, 112–114, 2002.
62. Wagner-Dobler, I., Beil, W., Lang, S., Meiners, M., Laatsch, H., Integrated approach to explore the potential of marine microorganisms for the production of bioactive metabolites. *Adv. Biochem. Eng./Biotechnol.*, 74, 207–238, 2002.

6

Nanoparticles and Phytoconstituents from Marine Mangrove Plant: Potential for Biochemical Applications

Divya Kumari and Pracheta Janmeda*

Department of Bioscience and Biotechnology, Banasthali Vidyapith, Rajasthan, India

Abstract

Considering that natural extracts have so many benefits, there is a growing body of research in medicine that looks into how they might be used pharmaceutically to treat various illnesses. These include practically limitless availability and compound sustainability because of the natural origin. They also help to promote the nations from which they are produced and extracted. Because of their advantageous qualities, wild active chemicals isolated from maritime mangrove plants are drawing more attention. These plants are now known to be possible sources of natural compounds having biologically active properties for a wide range of biological treatments. Many chemical substances, most of which have strong anticancer or antitumor effects, are beneficially supplied by these plants. These compounds include cyclic triterpenoids, flavonoids, iridoids, naphtaquinones, polyphenols, polysaccharides, and steroids. Additionally, by using enhanced selectivity toward cancer or tumorigenic cells to formulate the extracted chemicals from these plants into nanoparticles (NPs), researchers were able to improve the therapeutic outcome of therapies and lessen negative effects from nonspecific spread. This chapter primarily discusses the several bioactive substances that are extracted from marine mangrove plants and their uses in biochemistry and medicine.

Keywords: Anticancer, antimicrobial, biochemical, biomedical, mangrove plants, nanoparticles, phytocompounds

**Corresponding author*: pracheta@banasthali.in

Shaju S. S., Vipin. P. M. and Visakh. P.M. (eds.) Marine Biotechnology: Studies and Applications, (169–196) © 2025 Scrivener Publishing LLC

6.1 Introduction

There are 74 species of real mangroves among the 123 tropical and subtropical nations where mangrove plants are found [1]. Because of the variety of ecosystem services they offer, including improving fisheries, storing and sequestering carbon, and protecting coastlines from the many effects of natural disasters, mangroves rank among the most ecologically and economically significant ecosystems worldwide. Additionally, they neutralize contaminants from the land and ocean, preserve the marine ecology, and enhance the quality of coastal as well as nearshore waters via nutrient cycling. By giving them a home and food, they also support species of animals (mammals, reptiles, birds, etc.) and marine life (fish, shrimp, crabs, etc.). Anaerobic soil, high salinity, high temperatures, strong winds, and other extreme environmental conditions are all very difficult for the mangrove species *A. marina* to survive. Too much salt can hinder the growth of gray mangroves, but in environments with both fresh and saltwater, they flourish and grow to their maximum height. By excreting salts through its leaves, the species is able to withstand severe salinity [2, 3]. Mangroves are a group of tiny plants and shrubs that can withstand high salinity levels. Mangroves are rich in a variety of plants that are used for medicine, provide timber, and have exceptional CO_2 sequestration capabilities. Mangrove ecosystems are surrounded by a variety of anthropogenic agents and pollutants. Significant sources include industrial effluents, maritime activities, and wastewater runoff [4]. According to Choi *et al.* [5], Doshi *et al.* [6], and Hu *et al.* [7], improper management of wastewater treatment plants and the direct or indirect disposal of untreated effluent into aquatic systems are the main causes of environmental contamination. Human health, surface water quality, and ecosystems may all be harmed by the effluent concentration [8, 9]. Global human activity poses a serious threat to mangroves [10, 11], despite the fact that their ecological and economical significance is now well acknowledged [12]. Mangroves are distinct ecosystems that support a wide variety of animal and plant species. Mangrove species have a great deal of potential to enhance the quality of the environment. These plants can withstand the severe ecological stressors (nutrients and pollution from urban runoff) prevalent at these sites because of a variety of biological chemicals (phenol, alcohol, terpenoids, etc.) that they possess. This occurs as a result of these plants' successful adaptation to marshy and saline environments. These habitats are subject to anthropogenic loads in addition to natural environmental stresses.

However, this endangers the species that live in these locations and pollutes the mangrove neighborhood. Nonetheless, mangrove trees are utilized in many tropical nations as a bioremediation strategy for wastewater treatment because of their ability to absorb excess nutrient loads [13, 14]. The synthesis of bio-nanomaterials for environmental remediation and biomedical applications has drawn attention to mangrove ecosystems recently. Biomaterials produced from mangroves have distinct physical and chemical characteristics, including as a large surface area, high porosity, and the capacity to form stable complexes with a wide range of organic and inorganic substances. Because of their superior catalytic and adsorption qualities, synthesized nanomaterials enable efficient cleanup [15]. Another way to remediate untreated effluents is to use mangrove resources [16]. Mangrove-derived nanoparticle synthesis is a novel and environmentally friendly technique that may be able to get around some of the drawbacks of more conventional techniques. Their distinct characteristics render them perfect for the creation of nanoparticles (NPs). Mangroves have the ability to absorb and store a wide range of minerals and metals from saltwater, which can be utilized as building blocks for the creation of NPs. Mangrove tannins and other phytochemicals can also function as reducing agents, which makes it easier to synthesize NPs. Sonochemical synthesis, hydrothermal synthesis, and green synthesis are a few techniques for producing mangrove-derived NPs. The hydrothermal synthesis method of NP production necessitates high temperature and high pressure. The production of NPs is accomplished by the sonochemical technique, which uses ultrasonic vibrations to create cavitation bubbles. This chapter delves into the specifics of the morphological traits, biotechnological potential, and bioactive qualities of marine mangrove plants that have applications in the pharmaceutical sector. As a result, our work will speed and lay the groundwork for future studies on the biological applications of mangroves.

6.2 Marine Mangrove Plant

Due to their widespread industrial and medical applications, as illustrated in Figure 6.1, mangroves have long piqued the curiosity of both the general population and scientists [17]. According to Mitra *et al.* [18], they are mainly found in tropical and subtropical intertidal zones, where they are categorized as halophytes (salt tolerant). A taxonomically varied group of trees and shrubs known as "mangrove vegetation" predominates in salty

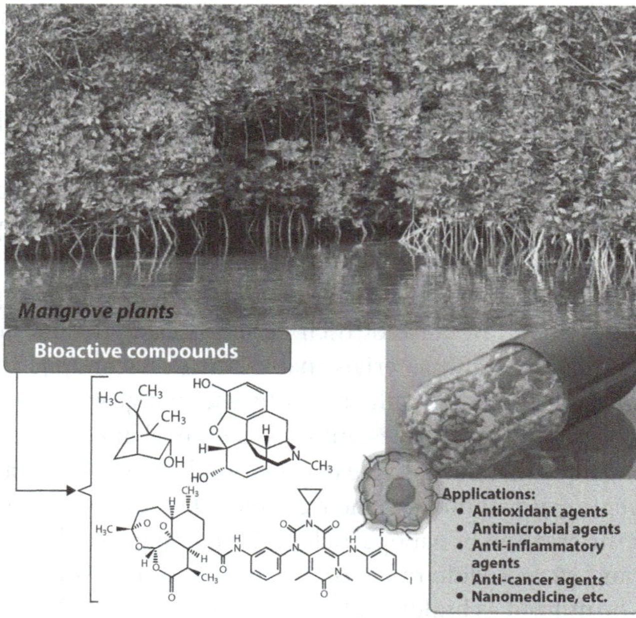

Figure 6.1 Diagrammatic representation of the marine mangroves showing biological benefits.

marshes and tidal areas along protected tropical and subtropical coastlines [19]. According to Setyawan *et al.* [20], mangroves play a significant role in the ecosystem processes and thrive in saline circumstances with daily flooding between mean sea level and the highest astronomical tides. They also provide vital habitats and food for similarly suited resident and migratory animals. Mangroves have developed a series of adaptations to withstand harsh climatic circumstances, including high salinity, strong winds, gradual tidal shifts, high temperatures, and anaerobic tidal swamps, as a result of their unique physical habitat [21]. Mangroves thus offer a favorable environment for a wide variety of micro- and macroorganisms with an abundance of bioactive substances and enzymes [22]. For survival and reproduction, everything has the special capacity to create bioactive metabolites, which serve as "chemical signals" for responding to, avoiding, or fending off environmental cues. The lengthy, sustaining roots of *Rhizophora*, the breathing roots of *Avicennia*, the leaves that release salt, and the water-dispersing, viviparous seedlings are a few examples of these. Mangroves have been used historically as a pesticidal, insecticidal, and medicinal plant [23]. They have also been used as a source of fuel and charcoal, for building houses, furniture, boats, and fishing gear, and to produce tannins for leather and dyeing [24]. The availability of essential nutrients, such as proteins, carbs, and amino acids, which are needed for the upkeep of biological processes and bioactive chemicals, was linked to the use of mangrove plants [25]. In the tropical Pacific, mangrove plants of the rhizophora variety are widely dispersed [26]. *Rhizophora* species (e.g., *Rhizophora apiculata*, *R. mucronata*, *R. mangle*, and *R. stylosa*) belonging to the Rhizophoraceae family contain a multitude of phytochemical substances that have great therapeutic potential. According to Wu *et al.* [27], these substances include hydrolyzable tannins, polyphenols, polymeric tannins, sesquiterpene, daucosterol, atranorin, palmitone, and triterpenoids and diterpenoids. Extracts from several species of *Rhizophora* have been demonstrated to possess a wide range of pharmacological effects, including antifungal, antibacterial, antiseptic, anti-inflammatory, and anti-ulcer characteristics [28].

R. stylosa Griff. is a common mangrove plant found in eastern Indo-West Pacific regions spanning from Australia to the western Pacific Ocean. It is a member of the Rhizophoraceae family [29]. One plant that may survive in salty conditions is *R. stylosa*, which can be found in river estuaries and coastal regions [30]. Another name for the main species in Indonesia's mangrove ecosystem is *R. stylosa* [31]. The physiological, morphological, and metabolic systems of higher plant orders vary as a result of the harsh climate [32]. According to Lalitha *et al.* [33], mangroves offer unique ecological environments and are a rich source of bioactive compounds. Mangrove trees can yield a variety of compounds, including phenolic, flavonoid, alkaloid, terpenoid, saponin, and tannin metabolites [34]. As a result, this plant is widely used in several Indonesian regions for traditional medicine, including anti-inflammatory and antipyretic purposes [35]. Many mangrove species have been used for generations, in many different countries, to treat ailments in accordance with local traditions. However, many of these species have not yet undergone comprehensive examination, and as a result, their medicinal properties have not been proven [36].

6.3 Nanoparticles from Marine Mangrove Plant

In light of its many uses, especially in the biomedical industry, nanotechnology has emerged as a significant scientific advance [37]. Silver NPs, which can be produced via a range of chemical, biological, and physical processes, are among the most researched materials [38]. In biological applications, silver NPs hold great potential, especially when used as antioxidants and antibacterials [39, 40]. The utilization of plant-derived phytochemicals presents an unexplored opportunity for the biological synthesis of NPs [41]. Plants can mediate synthesis called biosynthesis, which is more effective than physical and chemical methods. When synthesizing NPs, biosynthesis has the main benefit of not requiring any hazardous chemicals, high pressure, heat, or energy [42]. It is also possible that the negatively charged lipopolysaccharide coating on these bacteria accounts for their vulnerability to NPs. NP releases positively charged ions most of which are in close proximity to these negatively charged molecules, which increases intracellular damages and ion absorption. A multitude of mechanisms, including the generation of reactive oxygen species (ROS), changed gene expression, electrostatic interaction-mediated binding among NPs and microorganism membranes, cell wall invasion with the organelle in charge of synthesis of proteins, and binding of metabolites, combine to give NP their antibacterial activity [44].

6.3.1 Silver Nanoparticles of *Rhizophora stylosa* Griff. (Red Mangrove)

R. stylosa Griff. has the ability to synthesize NPs, which are utilized in biomedical applications and have lately been popular as alternatives in a number of industries [45]. Furthermore, the various biological activities of the phytoconstituents derived from mangrove plants (such as galactose, gallic acid, lupeol, catechins, carotenoids, and epicatechins) were studied. The nutraceuticals and pharmaceuticals use these molecules to create medications that have antibacterial, antioxidant, antidiabetic, and anticancer properties, among other benefits [46]. Combining silver NPs (AgNPs) with a sustainable supply generated from mangrove plants is recommended as an antibacterial agent. Researchers have been particularly concerned about AgNPs with a dimension of <100 nm due to their small size and huge surface area [47]. *Escherichia coli* and *Staphylococcus aureus*, which are representative of Gram-positive and -negative bacteria and were frequently used to assess the activity of NPs in previous studies, were the two types of bacteria against which the antimicrobial activity of RS–AgNps was tested. The agar diffusion method was used for the antibacterial tests. Amoxicillin and distilled water served as the positive and negative controls, respectively. The test was run twice, and the average of all the data that were provided was used. The results showed that the inhibition zone is significantly impacted by the three modifications in AgNP concentration.

For *E. coli* and *S. aureus* bacteria, the area of limitation for RS–AgNPs at concentrations of 0.001 M (5.5 mm), 0.005 M (7.2 mm), and

0.01 M (5.1 mm) was observed. The 0.005 M concentration of AgNPs demonstrated a greater zone of inhibition against *S. aureus* and *E. coli*. Additionally, Willian *et al.* found that AgNPs produced with 1, 5, and 10 mM precursor concentrations have higher antibacterial activity against *E. coli* than against *S. aureus*. These investigations show that RS–AgNPs are more selective than Gram-positive bacteria in their ability to kill Gram-negative bacteria. This could be because the bacteria's cell walls differ in thickness; Gram-negative bacteria have a thinner peptidoglycan layer (8 nm) with a lipopolysaccharide external membrane, while Gram-positive bacteria have a thick and complex peptidoglycan coating (80 nm) with covalent bonds to teichoic and teichuronic acids (1–3 m). In this case, the bacteria's cell walls are different. It is feasible to draw the conclusion that RS AgNPs could be created as an antibacterial agent in light of these findings.

6.3.2 Nanotechnology-Based Anticancer Formulation of *Avicennia marina*

Due to the medications' atypical dispersion into other organs and tissues and their low targeting ratio, many cancer treatments are ineffective and have adverse effects on other tissues. Thus, a great deal of research has gone into figuring out how to get therapeutic drugs to the tumor site in a way that minimizes toxicity and adverse effects [48]. One of the most promising methods in this situation for overcoming the obstacles of aspecific biodistribution, side effects, and minimal tumor accumulation is formulation based on nanotechnology [49, 50]. Drug delivery NPs (NPs) are particles that range in size from 1 to 1,000 nm and are made of a variety of materials, such as metals, polymers, lipids, and natural or synthesized compounds, that are both biocompatible and biodegradable. NPs are designed and developed to be loaded or covalently linked with bioactive compounds, such as proteins, peptides, antibodies, and nucleic acids, to (1) deal with issues with the molecules' dissolution and *in vivo* bioavailability; (2) prohibit the molecules from breaking down in the bloodstream; (3) improve the molecules' targeting, internalization, and accumulation in the desired cells and tissues; and (4) potentiate the drug's effect [51]. For these reasons, to achieve a therapeutic, synergistic impact, NPs are linked to plants and natural substances. A few studies have discussed the application of NPs to address the issues related to *A. marina*'s low tumor accumulation in an effort to enhance its therapeutic efficacy. Varunkumar *et al.* [52] created bioengineered AgNPs using an aqueous extract of *A. marina* leaves.

The resultant AgNPs induced ROS-mediated apoptosis in the A549 lung cancer cell line demonstrating dose-dependent cytotoxic action (IC_{50} = 50 μg/ml), which was further validated by RT-PCR and Western blotting analysis. It was shown that the caspase-independent and -dependent signaling pathways mediated by p53 were both implicated in the process. The anticancer property of produced AgNPs was confirmed in A549 lung cancer cells. Based on ROS activity, they reported a dose-dependent impact with an inhibition of 54% at 50 μg/ml and 94% at 80 μg/ml.

6.3.3 Metal-Based Nanoparticle Synthesis from Mangrove Plants

Mangrove species have a great potential to enhance the quality of the environment. These plants are able to withstand the harsh environmental stressors found at these places because of a variety of biological chemicals (phenol, alcohol, terpenoids, and so forth). This is as a result of the plants' successful adaptation to the salty marshlands. Numerous human burdens are added to these ecosystems in addition to the stressors posed by the natural environment. The mangrove neighborhood is contaminated as a result, but it also presents a challenge to these species. These two elements, along with a few others (seasonal fluctuations in pollutant adds, toxicity potentials of contaminants, amount of their addition, etc.), have allowed these plants to withstand the stressful conditions that have been forced upon them [53].

- ***Avicenna* sp.**
  AgNPs were synthesized from leaves, bark, and roots of *Avicenna marina*. *A. marina*, a kind of mangrove plant, was used in the biofabrication of AgNPs by Barnawi *et al.* [54]. The TEM (transmission electron microscopy) analysis result indicates that the NP form was spherical to rectangular, with a size range of 10–40 nm. AgNPs were reportedly produced from *A. marina*. The results of the SEM (scanning electron microscopy) and TEM analyses showed that the NPs' average size was 17.30 nm. In the study of Balakrishnan *et al.* [55], mangrove leaf extracts of *A. marina* plant species were used to manufacture AgNPs. The synthesized NPs were characterized by ultraviolet (UV)—Visible spectroscopy, FT-IR (Fourier transform infra-red), SEM, TEM, and DLS (dynamic light scattering). AgNPs were found to have a peak

in the UV–Visible spectra at 420 nm. During TEM investigation, the average size varied between 60 and 95 nm. The primary functional groups exhibiting phytochemicals (flavonoids, triterpenoids, and polyphenols) were determined by FT-IR analysis results. Das *et al.* [56] concentrated on employing mangrove leaf extract (*Avicennia officinalis*) to produce AgNPs in their study. Using the UV—Visible spectra, a maximum absorption peak at 420 nm was found to indicate the existence of AgNPs. Phenolic groups that could stabilize AgNPs were identified by the FT-IR data. The DLS investigation determined the average size of NPs, which was found to be between 10 and 15 nm.

- ***Excoecaria agallocha***
  In their study, Bhuvaneswari *et al.* [57] stated that *Excoecaria agallocha* leaf extract was used to produce AgNPs. The synthesized NPs were characterized by UV—Vis spectroscopy, FT-IR, SEM, TEM, and DLS. Micrographs from SEM, FT-IR, UV, energy-dispersive X-ray spectroscopy (EDAX), and X-ray diffraction (XRD) show hexagonal and spherical particles with sizes ranging from 23 to 42 nm. The results of the XRD study reveal a face-centered cubic AgNPs structure with a 20-nm mean size. FTIR examination has identified the presence of amides, carboxylates, flavonoids, and phenolic chemicals. In their research, Sangeetha *et al.* [58] stated that they used an aqueous leaf extract from *E. agallocha* to create artificial AgNPs. Characterization was carried out in the study utilizing UV—Visible, FT-IR, TEM, SEM, and DLS. The production of AgNPs was indicated by a maximal absorption peak that was detected at 434 nm, and TEM analysis was used to examine the spherical shape of the NPs.
- ***Acanthus ilicifolius***
  AgNPs were produced using *A. ilicifolius* leaf aqueous extracts. Characterization was carried out using FT-IR, XRD, SEM, UV, and EDAX. Additionally, FT-IR analysis was performed on phenolic and alcoholic groups. The findings demonstrate that the metal binds to produce AgNPs and confirm the presence of phenolic and alcoholic groups. According to Anuradha *et al.* [59], the size was also 50 nm. AgNPs are produced by the leaf extract of *A. ilicifolius* and are used to combat the larvae of *Aedes aegypti* and *Armigeres subalbatus. In vitro*, it demonstrates larvicidal action against

Aedes aegypti and *Armigeres subalbatus* larvae at lethal concentrations (LCs) greater than 50 and 90, respectively. Additionally, UV, SEM, EDAX, XRD, and FTIR were used to characterize the biogenic NPs that were produced. The study's findings showed that the average size of the UV—Visible spectrum is 180 nm, with a peak at 420 nm. Alcohols, phenols, carboxylic acids, primary amines, aromatic amines, ethers, and esters are revealed by FT-IR data, in contrast. This demonstrates that AgNPs produced from *A. ilicifolius* leaf aqueous extract have larvicidal action against mosquito larvae without endangering the environment.

- ***Ceriops* sp.**
  AgNP synthesis was carried out in an experiment by Dhas *et al.* [60] using a *Ceriops tagal* leaf extract. AgNPs showed a maximum UV–Vis absorption peak at 424 nm. In the UV–Vis spectrum, the NPs' highest absorbance was observed at 400 nm. AgNPs were created in a work by Ramteke *et al.* [61] using mangrove *C. tagal* stem extract. Additionally, UV, SEM, EDAX, XRD, and FT-IR were used to characterize the produced biogenic NPs. A maximal absorption peak, with spherical and ellipsoidal diameters ranging from 5 to 22 nm, has been discovered in this investigation at 455 nm. The technique was a greener substitute for chemical synthesis techniques. In a different experiment, *C. tagal* stem extract was used to create copper NPs. FT-IR, XRD, SEM, UV, and EDAX were used for the characterization. The crystalline size of 4 nm was ascertained by TEM, and the greatest absorption peak was found at 500 nm in the UV—Visible spectrum. AgNPs were produced from *Ceriops decandra* fruit by Maity *et al.* [62]. TEM examination revealed that biogenic AgNPs had an average size of 8 nm. It also showed significant antibacterial action against both Gram-positive and -negative microorganisms. The study came to the conclusion that a cost- and environmentally friendly technique of producing AgNPs was the green synthesis of polysaccharides (extracted from *C. decandra* fruits).
- ***Heritiera fomes***
  AgNPs and zinc oxide NPs (ZnONPs) were produced by *H. fomes* and *Sonneratia apetala* bark extracts. Using UV–Visible spectroscopy to characterize the NPs, the greatest absorption peak for AgNPs was found at 375 nm. AgNPs

formed by biosynthesis range in size from 70 to 100 nm, as confirmed by DLS analysis. The existence of oxime and other heterocyclic compounds is confirmed by FT-IR analysis. TEM was used to investigate NPs, which had a diameter of 40–50 nm. The investigation also showed the potential uses of biogenic NPs for antibacterial, anti-inflammatory, antidiabetic, and antioxidant properties. It may find use in the food, cosmetics, and medicinal sectors [63].

Mangrove-mediated AgNPs, which are produced from *H. fomes* species aqueous extracts, have been shown to be efficient against bacterial infections and mosquito vectors. TEM, FTIR, XRD, and UV–Visible spectroscopy were used in the characterization process. NPs ranged in size from 15 to 40 nm. However, the *H. fomes* extract and *H. fomes*-mediated AgNPs have been shown to be efficient against the vectors of dengue, zika, and malaria [64].

- ***Rhizophora* spp.**
  Using green biosynthesis, leaf extract from *Rhizophora lamarckii* was utilized to create antiviral AgNPs. In aqueous solutions, AgNPs can be readily reduced to metallic ions. This led to the high density production of crystalline AgNPs with mean diameters ranging from 12 to 28 nm. *Rhizophora mucronata* Lam leaf extract was employed by Nathan *et al.* [65]. Fe (III) oxide NPs (Fe_2O_3 NPs) were produced for the purpose of photocatalytically degrading synthetic dyes. The absorption peak was found to be at 368 nm based on the data. The resulting spherical NPs had an average size of 65 nm. Additionally, the photodegradation efficiency of these NPs was good. *Rhizophora mucronata* leaves were used by Jha *et al.* [66] to create selenium NPs (SeNPs) using the Taguchi design approach. SeNPs have a mean size of 31.82 nm and a spherical form, according to high resolution TEM data.

  The peak absorption of the NPs was detected at 265 nm, and the FT-IR and EDX spectra verified this observation. Furthermore, testing biogenic SeNPs against *A. hydrophila*, *E. coli*, *S. aureus*, and *P. aerogenosa* demonstrates good antibacterial and antibiofilm action.
- ***Sonneratia apetala***
  Nagababu and Rao [67] conducted a study to synthesize AgNPs using leaf extract from *Sonneratia apetala*. The TEM

investigation showed that the NPs had a spherical form. The absorption peak is located between 425 and 475 nm, according to the data. Additionally, they demonstrated significant antibiotic activity against every chosen bacterium. According to the study's findings, this report on the synthesized NPs from *S. apetala* was the first to do so [68].

6.4 Phytoconstituents from Marine Mangrove Plant

6.4.1 Phytochemistry of *Rhizophora stylosa* Griff.

A total of 46 chemicals were recovered from *R. stylosa's* leaves, flowers, fruits, stems, and twigs per the literature review. A variety of solvents were used to get the extract for the isolation procedure, including petroleum ether, n-hexane, ethyl acetate, chloroform, methanol, and n-butanol. To create a crude extract, the dehydrated material was macerated at room temperature using a 1:1 ratio of $CHCl_3$ to MeOH. It was then diluted with water and divided with additional solvents. The crude extract was purified using a variety of methods, such as semi-preparative high-performance liquid chromatography (HPLC) on the RP-18 column, Sephadex LH-20, and column chromatography on RP-18 silica gel. The chemicals were identified using nuclear magnetic resonance (NMR), mass spectrophotometry, UV, FT-IR, and polarimetry. These chemical reactions produced triterpenoid, steroid, lignan, megastigmadien, apocarotenoid, alanine derivatives, monoterpenoid, aromatic/phenolic, flavonoid, fatty acid, and secondary alcohol groups.

6.4.2 Phytochemistry of *Avicennia marina*

The plant (*Avicennia marina*) has been subjected to chemical analyses that have been reported in the literature. These analyses have revealed the presence of a wide range of compounds, such as amino acids, carbohydrates, alkaloids, carotenoids, fatty acids, hydrocarbons, iridoid glucosides, abietane diterpenoid glycosides, carboxylic acids, steroids, tannins, triterpenes, naphthalene derivatives, flavones, flavonoids, phorbol esters, phenolic and related compounds, pheromones, inorganic salts, minerals, phytoalexins, and vitamins [69–72]. In addition to avicenol A, avicenol B, avicequinone A, and avicenol C, which are recognized chemicals, Han *et al.* [73] reported the presence of seven unique naphthoquinone derivatives, known as avicennone, that were isolated from *A. marina* twigs that were gathered in Xiamen. These substances showed moderate cytotoxic, antibacterial, and strong antiproliferative properties. Two flavones, luteolin 7-O-methyl ether 30-O-β-D-glucoside and its galactoside counterpart, were identified by Sharaf *et al.* [74] from aerial portions of *A. marina* that were gathered in an area south of Hurghada. They discovered that, when tested against the BT-20 human cancer cell line, they showed considerable cytotoxicity (ED50 = 18 μg/ml). Jia *et al.* [43] found quercetin and kaemferol in the leaves, as well as 5-hydroxy-40,7-dimethoxyflavone, a novel flavone. Three hydroxylated flavones, including 40,5-dihydroxy-30,50,7-trimethoxyflavone, 40,5,7-trihydroxyflavone, and 30,40,5-trihydroxy-7-methoxyflavone, were identified by Feng *et al.* [75] from aerial portions of *A. marina* harvested in Hainan island. The DPPH radical-scavenging assay was utilized to assess the antioxidant activity of the material.

6.5 Biochemical Applications

6.5.1 Biological Activities of *Rhizophora stylosa*

For a considerable amount of time, traditional medicine has employed many *Rhizophora* species, most notably *R. stylosa*, as remedies for liver problems, rheumatism, hematuria, oral thrush, wound healing, muscular soreness, back pain, and bone pain. These uses have been documented in the literature, and various plant extracts (roots, leaves, fruits, and bark) have shown intriguing and noteworthy inhibitory activities in a range of assays, including cytotoxic, antibacterial, antidiabetic, anticancer, and antioxidant assays. With MIC (minimum inhibitory concentration) values of 0.1 mg/ml and MBC (minimum bactericidal concentration) values of 6.3 mg/ml against both Gram-positive and -negative bacteria, the chloroform extracts of *R. stylosa* bark demonstrated the strongest antibacterial capabilities when compared to the leaf extracts suggesting that they are more effective against specific bacteria. This effect is probably due to the fact that phytochemicals and active compounds found in bark extracts differ from those found in leaf extracts. Furthermore, the highest result inhibitory zone [ranging from 11 to 19 (mm)] against *E. coli* was seen in *R. stylosa* leaf extracts treated with ethyl acetate [76]. While the ciscareaborin compound inhibited the growth of BGC-823 and MCF-7 human cancer cells with IC_{50} values of 45.9 and 116.0 μmol/L, respectively, Yang *et al.* [77] reported that the taraxerol compound found in *R. stylosa* demonstrated cytotoxicity against HeLa and BGC-823 human cancer cells. Furthermore, three human cancer cell lines, KB (epidermoid carcinoma), LU-1 (lung adenocarcinoma), and SK-Mel-2 (melanoma) showed notable cytotoxicity against Rhizostyloside, a novel metabolite of cyclocartane glucoside, with IC_{50} values of 86.7 and 51.0 µg/ml, respectively [78]. This value implies that *R. stylosa* leaf extract is cytotoxic to HeLa cells. According to Suh *et al.* [79], the *R. stylosa* bark methanol extract exhibited a greater phenol content and an antioxidant activity that was 85.5% higher than that of the water extracts. According to Li *et al.* [80], proanthocyanidin, an isolated molecule from *R. stylosa*, had the highest level of antioxidant activity. Its IC_{50} of 4.3 g/ml was found to be four times higher than that of butyl hydroxyl toluene (BHT), the positive control (IC_{50} = 18.0 µg/ml). At a dose of 200 mg/kg of BB + 2 g/kg of glucose (group P-2), the ethanol extract of mangrove leaves, including *R. stylosa*, had the strongest antidiabetic activity and decreased blood glucose levels by 31.27% [81]. Furthermore, *Rhizophora* leaf extract was administered to the treatment group, and according to Efendi *et al.* [82], this resulted in a decrease in blood glucose levels compared to pre-treatment. A 500-mg/kg BW resulted in a 29% reduction in fasting blood sugar levels. These results validated the antidiabetic characteristics of *R. stylosa*.

6.5.2 Anticancer Activity of *Avicennia marina*

Two novel flavonoids were identified from the methanol extract of aerial parts of *A. marina*: luteolin 7-O-methylether 30-O-β-D-glucoside and its galactoside analog. These flavonoids were found to be cytotoxic against the human breast cancer cell line BT-20, with ED_{50} values of 16, and 18 µg/ml, respectively. Seven novel naphthoquinone derivatives, including avicennones and five other well-known natural compounds, were extracted from the twigs of *A. marina*. Avicennone A, stenocarpoquinone B, avicequinone C, avicenol A, avicenol C, and a combination of avicennones D and E were used in the tests. They were employed to show cytotoxic activity against the HeLa human cervix carcinoma cell line and an antiproliferative effect against K562 human chronic myeloid leukemia cells and L-929 murine fibroblasts. Stenocarpoquinone B, avicequinone C, and the combination of avicennone D and F demonstrated antiproliferative effects against L-929 (GI_{50} values of 1.2, 0.8, and 4.4 µg/ml, respectively) and K562 (GI_{50} values of 0.2, 1.1, and 7.5 µg/ml, respectively); however, these were less potent than standard paclitaxel (GI_{50} values of 0.1 and 0.01 µg/ml, respectively). These substances also show cytotoxicity against the HeLa cell line, as seen by their CC_{50} values, which were greater than those of the positive control (CC_{50} value: 0.01 µg/ml) at 4.3, 3.2, and 13.1 µg/ml, respectively.

One noteworthy observation is that the structural constituent p-dione of the naphthoquinone core is shared by all active molecules. Instead,

Behbahani and Sadeghi-aliabadi [83] state that the other chemicals have not demonstrated much efficacy against the previously stated cancer cell lines. With an IC_{50} value of 480 μg/ml, the methanolic extract of *A. marina* leaves demonstrated cytotoxic activity against human breast cancer cells MDA-MB 231; however, it had no discernible effect on the normal cell line L929. Furthermore, the extract demonstrated a time-dependent growth suppression impact of 40%, 44%, and 59% after 24, 48, and 72 h of treatment, respectively, in addition to inducing apoptosis in a dose-dependent manner. Human promyelocytic leukemia HL-60 cells were found to be susceptible to the cytotoxic effects of an ethanol extract from *A. marina* leaves, with IC_{50} values of 600, 400, and 280 μg/ml after 24, 48, and 72 h, respectively. These effects were found to be concentration and time dependent. The treated cells had more granular material, looked smaller, had less blebbing in the membrane, and were less refracted than the control cells. Additionally, flow cytometric examination demonstrated that a dose of 600 μg/ml produced 62% apoptosis at 24 h confirming that apoptosis was the mechanism of cell death generated by the extract [84]. *A. marina* leaf extracts, both methanolic and aqueous, were found to have cytotoxic effects similar to doxorubicin on human promyelocytic leukemia HL-60 cells (IC_{50} values of 277.129 and 291.773 μg/ml, respectively) and the human non-small cell lung cancer NCI-H23 cell line (IC_{50} values of 221.173 and 237.179 μg/ml, respectively). The stem bark of *A. marina* was shown to be cytotoxic to HL-60 and NCI-H23, when it was extracted in methanol (IC_{50} values of 297.934 and 210.987 μg/ml, respectively), and to be cytotoxic, when extracted in aqueous form (IC_{50} values of 281.175 and 220.127 μg/ml, respectively). The extracts showed minimal toxicity against the normal human embryonic kidney (HEK-293T) cell line and comparable cytotoxic IC_{50} values with conventional doxorubicin [85]. Anticancer activities of other mangrove plants are listed in Table 6.1.

6.5.3 Biomedical Applications of Mangrove-Synthesized Bionanoparticles

Mangroves are distinct ecosystems characterized by intricate physical, chemical, and biological interactions and a high level of biodiversity. According to recent research, these conditions are capable of producing NPs with possible uses in medicine. Bioactive substances discovered in mangroves have been used to treat a wide range of illnesses. Mangrove resources produce secondary metabolites (tannins, flavonoids, alkaloids, and saponins) that have therapeutic uses. Nonetheless, compared to mangrove-synthesized NPs, the majority of study has been on mangrove plant

Table 6.1 Anticancer profiles of isolated bioactive compounds from some marine mangrove plants.

S. no.	Mangrove plant species	Bioactive compounds	Activities	References
1.	*Acanthus* sp. (*A. ilicifolius*)	Benzoxaxoline	Anticancer activity against HeLa and KB cell lines, and antitumor growth	[86–88]
2.	*Avicennia* sp. (*A. officinalis*)	Triterpene, and betulinic acid	Anticancer activity toward Ehrlich ascites carcinoma (EAC) cells	[88, 89]
3.	*Avicennia germinans*	Tetracosane	Anticancer property toward MDA-MB-231, AGS, NIH 3T3, and HT-2918 cell lines	[90]
4.	*Avicennia* sp. (*A. marina*)	Stenocarpoquinone, naphthoquinones, iridoid glycosides, and avicequinones	Cytotoxic and antitumor, anticancer and antimutagenic activities on HL-60 cell lines	[91, 92]
5.	*Avicennia* sp. (*A. alba*)	Naphthoquinolines, avicequinones, and phenolic compounds	Anticancer and cytotoxicity on MCF-7 and HeLa cell lines	[93, 94]
6.	*Bruguiera* sp. (*B. gymnorrhiza*)	Brugin	Anticancer activity against lung carcinoma and antitumor activity toward sarcoma 180	[95]
7.	*Bruguiera* sp. (*B. sexangula*)	Tropine, brugin, and benzoic acid	Anticancer activity toward sarcoma 180 and lexis lung carcinoma	[96, 97]
8.	*Ceriops* sp. (*C. decandra*)	Quinine, triterpenes, clionasterol, stigmast-5-en-3-ol, lupeol, and diolein	Anticancer and antitumor activities toward malignant ulcers and buccal pouch carcinogenesis	[23]

(*Continued*)

Table 6.1 Anticancer profiles of isolated bioactive compounds from some marine mangrove plants. (*Continued*)

S. no.	Mangrove plant species	Bioactive compounds	Activities	References
9.	*Excoecaria* sp. (*E. agallocha*)	Tannins, diterpenes, and excoecarin	Antitumor activity, gastroprotective, antinociceptive, and also acted as central nervous system depressants	[98, 99]
10.	*Heritieria* sp. (*H. fomes*)	Phenolic compounds	Anticancer and cytotoxic effects	[51]
11.	*Rhizophora* sp. (*R. apiculata*)	Phenolic compounds	Anticancer activity (A549, lung cancer cells)	[100]
12.	*Rhizophora* sp. (*R. mucronata*)	Phytol	Cytotoxicity against human gastric adenocarcinoma (AGS) cells	[101]
13.	*Xylocarpus* sp. (*X. granatum*)	Flavonoids and phenolic compounds	Activity toward HeLa (cervical cancer) and MCF-7 (breast cancer) cell lines	[102]

extracts. Mangrove plants and microbes have been shown in certain investigations to possess antibacterial, antioxidant, antifungal, antiviral, and anticancer properties. NPs derived from mangroves have been assessed for their potential biomedical applications (radical scavenging, antifungal, antibacterial, parasitic, antiparasitic, antidiabetic, anti-inflammatory, and cytotoxic effects). Suresh *et al.* [103] carried out a study on the synthesis of insecticidal AgNPs through mangrove-mediated means utilizing *Suaeda maritima* as a stabilizing and reducing agent. The findings indicated that the tobacco cutworm *Spodoptera litura* and the dengue vector *Aedes aegypti* were both susceptible to larvicidal and pupicidal toxicity when exposed to an aqueous extract of *S. maritima* and AgNPs produced from mangroves.

6.5.4 Biogenic NPs in Industrial Applications

More than 800 products are available worldwide on the market for food, drinks, packaging, cheese, curd, frozen peas, and vegetables based on nanotechnology [104]. Nanotechnology is used in many consumer goods, such as paints, fabrics, cosmetics, and medications. In the industrial sector, nanomaterials are used for several purposes, including as dust, oxygen, and water vapor protection. Nowadays, NPs are made using both marine and terrestrial species. *Caesalpinia coriaria* leaf extract has been utilized by researchers to synthesize AgNPs that may have antibacterial effects [105]. Tea and cosmetics both include flavonoids and tannin NPs as antiaging ingredients. Usha *et al.* [106] have also shown how antimicrobial fabrics can be improved by using NPs made of Cu or ZnO that are produced by *Streptomyces* species, also known as actinomycetes. Research on biogenic NPs from mangroves is currently rare, but additional study could enhance their commercial applications. Nanotechnology has revolutionized many frontier fields and shown a number of potential applications. Fisheries and aquaculture are among the many industries using nanotechnology for effective nutrition delivery, illness diagnosis, medicine administration, and DNA vaccines [107]. Moreover, the seafood business heavily relies on bio-nanotechnology. According to the findings, fish growth is supplemented with SeNPs made from mangrove resources. Metal oxide NPs show antibacterial qualities against *Salmonella* spp., *E. coli*, *Staphylococcus* spp., and *Klebsiella* spp. Ag, Pt, and silica NPs can be used to make a variety of cosmetics and medical items. Sunscreen, toothpaste, mouthwash, shampoo, and perfume are among the goods [108, 109]. Mangrove fungus was also employed in the anaerobic synthesis of biogenic metallic NPs as a catalyst for the production of bioelectricity in the microbial fuel cell (MFC) technique. *Trichoderma* sp. was used to create silver and gold NPs (AuNPs) [110].

6.6 Conclusions

Mangrove plants are widely used in traditional medicine and have a long history of use because of their availability of potential natural chemical sources. Numerous bioactive chemicals have been isolated and discovered, and research on a range of metabolic processes has been done both *in vitro* and *in vivo*. Therefore, to fully realize the potential of the phytochemicals derived from these mangrove plants, more research is required. New biological substances that can be exploited in the pharmaceutical business will be produced by these study discoveries. The studies included in this chapter also demonstrated how recent and promising it is to do research into *A. marina*'s possible anticancer properties. Nanotechnology holds great promise for the removal of environmental pollutants from both man-made and natural sources. Several chemical and biological techniques can be used to synthesize NPs. Plants and microorganisms provide bioresources that are used in biological techniques.

However, mangrove plants thrive in difficult marshy and saline habitats, which makes them a good choice for the production of biogenic NPs. These biogenic NPs that are mediated by mangroves may find use in the breakdown of dyes. Mangroves offer a multitude of resources for the synthesis of nanomaterials and their possible uses in industry, agriculture, and medicine. Nonetheless, not much research has been done in this area. The importance of mangrove ecosystems in producing biological substrates and reducing agents for the synthesis of NPs is highlighted in this chapter. Marine resources were utilized in the synthesis of biogenic NPs to satisfy the demand for a secure, consistent, and environmentally benign method. A viable and sustainable solution for treating textile wastewater is to use bio-nanomaterials made from mangroves.

References

1. Kerry, R.G., Pradhan, P., Das, G., Gouda, S., Swamy, M.K., Patra, J.K., Anticancer potential of mangrove plants: Neglected plant species of the Marine ecosystem, in: *Anticancer Plants: Properties and Application*, vol. 1, pp. 303–305, Springer Nature, Singapore, 2018.
2. Cerri, F., Giustra, M., Anadol, Y., Tomaino, G., Galli, P., Labra, M., Campone, L., Colombo, M., Natural Products from Mangroves: An Overview of the Anticancer Potential of Avicennia marina. *Pharmaceutics*, 14, 2793, 2022. https://doi.org/10.3390/pharmaceutics14122793.
3. Sohaib, M., Al-Barakah, F., Migdadi, H., Husain, F., Comparative study among Avicennia marina, Phragmites australis, and Moringa oleifera based ethanolic-extracts for their antimicrobial, antioxidant, and cytotoxic activities. *Saudi J. Biol. Sci.*, 29, 111–122, 2022.
4. Vaish, S. and Pathak, B., Bionano technological approaches for degradation and decolorization of dye by mangrove plants, in: *Biotechnological Utilization of Mangrove Resources*, pp. 399–412, Academic Press, London, 2020.
5. Choi, S., Johnston, M., Wang, G.S., Huang, C.P., A seasonal observation on the distribution of engineered nanoparticles in municipal wastewater treatment systems exemplified by TiO2 and ZnO. *Sci. Total Environ.*, 625, 1321–1329, 2018.
6. Doshi, B., Sillanpaa, M., Kalliola, S., A review of bio-based materials for oil spill treatment. *Water Res.*, 135, 262–277, 2018.
7. Hu, X., Ren, C., Kang, W., Mu, L., Liu, X., Li, X., Wang, T., Zhou, Q., Characterization and toxicity of nanoscale fragments in wastewater treatment plant effluent. *Sci. Total Environ.*, 626, 1332–1341, 2018.

8. Ajaz, M., Shakeel, S., Rehman, A., Microbial use for azo dye degradation-a strategy for dye bioremediation. *Int. Microbiol.*, 23, 149–159, 2020.
9. Oon, Y.-L., Ong, S.-A., Ho, L.-N., Wong, Y.-S., Dahalan, F.A., Oon, Y.-S., Teoh, T.-P., Lehl, H.K., Thung, W.-E., Constructed wetland-microbial fuel cell for azo dyes degradation and energy recovery: Influence of molecular structure, kinetics, mechanisms and degradation pathways. *Sci. Total Environ.*, 720, 137370, 2020.
10. Duke, N.C., Meynecke, J.O., Dittmann, S., Ellison, A.M., Anger, K., Berger, U., Cannicci, S., Diele, K., Ewel, K.C., Field, C.D., Koedam, N., A world without mangroves? *Science*, 317, 5834, 41–42, 2007.
11. Polidoro, B.A., Carpenter, K.E., Collins, L., Duke, N.C., Ellison, A.M., Ellison, J.C., Farnsworth, E.J., Fernando, E.S., Kathiresan, K., Koedam, N.E., Livingstone, S.R., Miyagi, T., Moore, G.E., Ngoc Nam, V., Ong, J.E., Primavera, J.H., Salmo, S.G., Sanciangco, J.C., Sukardjo, S., Wang, Y., Yong, J.W.H., Hansen, D.M., The loss of species: mangrove extinction risk and geographic areas of global concern. *PLoS One*, 5, 4, e10095, 2010.
12. Lee, S.Y., Primavera, J.H., Dahdouh-Guebas, F., McKee, K., Bosire, J.O., Cannicci, S., Diele, K., Fromard, F., Koedam, N., Marchand, C., Mendelssohn, I., Mukherjee, N., Record, S., Ecological role and services of tropical mangrove ecosystems: a reassessment: Reassessment of mangrove ecosystem services. *Glob. Ecol. Biogeogr.*, 23, 7, 726–743, 2014.
13. Reef, R., Feller, I.C., Lovelock, C.E., Nutrition of mangroves. *Tree Physiol.*, 30, 9, 1148–1160, 2010.
14. Ouyang, X. and Guo, F., Intuitionistic fuzzy analytical hierarchical processes for selecting the paradigms of mangroves in municipal wastewater treatment. *Chemosphere*, 197, 634–642, 2018.
15. Chand, K., Cao, D., Fouad, D.E., Shah, A.H., Dayo, A.Q., Zhu, K., Lakhan, M.N., Mehdi, G., Dong, S., Green synthesis, characterization and photocatalytic application of silver nanoparticles synthesized by various plant extracts. *Arab. J. Chem.*, 13, 11, 8248–8261, 2020.
16. Vaish, S. and Pathak, B., Mangrove synthesized bio-nanomaterial and its applications: A review. *Environ. Nanotechnol. Monit. Manage.*, 20, 100866, 2023. https://doi.org/10.1016/j.enmm.2023.100866.
17. Kusmana, C., Sukristijiono. Mangrove resource uses by local community in Indonesia. *J. Nat. Res. Environ. Manag.*, 6, 217–224, 2016.
18. Mitra, S., Naskar, N., Chaudhuri, P.A., review on potential bioactive phytochemicals for novel therapeutic applications with special emphasis on mangrove species. *Phytomed. Plus*, 1, 100107, 2021.
19. Miranti, D.I., Ichiura, H., Ohtani, Y., The bioactive compounds and antioxidant activity of food products of Rhizophora stylosa fruit (coffee and tea mangrove). *J. For. Res.*, 2018, 2315329, 2018.
20. Setyawan, A.D., Ragavan, P., Basyuni, M., Sarno, S., Review: Rhizophora mucronata as source of foods and medicines. *Bonorowo Wetl.*, 9, 42–55, 2019.

21. Srikanth, S., Lum, S.K., Chen, Z., Mangrove root: Adaptations and ecological importance. *Trees*, 30, 451–465, 2015.
22. Thatoi, H., Behera, B.C., Mishra, R.R., Dutta, S.K., Biodiversity, and biotechnological potential of microorganisms from mangrove ecosystems: A review. *Ann. Microbiol.*, 63, 1–19, 2013.
23. Bandaranayake, W., Traditional and medicinal uses of mangroves. *Mangroves Salt Marshes*, 2, 3, 133–148, 1998.
24. Habeebula, M. and Velraj, M., Potential anti-diabetic mangroves in Kerala, India: A Review. *Int. J. Res. Ayuverda Pharm.*, 9, 194–198, 2018.
25. Rout, P., Singh, S., Kumar, N., Basak, U.C., Nutritional and antioxidant potential of some selected edible mangrove fruits of Odisha coast. *Int. J. Adv. Sci. Res.*, 1, 349–355, 2015.
26. Kainuma, M., Kezuka, M., Inoue, T., Chan, E.W., Tangah, J., Botany, uses, chemistry and bioactivities of mangrove plants I: Rhizophora stylosa. *ISME/GLOMIS Electron. J.*, 13, 12–17, 2015.
27. Wu, J., Xiao, Q., Xu, J., Li, M.-Y., Pan, J.-Y., Yang, M.-H., Natural products from true mangrove flora: Source, chemistry and bioactivities. *Nat. Prod. Rep.*, 25, 955–981, 2008.
28. Sormin, R.B., Nendissa, D.M., Mailoa, M.N., Rieuwpassa, F., Wenno, M.R., Antibacterial activity of Rhizophora apiculata extract originated from Inner Ambon Bay against selected pathogen bacteria, in: *Proceedings of the International Conference on Small Islands Community: Momentum of Transformation of Small Islands Community in New Normal Era*, p. 12017, Maluku, Indonesia, 12 December 2020.
29. Li, D., Li, X.-M., Wang, B.-G., Pentacyclic triterpenoids from the mangrove plant Rhizophora stylosa. *Nat. Prod. Rep.*, 22, 808–813, 2008.
30. Pramudji, P. and Dharmawan, I.W.E., Growth analysis of Rhizophora stylosa Griff. seedlings in mangrove rehabilitation area of Tanjung Pasir, Tangerang. *Oseanologi Limnol. Indones.*, 1, 91–100, 2016.
31. Setyawan, A.D. and Ulumuddin, Y.I., Species diversity of Rhizophora in Tambelan Islands, Natuna Sea, Indonesia. *Biodiversitas*, 13, 172–177, 2012.
32. Chanda, A., Akhand, A., Manna, S., Das, S., Mukhopadhyay, A., Das, I., Hazra, S., Choudhury, S.B., Dadhwal, V.K., Mangrove associates versus true mangroves: A comparative analysis of leaf litter decomposition in Sundarban. *Wetl. Ecol. Manag.*, 24, 293–315, 2015.
33. Lalitha, P., Parthiban, A., Sachithanandam, V., Purvaja, R., Ramesh, R., Antibacterial and antioxidant potential of GC-MS analysis of crude ethyl acetate extract from the tropical mangrove plant Avicennia officinalis L. *S. Afr. J. Bot.*, 142, 149–155, 2021.
34. Syahidah, and Subekti, N., Phytochemical Analysis of Mangrove Leaves (Rhizophora sp.), in: *Proceedings of the 14th Pacific Rim Bio-Based Composites Symposium*, pp. 1–7, Makassar, Indonesia, 29–31 October 2018.
35. Arbiastutie, Y., Diba, F., Masriani, M., Short Communication: Ethnobotanical and ecological studies of medicinal plants in a mangrove forest in Mempawah District, West Kalimantan, Indonesia. *Biodiversitas*, 22, 3164–3170, 2021.

36. Gopal, N., Ekegbu, J., Kaur, C.P., Paulraj, P., P, R., Bhavya, K.S., Evaluation of antibacterial properties of leaves and barks of Rhizophora stylosa against Gram-Positive and Gram-Negative organisms. *J. Pure. Appl. Microbiol.*, 13, 957–965, 2019.
37. Elahi, N., Kamali, M., Baghersad, M.H., Recent biomedical applications of gold nanoparticles: A review. *Talanta*, 184, 537–556, 2018.
38. Willian, N., Syukri, S., Zulhadjri, Z., Pardi, H., Arief, S., Marine plant mediated green synthesis of silver nanoparticle using mangrove Rhizophora stylosa: Effect of variable process and their antibacterial activity. *F1000Research*, 10, 768, 2022.
39. Ali, A., Akhtar, N., Khan, B., Khan, H., Rasul, A., Khalid, N., Waseem, K., Acacia nilotica: A plant of multipurpose medicinal uses. *J. Med. Plants Res.*, 6, 1492–1496, 2012.
40. Vasanth, K., Ilango, K., Mohankumar, R., Agrawal, A., Dubey, G., Anticancer activity of Moringa oleifera mediated silver nanoparticles on human cervical carcinoma cells by apoptosis induction. *Colloids Surf. B Biointerfaces.*, 117, 354–359, 2014.
41. Hembram, K.C., Kumar, R., Kandha, L., Parhi, P.K., Kundu, C.N., Bindhani, B.K., Therapeutic prospective of plant-induced silver nanoparticles: Application as antimicrobial and anticancer agent. *Artif. Cells Nanomed. Biotechnol.*, 46, S38–S51, 2018.
42. Srirangam, G. and Rao, K.P., Synthesis and characterization of silver nanoparticles from the leaf extract of Malachra capitata (L.). *Rasayan J. Chem.*, 10, 46–53, 2017.
43. Jia, R., Guo, Y., Hou, H., Studies on the chemical constituents from leaves of Avicennia marina. *Chin. J. Nat. Med.*, 2, 16–19, 2004.
44. Kalasuba, K., Miranti, M., Rahayuningsih, S.R., Safriansyah, W., Syamsuri, R.R.P., Farabi, K., Oktavia, D., Alhasnawi, A.N., Doni, F., Red Mangrove (Rhizophora stylosa Griff.)—A Review of Its Botany, Phytochemistry, Pharmacological Activities, and Prospects. *Plants*, 12, 2196, 2023. https://doi.org/10.3390/ plants12112196.
45. Willian, N., Syukri, Z., Labanni, A., Arief, S., Bio-Friendly synthesis of silver nanoparticles using mangrove Rhizophora stylosa leaf aqueous extract and its antibacterial and antioxidant activity. *Rasayan J. Chem.*, 13, 1478–1485, 2020.
46. Sadeer, N.B., Zengin, G., Mahomoodally, M.F., Biotechnological applications of mangrove plants and their isolated compounds in medicine-a mechanistic overview. *Crit. Rev. Biotechnol.*, 43, 393–414, 2022.
47. Singh, P., Kim, Y.J., Singh, H., Mathiyalagan, R., Wang, C., Yang, D.C., Biosynthesis of anisotropic silver nanoparticles by Bhargavaea indica and their synergistic effect with antibiotics against pathogenic microorganisms. *J. Nanomater.*, 2015, 4, 2015.
48. Kok-Yong, S. and Lawrence, L., Drug distribution and drug elimination, in: *Basic Pharmacokinetic Concepts and Some Clinical Applications*, T.A. Ahmed, (Ed.), IntechOpen, London, UK, 2015.

49. Rosenblum, D., Joshi, N., Tao, W., Karp, J.M., Peer, D., Progress and challenges towards targeted delivery of cancer therapeutics. *Nat. Commun.*, 9, 1410, 2018.
50. Mitchell, M.J., Billingsley, M.M., Haley, R.M., Wechsler, M.E., Peppas, N.A., Langer, R., Engineering precision nanoparticles for drug delivery. *Nat. Rev. Drug Discov.*, 20, 101–124, 2021.
51. Patra, J.K., Das, G., Fraceto, L.F., Campos, E.V.R., Rodriguez-Torres, M.d.P., Acosta-Torres, L.S., Diaz-Torres, L.A., Grillo, R., Swamy, M.K., Sharma, S., Nano based drug delivery systems: Recent developments and future prospects. *J. Nanobiotechnol.*, 16,71, 2018.
52. Varunkumar, K., Anusha, C., Saranya, T., Ramalingam, V., Raja, S., Ravikumar, V., Avicennia marina engineered nanoparticles induce apoptosis in adenocarcinoma lung cancer cell line through p53 mediated signaling pathways. *Process Biochem.*, 94, 349–358, 2020.
53. Soni, N. and Prakash, S., Factors affecting the geometry of silver nanoparticles synthesis in Chrysosporium tropicum and Fusarium oxysporum. *Am. J. Nanotechnol.*, 2, 1, 112–121, 2011.
54. Barnawi, A.A.B., Sharawi, S.E., Mahyoub, J.A., Al-Ghamdi, K.M., Larvicidal studies of Avicennia marina extracts against the dengue fever mosquito Aedes aegypti (Culicidae: Diptera). *Int. J. Mosq. Res.*, 6, 55–60, 2019.
55. Balakrishnan, S., Srinivasan, M., Mohanraj, J., Biosynthesis of silver nanoparticles from mangrove plant (Avicennia marina) extract and their potential mosquito larvicidal property. *J. Parasit. Dis.*, 40, 991–996, 2016.
56. Das, S.K., Behera, S., Patra, J.K., *et al.*, Green Synthesis of Sliver Nanoparticles Using Avicennia officinalis and Xylocarpus granatum Extracts and *In vitro* Evaluation of Antidiabetic and Anti-inflammatory Activities. *J. Clust. Sci.*, 30, 1103–1113, 2019. https://doi.org/10.1007/s10876-019-01571-2.
57. Bhuvaneswari, R., Xavier, R.J., Arumugam, M., Facile synthesis of multifunctional silver nanoparticles using mangrove plant Excoecaria agallocha L. for its antibacterial, antioxidant and cytotoxic effects. *J. Parasit. Dis.*, 41, 1, 180–187, 2017.
58. Sangeetha, A., Saraswathi, U., Singaravelu, G.J., Green synthesis of silver nanoparticles using a mangrove Excoecaria agallocha. *Int. J. Pharm. Sci. Invent.*, 3, 54–57, 2014.
59. Anuradha, V., Yogananth, N., Rajathilagam, R., Chanthuru, A., Marzook, S.M., Green synthesis of Silver nanoparticle by Acanthus ilicifolius mangrove plant against Armigeressubalbatus and Aedesaegypti mosquito larvae. *Int. J. Nano Dimens.*, 6, 2, 197–204, 2015.
60. Dhas, S.P., Mukerjhee, A.M., Chandrasekaran, N., Phytosynthesis of silver nanoparticles using Ceriops tagal and its antimicrobial potential against human pathogens. *Int. J. Pharm. Pharm. Sci.*, 5, 3, 349–352, 2013.
61. Ramteke, L., Jadhav, B., Gawali, P., Green synthesis of silver nanoparticles from aqueous stem extract of Ceriops tagal. *Int. Res. J. Pharm.*, 8, 12, 2017.

62. Maity, G.N., Sarkar, J., Khatua, S.O., Mondal, S.O., Acharya, K.R., Green synthesis of silver nanoparticles using mangrove fruit polysaccharide for bacterial growth inhibition. *Asian J. Pharm. Clin. Res.*, 12, 179–183, 2019.
63. Thatoi, P., Kerry, R.G., Gouda, S., Das, G., Pramanik, K., Thatoi, H., Patra, J.K., Photo-mediated green synthesis of silver and zinc oxide nanoparticles using aqueous extracts of two mangrove plant species, Heritiera fomes and Sonneratia apetala and investigation of their biomedical applications. *J. Photochem. Photobiol. B Biol.*, 163, 311–318, 2016.
64. Alshehri, M.A., Aziz, A.T., Trivedi, S., Alanazi, N.A., Panneerselvam, C., Baeshen, R., Alatawi, A., One-step synthesis of Ag nanoparticles using aqueous extracts from sundarbans mangroves revealed high toxicity on major mosquito vectors and microbial pathogens. *J. Clust. Sci.*, 31, 1, 177–184, 2020.
65. Nathan, V.K., Ammini, P., Vijayan, J., Photocatalytic degradation of synthetic dyes using iron (III) oxide nanoparticles (Fe2 O3-Nps) synthesised using Rhizophora mucronata Lam. *IET*, 13, 2, 120–123, 2019.
66. Jha, N., Esakkiraj, P., Annamalai, A., Lakra, A.K., Naik, S., Arul, V., Synthesis, optimization, and physicochemical characterization of selenium nanoparticles from polysaccharide of mangrove Rhizophora mucronata with potential bioactivities. *J. Trace Ele. Miner.*, 2, 100019, 2022.
67. Nagababu, P. and Rao, U., Costğeffective green synthesis and characterization of silver nanoparticles from avicennia alba blume leaves and their antibacterial activity. *Asian J. Pharm. Clin. Res.*, 9, 1, 301–303, 2016.
68. Murugan, K., Dinesh, D., Paulpandi, M., Subramaniam, J., Rakesh, R., Amuthavalli, P., Panneerselvam, C., Suresh, U., Vadivalagan, C., Alsalhi, M.S., Devanesan, S., Wei, H., Higuchi, A., Nicoletti, M., Canale, A., Benelli, G., Mangrove helps: Sonneratia alba-synthesized silver nanoparticles magnify guppy fish predation against Aedes aegypti young instars and down-regulate the expression of envelope (E) gene in dengue virus (serotype DEN-2). *J. Clust. Sci.*, 28, 1, 437–461, 2017.
69. Sun, Y., Ouyang, J., Deng, Z., Li, Q., Lin, W., Structure elucidation of five new iridoid glucosides from the leaves of Avicennia marina. *Magn. Reson. Chem.*, 46, 638–642, 2008.
70. Esau, L., Sagar, S., Bajic, V.B., Kaur, M., Autophagy inhibition enhances the mitochondrial-mediated apoptosis induced by mangrove (Avicennia marina) extract in human breast cancer cells. *Eur. J. Med. Plants*, 5, 304, 2015.
71. Eldohaji, L.M., Fayed, B., Hamoda, A.M., Ershaid, M., Abdin, S., Alhamidi, T.B., Mohammad, M.G., Omar, H.A., Soliman, S.S., Potential targeting of Hep3B liver cancer cells by lupeol isolated from Avicennia marina. *Arch. Pharm.*, 354, 2100120, 2021.
72. Ibrahim, H.A.H., Abdel-Latif, H.H., Zaghloul, E.H., Phytochemical composition of Avicennia marina leaf extract, its antioxidant, antimicrobial potentials and inhibitory properties on Pseudomonas fluorescens biofilm. *Egypt. J. Aquat. Res.*, 48, 29–35, 2022.

73. Han, L., Huang, X., Dahse, H.-M., Moellmann, U., Fu, H., Grabley, S., Sattler, I., Lin, W., Unusual naphthoquinone derivatives from the twigs of Avicennia marina. *J. Nat. Prod.*, 70, 923–927, 2007.
74. Sharaf, M., El-Ansari, M., Saleh, N., New flavonoids from Avicennia marina. *Fitoterapia*, 71, 274–277, 2000.
75. Feng, Y., Li, X.M., Duan, X.J., Wang, B.G., Iridoid glucosides and flavones from the aerial parts of Avicennia marina. *Chem. Biodivers.*, 3, 799–806, 2006.
76. Mouafi, F.E., Abdel-Aziz, S.M., Bashir, A.A., Fyiad, A.A., Phytochemical analysis and antimicrobial activity of mangrove leaves (Avicenna marina and Rhizophora stylosa) against some pathogens. *World Appl. Sci. J.*, 29, 547–554, 2014.
77. Azuma, H., Toyota, M., Asakawa, Y., Takaso, T., Tobe, H., Floral scent chemistry of mangrove plants. *J. Plant Res.*, 115, 47–53, 2002.
78. Huong, P.T., Diep, C.N., Thanh, N.V., Tu, V.A., Hanh, T.H., Cuong, N.T., Thao, N.P., Cuong, N.X., Thao, D.T., Thai, T.H., *et al.*, A new cycloartane glucoside from Rhizophora stylosa. *Nat. Prod. Comm.*, 9, 1255–1257, 2014.
79. Suh, S.-S., Hwang, J., Park, M., Park, H.-S., Lee, T.-K., Phenol content, antioxidant and tyrosinase inhibitory activity of mangrove plants in Micronesia. *Asian Pac. J. Trop. Med.*, 7, 531–535, 2014.
80. Li, D.-L., Li, X.-M., Peng, Z.-Y., Wang, B.-G., Flavanol derivates from Rhizophora stylosa and their DPPH radical scavenging activity. *Molecules*, 12, 1163–1169, 2007.
81. Usman, M.M.A., Erika, F., Nurdin, M., Kuncoro, H., Antidiabetic activity of leaft extract from three types of mangrove originating from Sambera Coastal Region Indonesia. *Res. J. Pharm. Technol.*, 12, 1707–1712, 2019.
82. Efendi, A., Halid, I., Ustiawaty, J., Effect of Rhizophora sp. mangrove leaf extract on mice blood glucose levels, in: *Proceedings of the 3rd International Conference on Bioscience and Biotechnology*, Lombok, Indonesia, 12–14 October 2020.
83. Behbahani, M. and Sadeghi-aliabadi, H., Evalution of cytotoxic effect of some extracts of Avicennia marina against MDA-MB 231 human breast cancer cell line. *Pharm. Sci.*, 16, 229–238, 2010.
84. Karami, L., Majd, A., Mehrabian, S., Nabiuni, M., Salehi, M., Irian, S., Antimutagenic and anticancer effects of Avicennia marina leaf extract on Salmonella typhimurium TA100 bacterium and human promyelocytic leukaemia human gastricadeno carcinoma (AGS) cells cells. *Sci. Asia*, 38, 4, 349–355, 2012.
85. Prakash, S. and Piyush, M., Biological screening of Avicennia marina for anticancer activity. *Der. Pharma. Sin.*, 4, 125–130, 2013.
86. Boopathy, N.S., Kandasamy, K., Subramanian, M., You-Jin, J., Effect of mangrove tea extract from Ceriops decandra (Griff.) Ding Hou. On salivary bacterial flora of DMBA induced Hamster buccal pouch carcinoma. *Ind. J. Appl. Microbiol.*, 51, 3, 338–344, 2011.

87. Khajure, P.V. and Rathod, J., Potential anticancer activity of Acan thus ilicifolius extracted from the mangroves forest of Karwar, West Coast of India. *World J. Sci. Technol.*, 1, 1–6, 2011.
88. Das, G., Gouda, S., Mohanta, Y.K., Patra, J.K., Mangrove plants: A potential source for anticancer drugs. *Ind. J. Geo-Mar. Sci.*, 44, 5, 666–672, 2015.
89. Sumithra, M., Anbu, J., Nithya, S., Ravichandiran, V., Anticancer activity of methanolic leaves extract of Avicennia officinalis on Ehrlich ascitis carcinoma cell lines in rodents. *Int. J. Pharmtechnol. Res.*, 3, 3, 1290–1292, 2011.
90. Subathra, M. and Mohideen, A.U., Phytochemical profiling of ethanolic leaves extract of Avicennia germinans. *Global J. Res. Anal.*, 6, 1–3, 2018.
91. Khafagi, I., Gab-Alla, A., Salama, W., Fouda, M., Biological activities and phytochemical constituents of the gray mangrove Avicennia marina (Forssk.) Vierh. *Egypt J. Exp. Biol.*, 5, 1, 62–69, 2003.
92. Sukhramani, P. and Patel, P., Biological screening of Avicennia marina for anticancer activity. *Der. Pharm. Sin.*, 4, 2, 125 –130, 2013.
93. Ito, C., Katsuno, S., Kondo, Y., Tan, H.T.W., Furukawa, H., Chemical constituents of Avicennia alba. Isolation and structural elucidation of new naphthoquinones and their analogues. *Chem. Pharm. Bull.*, 48, 3, 339–343, 2000.
94. Eswaraiah, G., AbrahamPeele, K., Krupanidhi, S., Bharath Kumar, R., Venkateswarulu, T.C., Identification of bioactive compounds in leaf extract of Avicennia alba by GC-MS analysis and evaluation of its *in-vitro* anticancer potential against MCF7 and HeLa cell lines. *J. King Saud Univ- Sci.*, 32, 1, 740–744, 2020.
95. Bunyapraphatsara, N., Jutiviboonsuk, A., Sornlek, P., Therathanathorn, W., Aksornkaew, S., Fong, H.H., Pezzuto, J.M., Kosmeder, J., Pharmacological studies of plants in the mangrove forest. *Thai J. Phytopharm.*, 10, 2, 2546, 2003.
96. Kathiresan, K., Boopathy, N.S., Kavitha, S., Coastal vegetationan underexplored source of anticancer drugs. *Nat. Prod. Radiance.*, 5, 2, 115–119, 2006.
97. Govindasamy, C. and Kannan, R., Pharmacognosy of mangrove plants in the system of unani medicine. *Asian Pac. J. Trop. Med*, 2, S38–S41, 2012.
98. Subhan, N., Alam, M.A., Ahmed, F., Shahid, I.J., Nahar, L., Sarker, S.D., Bioactivity of Excoecaria agallocha. *Rev. Bras. Farmacogn.*, 18, 4, 521–526, 2008b.
99. Subhan, N., Ashraful, A.M., Ahmed, F., Abdul, A.M., Nahar, L., Sarker, S.D., *In vitro* antioxidant property of the extract of Excoecaria agallocha (Euphorbiaceae). *DARU*, 16, 3, 149–154, 2008c.
100. Ramalingam, V. and Rajaram, R., Enhanced antimicrobial, antioxidant and anticancer activity of Rhizophora apiculate. *An Experim. Rep.*, 3, 8, 1–13, 2018.
101. Panjaitan, A.P. and Suprajitnob, E., Antibacterial compounds activity of mangrove leaf extract Srhizophora mucronata on aeromonas hydrophyla. *Russ. J. Agric. Socio-Econ. Sci.*, 73, 187–193, 2018.

102. Darmadi, J., Batubara, R.R., Himawan, S., *et al.*, Evaluation of Indonesian mangrove Xylocarpus granatum leaves ethyl acetate extract as potential anticancer drug. *Sci. Rep.*, 11, 6080, 2021. https://doi.org/10.1038/s41598-021-85383-3.
103. Suresh, U., Murugan, K., Panneerselvam, C., Rajaganesh, R., Roni, M., Aziz, A.T., Naji Al-Aoh, H.A., Trivedi, S., Rehman, H., Kumar, S., Higuchi, A., Canale, A., Benelli, G., Suaeda maritima-based herbal coils and green nanoparticles as potential biopesticides against the dengue vector Aedes aegypti and the tobacco cutworm Spodoptera litura. *Physiol. Mol. Plant Pathol.*, 101, 225–235, 2018.
104. Taha, R.H., Nanotechnology and its application in agriculture. *Adv. Plants Agri. Res.*, 3, 2, 00089, 2016.
105. Jeeva, K., Thiyagarajan, M., Elangovan, V., Geetha, N., Venkatachalam, P., Caesalpinia coriaria leaf extracts mediated biosynthesis of metallic silver nanoparticles and their antibacterial activity against clinically isolated pathogens. *Ind. Crop. Prod.*, 52, 714–720, 2014.
106. Usha, R., Prabu, E., Palaniswamy, M., Venil, C.K., Rajendran, R., Synthesis of metal oxide nano particles by Streptomyces sp. for development of antimicrobial textiles. *Global J. Biotechnol. Biochem.*, 5, 3, 153–160, 2010.
107. Ashraf, M., Aklakur, M., Sharma, R., Ahmad Sh, K.M., Nanotechnology as a novel tool in fisheries and aquaculture development: A review. *Iran J. Energy Environ.*, 2, 3, 258–261, 2011.
108. Pacitti, D., Lawan, M.M., Sweetman, J., Martin, S.A.M., Feldmann, J., Secombes, C.J., Chen, T.-Y., Selenium Supplementation in Fish: A Combined Chemical and Biomolecular Study to Understand Sel-Plex Assimilation and Impact on Selenoproteome Expression in Rainbow Trout (Oncorhynchus mykiss). *PLoS One*, 10, 5, e0127041, 2015.
109. Kumar, V. and Yadav, S.K., Plant-mediated synthesis of silver and gold nanoparticles and their applications. *J. Chem. Technol. Biotechnol.*, 84, 2, 151–157, 2009.
110. Saravanakumar, K., MubarakAli, D., Kathiresan, K., Thajuddin, N., Alharbi, N.S., Chen, J., Biogenic metallic nanoparticles as catalyst for bioelectricity production: a novel approach in microbial fuel cells. *Mater. Sci. Eng., B*, 203, 27–34, 2016.

7

Mesopelagic Fish Protein Hydrolysates: A Source of Antihypertensive and Antidiabetic Peptides

Hilary Kwesi Ketemepi[1], Sharifudin Md. Shaarani[2] and Nor Qhairul Izzreen Mohd Noor[1,3]*

[1]Food Security Research Laboratory, Faculty of Food Science and Nutrition, Universiti Malaysia Sabah, Kota Kinabalu, Sabah, Malaysia
[2]Food Biotechnology Programme, Faculty of Science and Technology, Universiti Sains Islam Malaysia, Nilai, Negeri Sembilan, Malaysia
[3]Halal Services and Research Centre, Faculty of Food Science and Nutrition, Universiti Malaysia Sabah, Kota Kinabalu, Sabah, Malaysia

Abstract

MFPHs contain bioactive peptides that can inhibit angiotensin-converting enzyme (ACE) activity from converting angiotensin I to II, promote vasodilation, enhance insulin sensitivity, inhibit alpha-glucosidase activity, and regulate glucose uptake. All these are crucial in tackling hypertension and diabetes. In this chapter, compelling evidence supporting the effectiveness of MFPH peptides in addressing hypertension and diabetes is brought to bear. The findings have the potential to improve cardiovascular health and control blood sugar levels for those battling with hypertension and diabetes. Additionally, researchers, industry professionals, and policymakers concerned about hypertension and diabetes can use this information to find ways.

Keywords: Mesopelagic fish, protein hydrolysates, Mesopelagic zone, Mesopelagic fish, hydrolysis and hydrolysates, antihypertensive, peptides of pelagic, antidiabetic peptides

**Corresponding author*: qhairul@ums.edu.my

Shaju S. S., Vipin. P. M. and Visakh. P.M. (eds.) Marine Biotechnology: Studies and Applications, (197–238)

General overview

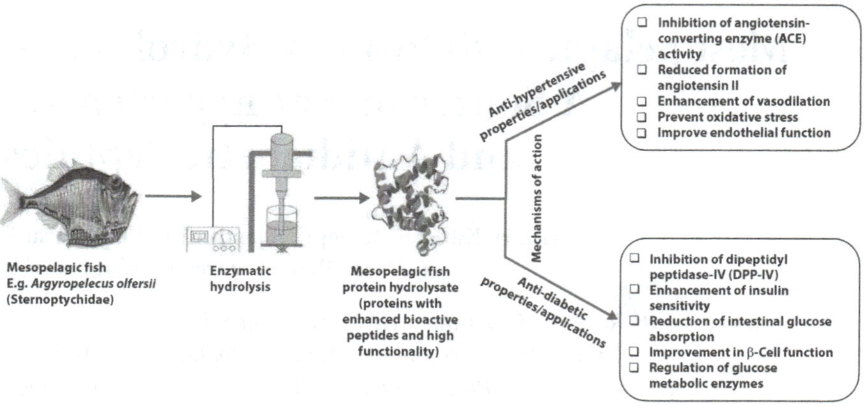

7.1 Introduction

The world's population is projected to grow to approximately 9 billion people by 2050, and as such, a constant supply of nutritional food would be required, and this has the potential to put significant pressure on the demand for protein [1] (Scott, 2017). Despondently, increasing rates of communicable and non-communicable diseases coupled with globalization in this 21st century make it impossible to achieve this sustainable future. According to Ridoutt *et al.* [2], food mandates are foreseen to rise by approximately 70% in the conceivable future compared to current existing levels. Nevertheless, 75% of global food consumption is now derived from a limited range of food sources. In addition to rice, maize, wheat, soybeans, palm oil, potatoes, sweet potatoes, barley, cassava, sorghum, sugar cane, and sugar beet, there are five animal species that contribute to this supply chain, including chicken, cow, sheep, goat, and pig [3]. Alarmingly, the reliance on a limited variety of crops (e.g., rice, maize, and wheat) and animals underscores the need for dietary diversity [4]. In light of socioeconomic shifts involving urbanization, rising income, and an aging population, protein-rich food is in high demand [5]. Protein, being an indispensable macronutrient, is greatly needed by the body for a number of structural and functional purposes like muscle mass maintenance, immunological responses, cell signaling, and cell repair [6]. Since time immemorial, animal-derived proteins have been the traditional sources of protein; nevertheless, livestock production is reported to contribute to approximately 12% of greenhouse gas (GHG) emissions.

Amid the anticipated growth in population, it will be impossible to meet the projected demand of protein from conventional sources alone taking a closer look at the projected demand of protein in the future. Alternate protein sources are therefore highly recommended. This may well involve exploring underexploited protein sources that possess the likelihood to bridge the gap identified between protein demand and supply. The marine ecosystem is renowned for its enormous and varied oceanic environments, which have the potential to provide food and bioactive compounds. This ecosystem integrates multiple regions that contribute to the complex dynamics exhibited [7]. The mesopelagic zone is usually sited between 200 and 1,000 m below the marine surface (Figure 7.1) and is one of the critical regions in the marine ecosystem. This zone is also referred to as the twilight zone because limited light penetrates in there, with cold temperatures and high pressures. These occurrences impede photosynthesis and thus lead to the creation of a unique habitat that is characterized by distinct biological and ecological processes. Notably, a key role is played by the mesopelagic zone in retaining ecological balance and biogeochemical cycles of the marine ecosystem. The zone serves as a haven for different

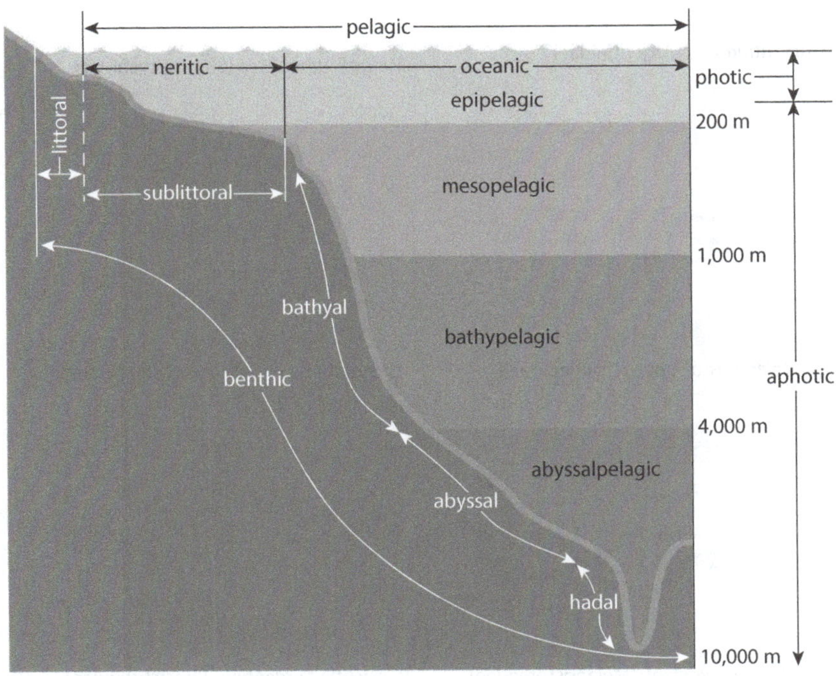

Figure 7.1 A view of the mesopelagic zone of the ocean, just below the pelagic and epipelagic zones, with a depth of 200 to 1,000 m. Adapted from ref. [14].

organisms, including zooplankton (e.g., medusae, siphonophores, salps, larvaceans, and ctenophores), crustaceans (e.g., copepods and euphausiids), predatory worms (e.g., Chaetognatha and osedax), midwater fish (e.g., Gonostomatidae and Myctophidae—the two dominant families

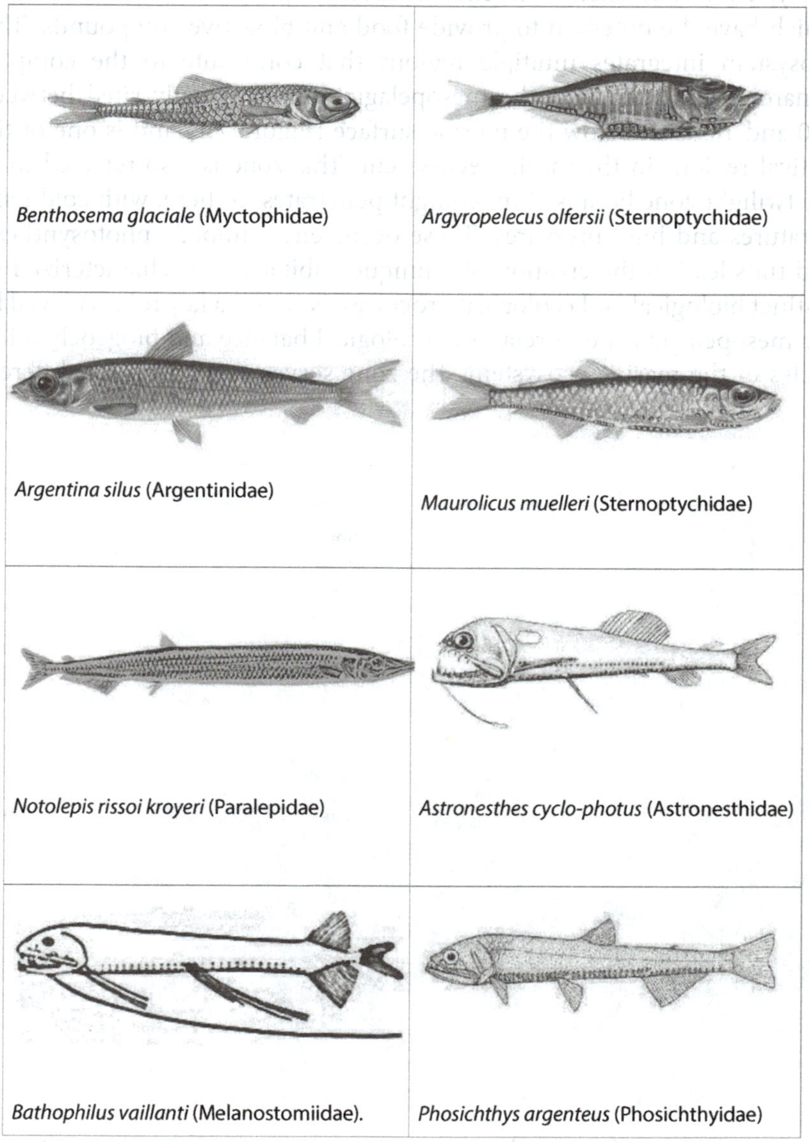

Figure 7.2 Images of some common mesopelagic fish species.

worldwide), and cephalopods (e.g., nautilus and octopus), among others. Present in the zone are also squid, gelatinous animals, and fish [8].

The contribution of siphonophores to mesopelagic biomass is particularly challenging to quantify given that these animals are gelatinous and delicate, breaking up when trawled. As such, fishing them (siphonophores) has no commercial value. The fish usually present at the mesopelagic zone are called mesopelagic fish or deep-sea fish. Myctophidae, Argentinidae, Paralepidea, Melanostomiidae, Sternoptychidae, Paralepidae, Phosichthyidae, Sternoptychidae, Astronesthidae, and Melanostomiidae are all examples of such fish species. All these make up the mesopelagic fish biomass [9]. Figure 7.2 displays images of some common mesopelagic fish species. Oceanic food webs also depend on these organisms (zooplankton, squid, gelatinous animals, and fish) because they connect surface waters (where primary production occurs) with deeper ocean layers. Through the feeding and reproductive activities of mesopelagic species, the vertical transfer of nutrients and energies is facilitated [10]. In recent times, focus has been concentrated on the potential health benefits of mesopelagic fish beyond their ecological significance. Several studies have reported that mesopelagic species provide proteins of high quality, essential fatty acids (e.g., omega-3 fatty acids), vitamins (e.g., vitamin A1 and D), minerals (e.g., calcium, phosphorus, selenium, and iodine), and polyunsaturated fatty acids (PUFAs) like eicosapentaenoic acid (EPA) (C20:5n3) and

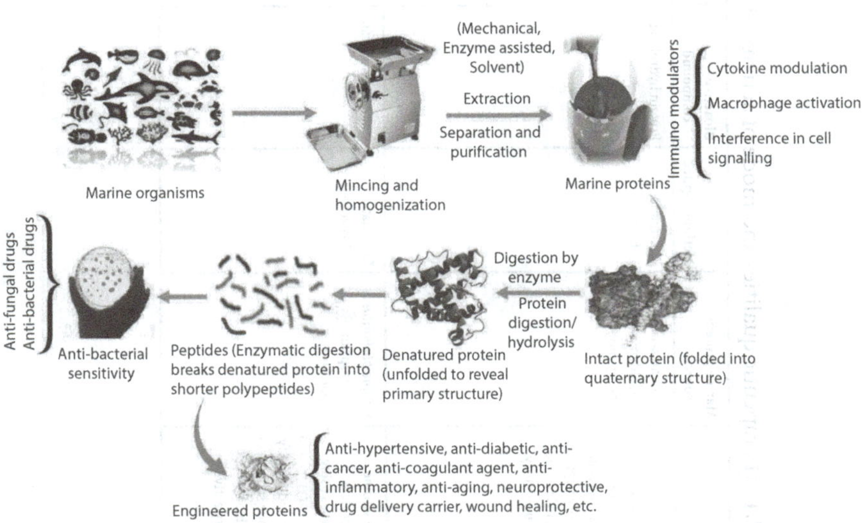

Figure 7.3 Extraction or hydrolysis processes of protein source and applications of protein extracts or hydrolysates for therapeutic purposes.

Table 7.1 Therapeutic qualities and mode of activity of marine protein sources.

Objectives	Marine protein source	Method applied	Bioactive compound	Therapeutic property	Test conducted for therapeutic effect	Key findings	Limitations	Implications	References
To investigate the antidiabetic effects of *Trachinotus ovatus* protein hydrolysates and identify peptides with α-amylase and DPP-IV inhibitory activities	*Trachinotus ovatus* (Golden/ silver pompano)	Enzymatic hydrolysis	Protein hydrolysates	Antidiabetic potential effects	α-amylase (AAM) inhibitory activity assay and dipeptidyl peptidase IV (DPP-IV) inhibitory activity assay—(*in vivo*)	Bioactive peptides with potential antidiabetic effects were noticed. These peptides exhibited significant reduction in blood glucose levels, and protective effects on organs	Peptide identification and screening process were complex and time consuming	Promising natural ingredients for diabetes treatment or prevention is provided	[15]
To produce sardine muscle protein hydrolysates using different enzymes and to purify and identify the peptide sequences responsible for inhibitory activity	*Sardine pilchardus* (European pilchard)	Enzymatic hydrolysis	Protein hydrolysates	Antidiabetic potential effects	Simulated gastrointestinal digestion (SGID) and DPP-IV inhibition assay—(*in vitro*)	Peptides <1,400 Da fractions exhibited the highest DPP-IV inhibitory activity, with CGGWLF identified as a potentially bioactive peptide	There is the need for further verification of bioactivity through chemical synthesis of peptides like CGGWLF	Discarded fish protein sources could be used to produce bioactive peptides with therapeutic properties, particularly for antidiabetic effects	[16]

(*Continued*)

Table 7.1 Therapeutic qualities and mode of activity of marine protein sources. (*Continued*)

Objectives	Marine protein source	Method applied	Bioactive compound	Therapeutic property	Test conducted for therapeutic effect	Key findings	Limitations	Implications	References
To assess the antidiabetic potential of salmon co-product hydrolysates *in vitro* and determine the mechanism of their antidiabetic activity	*Salmo salar* (Atlantic salmon)	Enzymatic hydrolysis	Gelatin hydrolysates	Antidiabetic potential effects	Insulin secretion in BRIN-BD11 cells, determination of dipeptidyl peptidase-IV (DPP-IV) inhibitory activity, and determination of GLP-1 secretion from GLUTag cells—(*in vitro*)	The hydrolysates showed promising *in vitro* antidiabetic activity, with specific pathways identified for their insulinotropic activity	Further research is needed to identify the most potent bioactive peptides and amino acids responsible for the observed antidiabetic activity	Salmon co-products can be valuable sources of biologically active hydrolysates offering potential therapeutic benefits for managing metabolic disorders, e.g., diabetes	[17]
To identify and characterize NK-lysin-like transcripts from *Salmo salar* based on expressed sequence tag (EST) reported sequences	*Salmo salar* (Atlantic salmon)	Chemically synthesized	Lysine-derived peptide	Immunomodulatory potential activities	Effects of NK-lysin and cathelicidin-derived peptides on cytokine expression in head kidney leukocytes—(*in vitro*)	The identified NK-lysin-like peptides induced the expression of key immune response markers like IL-1β, IL-8, and IFN-γ, in *Salmo salar* head kidney leukocytes exhibiting their immunomodulatory potential	The study focus was only on *in vitro* assays. Hence, studies are required for *in vivo* assays to validate the therapeutic potential of the identified peptides in *Salmo salar*	Aquaculture could be enhanced by developing novel therapeutic products with immunomodulatory potentials derived from NK-lysin and cathelicidin peptides, thus reducing the need for chemicals	[18]

(*Continued*)

Table 7.1 Therapeutic qualities and mode of activity of marine protein sources. (*Continued*)

Objectives	Marine protein source	Method applied	Bioactive compound	Therapeutic property	Test conducted for therapeutic effect	Key findings	Limitations	Implications	References
To isolate a peptide from marine crab leg muscle and evaluate its anti-inflammatory effects on macrophage cells	*Charybdis natator* (marine crab)	Enzymatic hydrolysis	Muscle peptide	Anti-inflammatory potential effects	Cell proliferation assay and enzyme-linked immunosorbent assay (ELISA)—(*in vitro*)	The low-molecular weight peptide sequence L–G–L–G–A–A–V–L (713.456 Da) showed anti-inflammatory effects on macrophage cells	The study focused on the anti-inflammatory effects of the peptide on macrophage cells, potentially limiting broader applications	The potential of marine crab peptides for developing anti-inflammatory therapeutics are brought to light	[19]
To characterize mesopelagic protein hydrolysates (MPHs) and extracts of mesopelagic fish species captured during different commercial trawls	*Maurolicus muelleri* (silvery light fish) and *Meganyctiphanes norvegica* (Northern krill)	Enzymatic hydrolysis	Protein hydrolysates	Antidiabetic and antihypertensive potential effects	ACE-1 inhibition assay, AChE inhibition by mesopelagic samples, and DPP-IV inhibition by mesopelagic samples—(*in vitro*)	Hydrolysates and aqueous extracts of mesopelagic fish exhibited promising bioactivities by inhibiting enzymes linked to common diseases. The hydrolysates and extracts also contained high protein levels and amino acid composition compared to egg white	The protein contents determined were less than standard for fish protein hydrolysates. Also, the ash content was higher than the recommended limit for food applications, probably limiting their utilization in some products	Mesopelagic fish could potentially be a source of sustainable bioactive compounds with therapeutic properties paving the way for the development of nutraceuticals and functional food products to improve human health	[20]

(*Continued*)

Table 7.1 Therapeutic qualities and mode of activity of marine protein sources. (*Continued*)

Objectives	Marine protein source	Method applied	Bioactive compound	Therapeutic property	Test conducted for therapeutic effect	Key findings	Limitations	Implications	References
To determine the antioxidant and anti-inflammatory activities of peptide fractions obtained from oyster soft tissue using enzymatic hydrolysis	*Crassostrea talienwhanensis* (oyster)	Enzyme hydrolysis	Peptide fraction	Antioxidant and anti-inflammatory potential effects	DPPH scavenging activity test, ferric-reducing antioxidant power (FRAP) assay, and oxygen radical absorbance capacity (ORAC) assay—(*in vitro*)	Bioactive peptide fractions exhibited antioxidant properties, with higher scavenging activity than other fractions, which suggests their potential therapeutic effects	Ethanol was utilized for peptide fractionation, and this might have introduced some bias, as certain peptides might be more soluble in ethanol, hence more likely to be included in the fractions	Oyster-derived bioactive peptides have the potential to be utilized as natural antioxidants and anti-inflammatory agents	[21]
To examine the anti-inflammatory activity of oyster β-thymosin derived from *Crassostrea gigas* in lipopolysaccharide (LPS)-induced RAW264.7 cells	*Crassostrea gigas* (Pacific oyster)	Enzyme hydrolysis	β-thymosin peptides	Anti-inflammatory potential effects	Inhibition of nitric oxide (NO) production, prostaglandin E_2 (PGE_2) production, and inflammatory cytokine expression—(*in vitro*)	Similar to human β-thymosin, oyster β-thymosin suppressed NO generation, PGE_2 synthesis, and the expression of inflammatory cytokines through nuclear factor kappa-light-chain-enhancer of activated B cells (NF-κB) in LPS-induced RAW264.7 cells	Although the study showed that oyster β-thymosin inhibits inflammatory indicators *via* the NF-κB pathway, it did not fully explore other potential signaling pathways involved in its anti-inflammatory effects	Oyster β-thymosin obtained from *Crassostrea gigas* might be a potential therapeutic agent for reducing inflammation by targeting NF-κB in macrophage cells suggesting the prospect of being utilized in anti-inflammatory treatment	[22]

(*Continued*)

Table 7.1 Therapeutic qualities and mode of activity of marine protein sources. (*Continued*)

Objectives	Marine protein source	Method applied	Bioactive compound	Therapeutic property	Test conducted for therapeutic effect	Key findings	Limitations	Implications	References
Investigate the hepatoprotective effects of krill peptide hydrolysate fractions against ethanol-induced liver damage in mice	*Euphausia superba* (krill fish) was used as the marine protein source in the study	Enzyme hydrolysis	Peptide fraction	Hepatoprotective potential effects	Antioxidant enzyme activity for superoxide dismutase (SOD), catalase (CAT), and glutathione peroxidase (GPx) in liver homogenate mice—(*in vivo*)	Hepatoprotective effects of 1–3 kDa fraction (KPF) resulted in increased weight and survival rate, reduction in liver damage markers, and an increase in antioxidant enzymes	Animal models (mice) were the subject utilized in this study. Thus, further research is required to investigate the effects in human subjects	Functional foods containing krill-derived peptides could improve liver health and function potentially leading to new therapeutic interventions for liver diseases	[23]
To investigate the anti-inflammatory, anti-ulcerative colitis (UC), and regulatory effects on the intestinal flora of Skipjack enzymatic peptide (SEP)	*Katsuwonus pelamis* (skipjack)	Enzyme hydrolysis	Enzymatic peptide	Anti-inflammatory potential effects	Mouse ulcerative colitis model and a transgenic zebrafish inflammation model—(*in vivo*)	SEP significantly reduced inflammatory markers, repaired intestinal damage, reduced neutrophil granulocyte aggregation, alleviated symptoms in mice, and modulated the intestinal flora composition	The specific anti-inflammatory mechanisms of SEP, signaling pathways or detailed cellular interactions were not delved into	It is possible to develop SEP as an anti-inflammatory drug for application as an auxiliary treatment for senescence-associated secretory phenotype (SASP)	[24]
To examine the anticancer, antidiabetic, and wound healing properties of collagen peptides derived from unicorn leatherjacket skin	*Aluterus monoceros* (unicorn leatherjacket)	Enzyme hydrolysis	Collagen hydrolysate	Antidiabetic, anticancer, and wound healing potential effects	Anticancer assay, antidiabetic assay, anti-inflammation assay, and wound healing assay—(*in vitro*)	CP-5 exhibited the maximum *in vitro* anticancer, antidiabetic, and wound healing activities than CP-25 and CP-50	Focus was more on *in vitro* experiments. Thus, further research using animal models is necessary to validate the bioactive properties observed	Fish skin collagen peptides can potentially be utilized for various biomedical applications owing to their bio-functional properties	[25]

docosahexaenoic acid DHA (C22:6n3). Additionally, mesopelagic fishes often accumulate bioactive compounds, including carotenoids, peptides, and polyphenols, offering possible health benefits that go beyond basic nutrition [11, 12]. Researchers, such as Leone *et al.* [13], found that the mesopelagic zone harbors a variety of gelatinous organisms that have significant pharmaceutical and commercial potentials, which could be exploited for medicine, food, and other uses as indicated in Figure 7.3. Also, in Table 7.1 is the display of the therapeutic qualities and the mode of activities of marine protein sources. Bioactive peptides that are derived from mesopelagic fish are short chains of amino acids that display physiological characteristics beyond basic nutrition. Just as the therapeutic applications of biologically active peptides obtained from protein sources, namely, plants, animals, and some marine organisms, have been well explored, mesopelagic fish species, which are also rich protein sources, can as well be hydrolyzed to produce hydrolysates with enhanced bioactive peptides for therapeutic applications. Gloomily, such sources (i.e., mesopelagic fish species) are relatively underexploited with regard to their utilization to produce bioactive peptides from their hydrolysates for therapeutic applications. More so, one identified promising area of research, with the potential of mitigating the pressures associated with protein demands and healthy living in the future, is commercially exploitable marine resources, most especially mesopelagic fish species. Hence, this chapter discusses how mesopelagic fish species can be employed as protein sources to produce protein hydrolysates (PHs) that contain bioactive peptides that may possibly be utilized to manage two common and interconnected chronic diseases such as diabetes and hypertension.

7.2 Mesopelagic Fish Protein Hydrolysates

7.2.1 Mesopelagic Zone

Occupying approximately 60% of land area in the world and filling approximately 20% of the world's ocean volume is the mesopelagic zone, which makes it one of the largest parts of the biosphere and perhaps one of the least explored. This zone serves as a home that hosts remarkable marine biodiversity, with some species of fish, namely, lanternfish (Myctophidae), Hatchetfishes (Sternoptychidae), *Benthosema glaciale*, *Maurolicus muelleri*, krill, bristlemouths, and deep-sea smelts, being typical examples. Furthermore, the mesopelagic zone has also been identified as hosting the largest biomass of fish on earth. However, it is believed that the fish in this zone (i.e., mesopelagic fishes) are the Earth's topmost untapped resource. Fish biomass in the mesopelagic zone is estimated to range between 8 and 16 gigatons (Gt) [26, 27].

7.2.2 Mesopelagic Fish

Multiple commercial attempts to exploit mesopelagic fish species have taken place since the 1970s [28]. By and large, the fish are planktivorous in nature, small (2–15 cm), and conduct diel vertical migration (DVM) at night to the water surface (a more nutrient-rich environment) daily to feed. To stay ahead of the sun and avoid predators, mesopelagic fish return to the dark depths of the twilight zone before dawn. Energy and materials are transferred between the surface and the deep sea all through these movement processes. In addition, carbon is transferred into the deep ocean throughout the processes [29]. Despite the potential uses of mesopelagic fish for food, nutraceuticals, and personal care products, most of them are reduced by industrial usage into the formulation of fish meal and fish oil. The perceived relatively low value of this fish is demonstrated by its average price of 329 euros per ton, according to Standal and Grimaldo [30]. Even in present times, mesopelagic fish are still regarded as low-value fish; thus, they need huge catches to be profitable. This is evident in a report by Fjeld *et al.*, which establishes that only 1,895 tons of mesopelagic fish were captured by Norwegian vessels between 2018 and 2021 (2018: 31 tons; 2019: 1,693 tons; 2020: 50 tons; and 2021: 121 tons) [31]. Although mesopelagic fish on the regular are underappreciated and overlooked owing to their small sizes, oiliness, and boniness, they could have significant applicability in the food and pharmaceutical (nutraceutical) industries. By processing the fish into hydrolysates, it is possible to enhance the functionalities of its protein for varied applications. Hence, mesopelagics can be repurposed into other forms to develop other products instead of their exclusive utilization as fish feeds and fish oils. This will help avoid the intermediary of feeding them to other fish.

7.2.3 Protein Hydrolysis and Protein Hydrolysates

7.2.3.1 Chemical Hydrolysis (Acid and Alkali)

Fish protein can be extracted in various ways; however, parameters like agitation time, pH, homogenization, weight-to-volume ratios, temperature, and number of sequential extractions are factors greatly considered to be dependent on the designed utilization of protein extract or hydrolysate.

Hydrolysates consist of a mixture of modified peptides, oligopeptides, polypeptides, and free amino acids produced *via* protein hydrolysis—the breakdown of protein molecules into smaller peptides and amino acids. The functionality and bioavailability of proteins are enhanced or modified during the hydrolysis process. In 1820, Braconnot was reported to have used the acid hydrolysis method, and this has since been recorded as the first chemical hydrolysis method applied [32]. In present times, this method remains a common method employed, although it has been in existence and utilized over years. Acids, like hydrochloric acid (HCl) and sulfuric acid (H_2SO_4), at high concentrations (e.g., 6 M HCl) operating at high temperatures of approximately 110°C–121°C in combination with pressure of approximately 100 kPa to hydrolyze fish proteins into peptides of varying molecular weights, are usually made use of in this method. Whereas some experiments may use shorter hydrolysis time (approximately 2 to 6 h), others can use longer hydrolysis time (approximately 24 h) [33, 34]. In some situations, the pH-shift approach may be employed for protein hydrolysis of fish, and when this method is undertaken, protein may precipitate from fish depending on the pH and isoelectric point. Nonetheless, the surimi process, which is primarily a water-based approach, can be utilized in a stepwise manner to extract protein from fish; minced fish is washed and dewatered to remove impurities before drying. However, the former is more effective and preferable compared to the latter because the latter results in lower protein yield, reduced gelation ability and product quality, and loss of sarcoplasmic protein during the process [35].

Furthermore, alkali, such as sodium hydroxide (NaOH), calcium hydroxide ($Ca(OH)_2$), and potassium hydroxide (KOH), are the common chemicals employed under the chemical hydrolysis approach. Normally, for alkali treatment, hydrolysis is performed with the use of alkali of 4 mol/L from 105°C to 180°C for approximately 20 h [36]. In a study conducted by Oliyaei *et al.* [37] to evaluate the effect that temperature and alkaline pH have on the physicochemical properties of protein isolates extracted from the whole ungutted lanternfish (*Benthosema pterotum*), the experiment was conducted in alkali pHs 10 and 12 with extraction temperatures of 0°C, 4°C, 11°C, 25°C, and 30°C. From the results, some functional properties, including water holding capacity, oil holding capacity, emulsifying capacity, foaming capacity, and solubility of the protein isolated at pH 12, were higher than those isolated at pH 10 ($p < 0.05$). In addition to the above, the protein recovery yield was higher when extracted at pH 12 compared to pH 10. Protein recovery yield and functional properties improved as temperature increased.

In whichever way one may look at the chemical process (e.g., acid and alkali) utilized for protein hydrolysis, the food and pharmaceutical industries rarely apply chemical process owing to the dangers that the chemicals utilized can cause. In addition, they are associated with harsh conditions that are unfavorable because they destroy some functionalities of hydrolysates that may be valuable and intended for functional foods and nutraceuticals. An alternative approach is hence needed to produce high-quality protein hydrolysates with wide commercial applications. Numerous studies have suggested enzyme-assisted extraction techniques to be a safer substitute for protein hydrolysis, as enzymes commonly employed are predominantly food-grade proteolytic enzymes. Moreover, enzymes possess intrinsic attributes, namely, selectivity and specificity, which enable them to successfully hydrolyze proteins to produce hydrolysates with enhanced yield and desired characteristics [38, 39].

7.2.3.2 Enzymatic Hydrolysis

Regarding enzymatic hydrolysis, depicted in Figure 7.4, enzymes with either endogenous or exogenous activity, or those exhibiting both activities are employed in the hydrolysis of proteins. Likewise, enzymes are predominantly applied due to their proteolytic activity, temperature tolerance, specificity, selectivity, mild action, pH adjustability, anti-inflammatory effects, and digestive aid. The proteolytic activity of enzymes is exhibited in their cleaving activity; however, enzymes are classified depending on the activity they display, be it exo- or endo-acting peptidases or proteases. Exo-peptidases cleave terminal peptide bonds to produce smaller peptides or single amino acids from the C-terminus (carboxypeptidases) or N-terminus (aminopeptidases). Contrarily, endo-proteases cleave peptide bonds within proteins to produce short fragments or peptides [40].

Moreover, the food and pharmaceutical industries greatly make use of enzymes for most of their processes due to the safety associated with enzyme usage in general (particularly food-grade enzymes) and the ability of enzymes to enhance functional and nutritional properties of hydrolysates. The mild processing parameters of enzymes usually operating at pH 6.0–8.0 and temperatures of 40°C–60°C make all these possible, as these are the parameters that the food and pharmaceutical industries desire. Another significant feature associated with the use of enzymes is that there are no residuals of toxic chemicals or organic solvents present after hydrolysis [41].

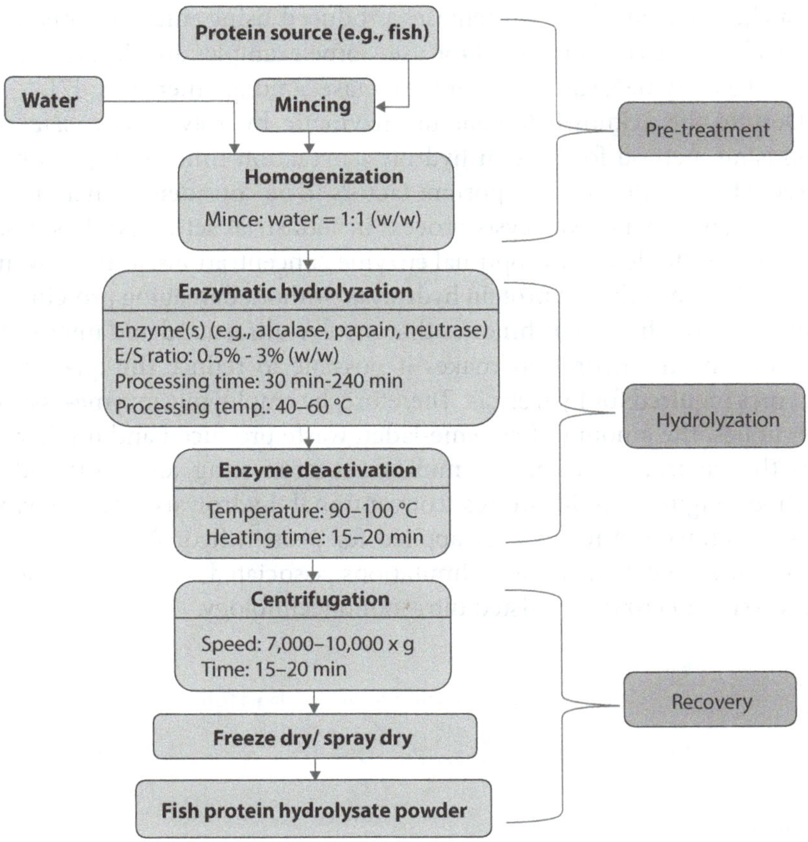

Figure 7.4 Schematic diagram depicting the steps involved in producing fish protein hydrolysates by enzymatic processing.

Due to the varying nature of enzymes, they possess different proteolytic characteristics, and they are obtained from diverse sources, including microbes (e.g., flavorzyme, alcalase, protamex, metalloproteases, serine protease, proteinase K) [42]; plants (e.g., bromelain, papain, ficin) [43]; and animals (e.g., trypsin, neutrase, α-chymotrypsin) [44]. A wide range of food-grade (e.g., papain, flavorzyme, bromelain, alcalase, proteinase K) and non-food-grade (e.g., serine protease, metalloproteases) enzymes are used by the food and pharmaceutical industries because they enhance the functional characteristics of resultant hydrolysates making them desirable [45]. However, the cost of enzyme(s) could be a significant factor that might impact the application of specific enzyme(s) for production.

As such, relatively cheap proteinases produced using microorganisms or by-products are recommended for use. Some examples are alcalase®, neutrase®, flavorzyme®, subtilisin, and orientase, among others [46, 47].

Despite the acknowledgment of enzymatic hydrolysis as a safe and promising method for protein hydrolysis, reaction time, and quantity of enzyme to be utilized are important factors to be considered when implementing an enzyme hydrolysis process in industrial setting(s). Besides, it is imperative to determine optimal enzyme concentrations, as these would aid in achieving efficient protein hydrolysis without damaging protein. The application of physical techniques that involve ultrasound and microwave to assist enzyme hydrolysis makes it possible to reduce the quantity of enzymes required for hydrolysis. Therefore, by employing enzyme-assisted techniques, the amount of enzyme-laden waste produced and discharged into the environment can be minimized promoting an eco-friendlier method. Figure 7.5 illustrates conventional hydrolysis and enzyme-assisted ultrasound technology applied for protein hydrolysis, and Table 7.2 outlines the benefits and limitations associated with conventional hydrolysis and enzyme-assisted ultrasound technology.

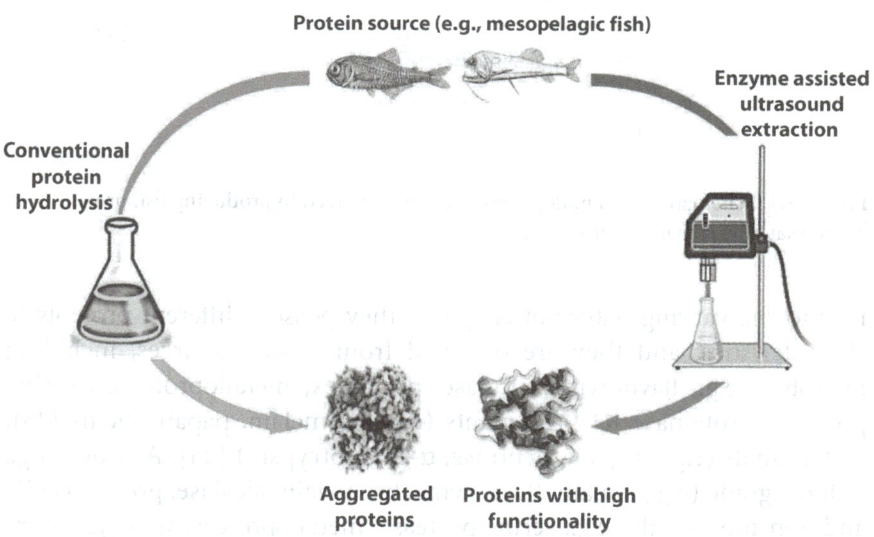

Figure 7.5 Graphic representation of conventional hydrolysis and enzyme-assisted ultrasound technology applied for protein hydrolysis. Adapted from ref. [39].

Table 7.2 Benefits and limitations associated with conventional hydrolysis and enzyme-assisted ultrasound technology.

Conventional hydrolysis	**Benefits**	**Limitations**
	Relatively cheaper chemicals	High extraction temperature
	No extra cost for equipment	Longer extraction time
		Poor energy efficiency
		Harsh extraction conditions
		Poor selectivity
		Higher environmental pollution
Enzyme-assisted ultrasound technology	**Benefits**	**Limitations**
	Low extraction temperature	Relatively expensive enzymes
	Shorter extraction time	High cost of equipment
	Higher energy efficiency	
	Mild extraction conditions	
	Higher selectivity	
	Low environmental pollution	

7.3 Mesopelagic Fish Protein Hydrolysates: Antihypertensive Peptides of Pelagic Fish Protein Hydrolysates

Hypertension, often described as high blood pressure, is a long-term health disease/condition marked by consistently elevated blood pressure levels that surpass the normal range. Blood pressure, in general, measured in millimeters of mercury (mmHg), signifies the force applied by blood onto artery walls during circulation. This pressure is represented by two values as follows: systolic pressure, which reflects arterial pressure during heartbeats, and diastolic pressure, which indicates arterial pressure when the heart is in its resting phase between beats. Usually, approximately 120/80 mmHg is considered the normal blood pressure range [48]. When a series of measurements consistently register blood pressure of 130/80 mmHg or above, hypertension is thus diagnosed. Hypertension is classified into different stages based on the severity of elevation, ranging from stage 1 hypertension (130–139/80–89 mmHg) to stage 2 hypertension (140/90 mmHg or higher). Additionally, individuals with blood pressure readings of 120–129/<80 mmHg are considered to have elevated blood pressure, which may advance to hypertension if left untreated [49]. Some of the health implications associated with this condition are as follows:

Increased Risk of Cardiovascular Disease: Hypertension is the key risk factor for cardiovascular diseases (CVD), which encompass myocardial infarction (heart attack), coronary artery disease, cardiac failure (heart failure), and arrhythmias (irregular heartbeats). When blood pressure rises, additional strain is put on the heart and blood vessels causing atherosclerosis (hardening and narrowing of the arteries). This phenomenon impedes blood flow and results in an increase in the risk of heart-related complications [50].

- ***Stroke:*** "Hypertension is the leading cause of stroke—a medical emergency that ensues when blood flow to the brain is interrupted or reduced leading to brain damage" [51].

  The brain's blood vessels can become weaker and damaged owing to high blood pressure, which increases the risk of a rupture or blockage that might cause ischemic stroke (caused by blockage in blood vessel) or hemorrhagic stroke (caused by bleeding in the brain) [52].
- ***Kidney Failure:*** The blood vessels in kidneys can be damaged as a result of chronic hypertension. When this happens, the efficiency with which the kidney filters waste products from the blood is impaired. As hypertension-related kidney disease progresses, it could result in chronic kidney disease (CKD), which eventually causes kidney failure. This condition can only be managed through dialysis or a kidney transplant. It is significant to highlight therefore that end-stage renal disease (ESRD) is one of the leading causes of death globally due to kidney failure [53].

- ***Other Health Complications:*** Hypertension is not only about high blood pressure but also a condition that could open doors for diverse health issues. For instance, hypertension can lead to problems associated with vision such as hypertensive retinopathy, where the retina suffers due to high pressure [54]. Peripheral artery disease is another known risk associated with the condition. This disease causes reduced blood flow to the limbs [55]. In some cases, erectile dysfunction may be experienced by some men who suffer from hypertension [56].

Furthermore, it is highly possible that cognitive decline can occur with patients suffering from hypertension, as this could possibly progress to other health conditions like dementia and Alzheimer's disease [57].

7.3.1 Mechanisms of Action

The antihypertensive properties possessed by peptides obtained from resultant mesopelagic fish protein hydrolysates (MFPHs) are suggested to control blood pressure *via* numerous approaches. In so doing, they decrease the risks associated with hypertension and its associated issues. Figure 7.6 displays a simplified schematic diagram illustrating the mechanisms of action of MFPHs as antihypertensive peptides. To begin with, peptides from MFPHs inhibit angiotensin-converting enzyme (ACE) activity. Meanwhile, ACE is significant in regulating blood pressure and fluid balance in the body. When mesopelagic fish protein hydrolysate (MFPH) peptides impede ACE activity, they block the conversion of angiotensin I to II. Blood vessels can be constricted by angiotensin II, and this could result in an elevation of blood pressure [58]. In conjunction with the mechanism of MFPH peptides closely mimicking the activity/action of ACE inhibitors, these peptides are naturally sourced from mesopelagic fish. Hence, they present an added advantage as the likelihood of any possible side effects may be relatively minimal. Second, vasodilation, which is the process by which blood vessels widen or dilate is boosted by MFPH peptides [59]. This phenomenon is crucial to the body because through such a process, the peripheral area resistance is reduced, which in turn improves blood flow in the body. In endothelial cells lining the blood vessels, some peptides stimulate the release of nitric oxide (NO). NO is a naturally occurring vasodilator, which regulates blood pressure by controlling and regulating vascular tone [60]. Finally, MFPH peptides help prevent oxidative stress.

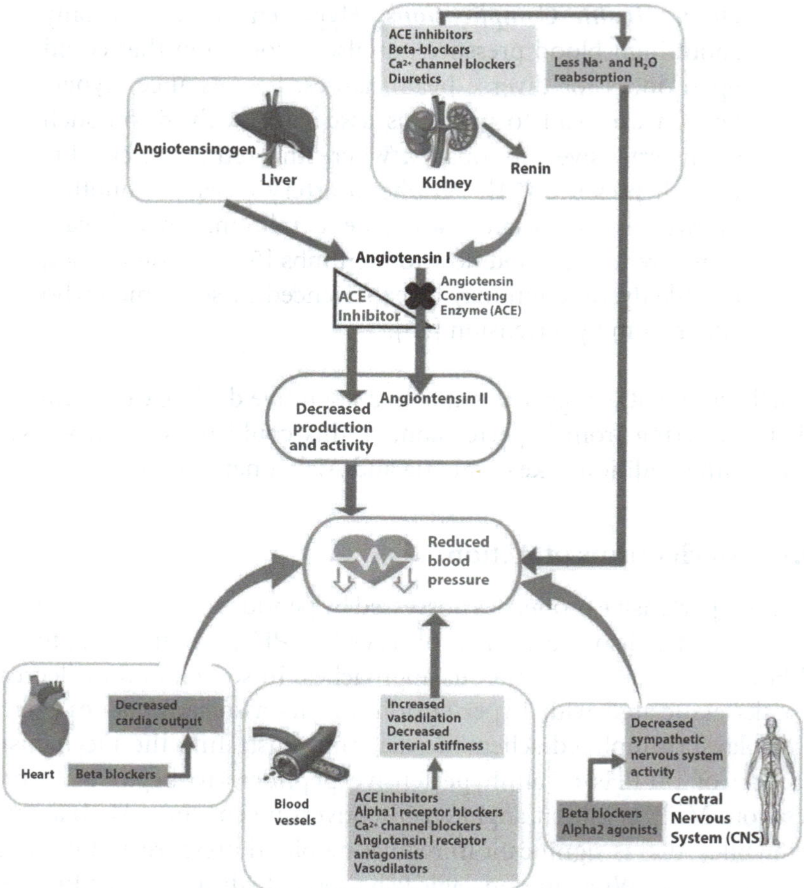

Figure 7.6 Schematic diagram illustrating the mechanisms of action of mesopelagic fish protein hydrolysates as antihypertensive peptides.

This happens when the balance between destructive reactive oxygen species (ROS) and protective antioxidants varies [61]. A major contributing factor to hypertension is the circumstance where oxidative stress damages blood vessels, which then promotes inflammation. The antihypertensive peptides of MFPH possess antioxidant properties that help lower ROS production, neutralize free radicals, and enhance the activity of antioxidant enzymes. MFPH peptides aid in maintaining endothelial function and vascular health and halt the advancement of hypertension and its related cardiovascular problems through the reduction of oxidative stress [62].

7.3.2 Research Evidence

The possibility of using MFPHs in managing hypertension using laboratory (*in vitro*) and animal (*in vivo*) models has been explored by several research activities. These studies provide insightful information about the mechanisms and efficiency of these bioactive peptides in managing high blood pressure. Below is a summary of key findings from selected studies:

In Vitro and In Vivo Studies

Findings from various research studies conducted demonstrated that peptides from MFPHs can inhibit angiotensin-converting enzyme (ACE), which is crucial for blood pressure regulation.

Hayes *et al.* [63] undertook a study with the aims of recovering proteins from boarfish (*Capros aper* Linnaeus) using the pH-shift process, generating protein hydrolysates with ACE-I inhibitory and antihypertensive properties, and evaluating their potential for maintaining normal blood pressure in individuals with hypertension. Hayes *et al.*'s study reported for the first time angiotensin-I-converting enzyme (ACE-I; EC 3.4.15.1) inhibitory and antihypertensive peptides produced from boarfish proteins. Alcalase CLEA, papain, and protease AP, which are commercial proteolytic enzymes, were utilized to produce boarfish protein hydrolysates (BPHs). Hayes *et al.* employed mass spectrometry to analyze protease AP-generated hydrolysate. The results they acquired emphasized the potential of the hydrolysate as a source of ACE-I inhibitory peptides. The peptides less than 10- and 3-kDa were subsequently enriched using the molecular weight cutoff (MWCO) filtering techniques. The inhibitory activity of the resulting hydrolysates and MWCO fractions against the angiotensin-I-converting enzyme (ACE-I; EC 3.4.15.1) was evaluated *in vitro*. Then, the hydrolysates were tested in triplicate using an ACE-I inhibitor kit at a concentration of 1 mg/ml following the directions of the manufacturer. In comparison to positive control captopril, the findings from the test conducted revealed that BPHs of 3-kDa hydrolysate produced using protease AP demonstrated significant ACE-I inhibitory activity of 85.8% (±1.81) when tested at 1 mg/ml. Hayes *et al.* conducted further *in vivo* experiments on spontaneously hypertensive rats (SHRs) using the resultant hydrolysates produced. Boarfish protein hydrolysate (BPH) generated with protease AP had antihypertensive effects on the hypertensive rats. This was observed through the rapid reduction in systolic blood pressure (SBP) indicating its potential as a functional food ingredient for managing hypertension [63].

Antioxidative Studies

Mesopelagic species, like Myctophidae family (e.g., lanternfish), Gonostomatidae family (e.g., bristlemouths), Sternoptychidae family (e.g., hatchetfish), Scopelarchidae family (e.g., pearleye), and Calanidae family (e.g., *Calanus finmarchicus*), just to mention a few, are great sources of marine lipids [64]. Lipids can be extracted, and the remaining peptide fractions are by-products with bioactive characteristics during enzymatic protein hydrolysis. Matić *et al.* [65] carried out research to investigate the potential health-promoting bioactive peptides derived from *Calanus finmarchicus* through enzymatic hydrolysis with the aim of evaluating the bioactivity of the hydrolysates, particularly in terms of antioxidant activity and DPP-IV inhibition. From the research, size exclusion chromatography (SEC) and nuclear magnetic resonance (NMR) spectroscopy were applied to evaluate the antioxidative activities of 63 resultant hydrolysates produced from *Calanus finmarchicus*. A significant increase in bioactivity of the hydrolysates was observed compared to the unhydrolyzed control, which is indicative of the strong antioxidative properties of hydrolysates. DPP-IV inhibition and antioxidant activity analyses were employed to assess the bioactive properties of the resultant hydrolysates. Trolox equivalent antioxidant capacity (TEAC) values were utilized to assess the antioxidant activity of hydrolysates. Findings disclosed promising antioxidative activity in hydrolysates highlighting their potential health benefits as bioactive peptide sources. Therefore, bioactive peptides from mesopelagic species could be helpful in the reduction of oxidative stress in blood vessels.

From the studies assessed, it is pertinent to note that there are great economic opportunities for marine processors, researchers, industry professionals, and policy makers in recovering fish proteins and generating high-value bioactive peptides from underutilized mesopelagic species for therapeutic applications.

7.3.3 Bioavailability and Stability

Mesopelagic fish protein hydrolysates contain multiple antihypertensive peptides that are bioavailable and stable, but their efficacy and therapeutic potential in the management of hypertension are strongly impacted by numerous factors [66]. When a compound has high bioavailability, it means that it is available and rapidly absorbed for biological action in the body, while its stability indicates how stable it is regarding degradation and maintenance of its activity. Below are some of the factors discussed:

- ***Digestibility and Absorption:*** Peptides must be ingested and then absorbed into the bloodstream before they can work efficiently and effectively. Notwithstanding, the degree to which they withstand stomach digestion is dependent on their structural makeup. There are some amino acid sequences in peptides that make it somehow difficult for the stomach and intestinal enzymes (digestive enzymes) to break down. However, once broken down and absorbed into the bloodstream, antihypertensive peptides may exert their effects directly on target tissues or undergo further metabolism to activate or inactivate metabolites [67].
- ***Interactions with Carrier Molecules:*** Antihypertensive peptides may interact with carrier molecules, such as transport proteins or lipids, which can affect their absorption and distribution in the body. Carrier-mediated transport systems can facilitate the uptake of peptides across cell membranes and enhance their bioavailability [68].

  Peptide absorption may also be influenced by the presence of other dietary components involving fats, fibers, and minerals. These components can influence gastrointestinal transit time, absorption kinetics, and peptide stability [69].
- ***Metabolic Stability:*** For antihypertensive peptides to exert their desired physiological effects, they must remain stable in the bloodstream [70].

  Peptides nevertheless may be disposed to degradation by circulating enzymes, for instance, peptidases and proteases. These circulating enzymes can cleave peptide bonds and render them inactive. To improve the metabolic stability of antihypertensive peptides, crucial strategies, including modifying peptide sequences to enhance resistance to enzymatic degradation, encapsulating peptides in liposomes or nanoparticles to protect them from enzymatic attack, or co-administering peptides with enzyme inhibitors to prolong their half-life during circulation, can be employed [71].
- ***pH and Temperature Sensitivity:*** Environmental factors, such as temperature and pH, have the potential to influence the stability of antihypertensive peptides [72].

  Peptides might experience conformational changes or denaturation under acidic or alkaline conditions affecting their bioactivity [73]. To determine their suitability for oral administration and systemic absorption, it is imperative to conduct a study of the stability of antihypertensive peptides under physiological conditions (pH 7.4, body temperature).

- ***Storage Conditions:*** Appropriate storage conditions are fundamental for the preservation of the stability of antihypertensive peptides during formulation, storage, and distribution. Hence, to avoid degradation and preserve peptides' bioactivity, it is recommended that peptides are stored in a cool, dry environment away from or protected from light and oxygen. Furthermore, packaging materials and storage containers that can minimize moisture, heat, and other conditions that may compromise the stability of peptides over prolonged periods should be chosen and applied/implemented.

A better understanding of these factors by scientists can empower them to address the worrisome situation through the development of antihypertensive peptides that would be more effective and reliable to better manage hypertension.

7.3.4 Applications and Future Directions

MFPHs are being investigated for their potential antihypertensive properties across varied applications, including functional foods, dietary supplements, and pharmaceuticals. The unique properties of these peptides make them capable of effectively reducing hypertension and improving cardiovascular health. Below are some potential applications and what could be considered in the future to improve the peptides:

Applications

- ***Functional Foods:*** Consideration can be given by the food industry when developing new product lines by applying MFPH peptides as antihypertensive peptides for incorporation into products to develop functional foods, such as beverages, snacks, and dairy products, which can provide consumers with convenient ways of managing blood pressure while enjoying their diets.
- ***Dietary Supplements:*** MFPH peptides can be formulated into capsules, tablets, or powders to offer a concentrated

and controlled dose for those seeking targeted hypertension management through supplementation. Consequently, consumption of these doses should be by prescription mainly as this would set the course for monitoring dosage consumption.

- ***Pharmaceuticals:*** Peptides of MFPHs create new avenues for significant marketing chain or line for the pharmaceutical industry as new pharmaceutical products that have the potential to manage the treatment of hypertension can be developed. In addition, the peptides used for developing the products are natural peptides obtained from a natural source (mesopelagic fish). This potentially offers a safer alternative to synthetic drugs with minimal side effects.

Future Directions

- ***Optimization of Extraction Methods:*** Researchers are still fine-tuning the extraction process. Also, the means of getting the most out of peptides from MFPHs are being sought. This involves exploring alternative approaches for extracting/producing high-quality protein hydrolysates with enhanced and desired characteristics, modifying/optimizing processing conditions, and exploring/utilizing new fish source(s) with desirable properties.
- ***Assessment of Long-Term Effects:*** The understanding of the safety and long-term effects of MFPH peptides in humans is crucial. Large-scale studies over extended periods can shed light on how peptide supplementation affects blood pressure control, overall heart health, and general well-being.
- ***Investigation of Synergistic Interactions:*** In addition to MFPH peptides known to have potential medicinal value, researchers are also interested in how other beneficial compounds interact with these peptides. Pairing the peptides with antioxidants, polyphenols, or omega-3 fatty acids could lead to even greater cardiovascular benefits [74].
- ***Development of Novel Delivery Systems:*** A very vital factor to be considered also is discovering techniques to improve and enhance the delivery of MFPH peptides in the body. Hence, investigations are being undertaken in a number of advanced delivery systems, such as nanoparticles and

liposomes, to improve absorption, extend their presence in the bloodstream, and protect them from degradation. This is to maximize the potentials of the peptides [75, 76].

The full potential of antihypertensive peptides therefore holds the possibility of being unlocked due to the continued research efforts of scientists. Achieving this feat in science will pave the path for more effective, accessible, and targeted treatments for hypertension and cardiovascular health diseases.

7.4 Mesopelagic Fish Protein Hydrolysates: Antidiabetic Peptides

Diabetes mellitus (DM), so often known as diabetes, is "a chronic disease or condition characterized by elevated sugar/glucose levels in the blood, with symptoms such as increased hunger, increased thirst, and frequent urination" [77]. If unchecked and managed poorly, it may result in various health complications involving nephropathy (kidney damage), cardiovascular disease, retinopathy (eye damage), neuropathy (nerve damage), skin conditions, foot complications, hearing impairment, gastrointestinal issues, dental problems, and Alzheimer's disease, among others [78, 79]. It is possible that in acute cases, diabetes could result in hyperglycemic state, diabetic ketoacidosis, or even death [80]. Largely, diabetes is triggered by either the insufficient production of insulin by the pancreas or inappropriate insulin cellular response [81]. As far as the current information available, the cure for this condition is unknown and has thus become a big concern globally because more and more people get affected each year. Below are some issues concerning diabetes:

- ***Epidemiology:*** Diabetes is on the rise worldwide. Approximately 422 million adults in 2014 suffered from diabetes, and it is despondently projected that over 592 million people might be affected by diabetes by 2035 [82].

  Most of the cases reported concerning diabetes are type 2 diabetes, which makes up 90%–95% of diagnoses. Type 1 diabetes, which is traditionally observed in kids, is also increasing, especially among young adults [83, 84]. The effects of diabetes on people differ from person to person. Older adults, certain ethnic groups, individuals with weight issues or inactive lifestyles, and those with family histories of diabetes are considered more vulnerable and are at more risk. The financial worth of individuals and their accessibility to healthcare also play vital roles in the management of diabetes [85].

- ***Pathophysiology:*** Type 1 diabetes occurs when the immune system is attacked by insulin-producing cells in the pancreas and destroys them. The insufficiency of insulin hinders the entrance of sugar into cells to generate energy for the body. This leads to high blood sugar levels and the occasional production of ketones [86] (Roep *et al.*, 2021).

  As the cells of the body become less responsive to insulin, and the pancreas continually produces less insulin, type 2 diabetes develops. The consequence of this combination is the high levels of sugar in the blood. Lifestyle habits, genetics, overweight, and inflammation are common precursors that can set the stage for type 2 diabetes [87]. Diabetes from certain gene mutations [88], gestational diabetes during pregnancy [89], and secondary diabetes linked to other health issues or medications [90] are some other forms of diabetes that are known.
- ***Current Treatment Options:*** In the management of diabetes, lifestyles/habits must be checked, complications must be prevented, and the constant monitoring of blood sugar levels is key. Typically, this requires a combination of lifestyle modifications, medications, and, in some cases, insulin therapy. The adoption of healthier diets, engagement in regular physical activities, management of weight, and quitting smoking are fundamental approaches for managing diabetes and reducing the risk of cardiovascular complications [91].

Metformin and injectable insulin are key examples of medications that help increase insulin sensitivity, stimulate insulin production, reduce hepatic glucose production, and support weight loss [92]. With regard to individuals suffering from type 1 diabetes, insulin therapy is very critical for the maintenance of blood sugar levels and preventing diabetic ketoacidosis [93, 94]. Insulin can be administered by injecting several times a day or through an insulin pump that provides continuous delivery. The adjustment of the dosage of insulin largely relies on blood sugar levels, dietary intake, and physical activity levels.

7.4.1 Mechanisms of Action

Mesopelagic fish protein hydrolysates offer a mixture of biologically active peptides that possess antidiabetic characteristics, which present promising therapeutic benefits for diabetes mellitus management. Foremost, insulin sensitivity can be enhanced by MFPH peptides. The effective response of cells to insulin's signal to absorb glucose from the bloodstream is referred to as insulin sensitivity. Blood sugar levels begin to elevate in circumstances where cells fail to respond appropriately to insulin's signal (i.e., insulin resistance), until a point where the resultant effect is type 2 diabetes. The presence of MFPH peptides leads to the enhancement of insulin sensitivity by activating the signaling pathways that insulin uses. This enhances the interaction between insulin and its receptors on cell surfaces and, in that way, facilitates more effective signal transmissions. MFPH peptides help muscle and fat cells absorb glucose more efficiently and reduce glucose production by the liver when insulin sensitivity is boosted. This causes blood sugar levels to be lowered aiding in diabetes management [95]. Additionally, MFPH peptides modulate glucose uptake. Glucose uptake into cells is made possible by specialized transporters such as glucose transporter type 4 (GLUT4). GLUT4 is essential in the maintenance of glucose homeostasis. The impairment of GLUT4 may possibly result in poor blood sugar control and reduced insulin sensitivity (i.e., glucose uptake process). Glucose uptake can be enhanced by the presence of MFPH peptides by directing glucose transporters to the cell membrane or increasing their activity. Consequently, the functions of these transporters (GLUT 4) are optimized by these peptides, which in turn help cells absorb more glucose. By boosting glucose uptake also, MFPH peptides are able to increase insulin sensitivity and reduce blood sugar levels, thereby offering significant support in managing diabetes [96].

Finally, MFPH peptides play a significant role in the inhibition of alpha-glucosidase activity. Alpha-glucosidase breaks down carbohydrates into simple sugars in the small intestines. In certain scenarios where the enzyme is overly active, it can cause post-meal glucose levels to surge. MFPH peptides may inhibit alpha-glucosidase activity, thereby effectively slowing down the digestion of carbohydrates. MFPH peptides help prevent sharp spikes in blood glucose levels after meals in the moderation of this process. Moreover, a more stable glucose profile, which enables diabetes management, is attained when MFPH peptides are able to control alpha-glucosidase activity and help smooth out the fluctuations in blood sugar levels [97].

7.4.2 Research Evidence

Researchers have carried out extensive studies and explored the potential of MFPHs in managing diabetes and combining laboratory experiments with clinical trials. Here are some key findings:

Exhibition of Alpha-Glucosidase Inhibitory Activity

Laboratory studies have revealed that MFPHs can inhibit the activity of alpha-glucosidase, an enzyme crucial for carbohydrate digestion.

According to Cavan *et al.* [98], the Antarctic krill (*Euphausia superba*) inhibits the Southern Ocean's mesopelagic zone. They contribute significantly to the availability of nutrients for the marine food chain and to global biogeochemical cycles in the ocean while in the zone. Because mesopelagic species often accumulate bioactive compounds, a study was undertaken to investigate the possibility of Antarctic krill-derived peptides as α-glucosidase inhibitors for the treatment of type 2 diabetes. Response surface methodology (RSM) was utilized for hydrolysis to optimize enzymolysis conditions using hydrolysis temperature and time of 45°C and 5.4 h, respectively. Enzyme concentrations of 6×10^3 U/g were the optimal conditions employed for hydrolysis. To determine the molecular weight (Mw) distribution and sequences of the hydrolysates, gel chromatography and liquid chromatography-tandem mass spectrometry (LC-MS/MS) were utilized. Both computer-aided techniques and traditional techniques were used to identify and analyze the mechanism behind α-glucosidase inhibitory peptides. To further validate the binding affinities between peptides and α-glucosidase, biolayer interferometry (BLI) assay was applied. It was observed from the findings that the greatest rate of α-glucosidase inhibition was detected in hydrolysates produced by Neutrase with four peptides (LPFQR, PSFD, PSFDF, and VPFPR), which exhibit strong affinity for binding to the active site of α-glucosidase. Hydrogen bonding and hydrophobic interactions made all these possible [99]. It is understood from Zheng *et al.*'s study that peptides obtained from mesopelagic species (e.g., Antarctic krill) have the potential of being utilized to curb α-glucosidase activity, which may aid in regulating blood sugar levels.

Enhancing Insulin Sensitivity

In 2019, Parthsarathy *et al.* conducted a study with the aim of examining the *in vitro* insulinotropic and glucagon-like peptide-1 (GLP-1) stimulatory effects of a boarfish (*Capros aper*) protein hydrolysate (BPH). They attempted further to examine the effects of oral administration of BPH on blood glucose control in healthy mice. The enzymes utilized to produce BPHs were alcalase and flavorzyme. After the production of BPHs, they were subjected to simulated gastrointestinal digestion (SGID). Tests were conducted using BPH and SGID samples (0.01–2.5 mg/ml) to determine *in vitro* dipeptidyl peptidase-4 (DPP-IV) inhibition, insulin secretion, and GLP-1 secretion from BRIN-BD11 (a cell line for rat insulinoma generated from beta cells in the pancreas) and GLUTag cells (obtained from entero-endocrine cells found in the lower intestine—ileum), respectively. Both BPH and SGID samples significantly increased insulin secretion from BRIN-BD11 cells and inhibited DPP-IV activity. Some selected healthy mice were orally administered BPH, while others were not. The BPH dose administered was 50 mg/kg bodyweight. Thereafter, an oral glucose tolerance test (OGTT) was done in healthy mice (i.e., both those administered with BPH and those without BPH) for a total duration of 120 min. Changes in plasma insulin levels and blood glucose concentration were monitored in the mice during the period of testing. Results obtained after OGTT indicated that BPH administration resulted in the increment of plasma insulin levels and reduced blood glucose concentration compared to controls. Remarkably, peptides of MFPHs have promising antidiabetic characteristics, which suggests the possibility of MFPHs being employed as functional ingredients that could be applicable to help manage pre-diabetes or type 2 diabetes from the findings of the various research conducted.

7.4.3 Challenges and Opportunities

A couple of challenges and opportunities have been identified to be associated with the research innovations and practical applications of antidiabetic peptides obtained from MFPHs. These peptides have the potential to possibly improve the health and lives of millions of people suffering from diabetes. Nonetheless, additional study and clinical trials are recommended to prove their effectiveness and safety. In addition, to ensure their affordability and accessibility, the cost of producing these peptides needs to be subsidized. Underneath are further examples of some challenges and opportunities:

Challenges

- ***Safety Concerns:*** Ensuring the safety of peptides is paramount prior to the release of MFPH peptides into the market. Further studies that ensure the toxicity, potential allergenicity, and long-term effects of these peptides are commended to certify their safety, as this could help broaden application and consumption.

- ***Optimization of Dosage and Formulation:*** To determine the optimal dose and formulation for MFPH peptides is not straightforward. The rate of absorption, stability, and efficiency are examples of some factors that play key roles in the optimization and formulation. To enhance peptide efficacy, innovative methods, such as encapsulation, or specialized delivery systems can be employed.
- ***Regulatory Considerations:*** A complex web of regulations must be adhered to when transitioning MFPH peptides from the laboratory to the market. However, these stringent quality control standards can also pose significant challenges in an attempt to meet all these regulatory requirements, as obtaining necessary approvals and ensuring compliance cannot be disregarded.
- ***Standardization and Quality Control:*** The quality of MFPH peptides is critical and, hence, must be consistently ensured across each batch during the production. Thus, reliable production methods, robust testing protocols, and rigorous quality control measures must be established to meet the required regulatory standards and guarantee the safety of patients.
- ***Cost and Accessibility:*** Affordability and accessibility may pose challenges in ensuring that these therapies are affordable and can be accessed by those who might need them, particularly in regions with limited resources. Overcoming these hurdles to make the treatments affordable and widely available might be a huge challenge.

Opportunities

- ***Synergistic Effects with Other Bioactive Compounds:*** When MFPH peptides are combined with other beneficial compounds, like antioxidants or omega-3 fatty acids, their effectiveness against diabetes could be enhanced. An opportunity is then presented to address the condition from diverse angles.
- ***Therapeutic Agents for Diabetic Complications:*** MEPH peptides have the potential not only to aid in the management of diabetes but also to assist in the combat to mitigate

the complications linked with diabetes such as kidney damage, nerve damage, heart disease, eye damage, skin and mouth conditions, and foot damage. The exploration of this potential can lead to new avenues for treatment and improved patient outcomes.

- ***Personalized Medicine Approaches:*** Everyone is unique; hence, treatments should be tailored accordingly. The customization of these peptides based on an individual's genetic profile, metabolism, and dietary needs possibly will potentially increase their effectiveness and reduce risks.
- ***Advanced Delivery Systems:*** Employing innovative delivery approaches, such as targeted tissue delivery or sustained-release formulations, may possibly improve the efficacy and convenience of these peptides for patients.

We inch closer to transforming the promising peptides of MFPHs into practical solutions for the management of diabetes as we address the challenges that confront us and capitalize on the opportunities presented.

7.4.4 Implications

This study is relevant and holds significant implications for the public health, food industry practices, and future research initiatives. When a closer look is taken regarding public health perspective, incorporating MFPHs into functional foods, dietary supplements, and pharmaceutical formulations could present innovative approaches that can aid in the prevention and management of hypertension and diabetes. With respect to the growing interest in holistic approaches to health and wellness, there has been the emergence of novel therapeutic interventions from natural sources. The food industry can also leverage on MFPHs to produce new market lines and product innovation. This can be attained when bioactive peptides obtained from MFPHs are incorporated into food and beverage products. In doing so, the demand from consumers for functional foods with specific health outcomes in regulating blood pressure and controlling blood sugar levels can be met. Collaborative efforts between the food industry and research institutions could aid in developing new formulations and delivery systems that enhance bioavailability and stability of antihypertensive and antidiabetic peptides.

Toward addressing knowledge gaps and improving extraction and production methods, future research studies should be tailored to the exploration of innovative ways (e.g., ultrasound assisted extraction) of producing MFPHs, so that the resultant hydrolysates would have enhanced functionalities, which are desirable by the food and pharmaceutical industries for product development. Continual studies are recommended to help elucidate specific activity mechanisms of bioactive peptides, evaluate their efficacy and safety through human clinical trials, and investigate potential synergies with other bioactive compounds. Research activities should be designed such that attention is given to standardizing production methods, ensuring product quality and consistency, and the adherence to regulatory requirements to facilitate the transformation of findings from research into practical implementations.

7.5 Conclusion

This chapter primarily explored the potential of the mesopelagic fish inhabiting the mesopelagic zone as a significant source of protein to produce MFPHs as valuable sources of bioactive peptides. These peptides possess enhanced characteristics that make them applicable in the management of hypertension and diabetes. Globally, these diseases, owing to their association with serious complications and high prevalence, have significant impact on healthcare systems not forgetting the burden they put on economies. Moreover, these peptides are naturally sourced; hence, they present an added advantage of possibly marginal or no side effects. MFPHs contain bioactive peptides that can inhibit angiotensin-converting enzyme (ACE) activity from converting angiotensin I to II, promote vasodilation, enhance insulin sensitivity, inhibit alpha-glucosidase activity, and regulate glucose uptake. All these are crucial in tackling hypertension and diabetes. From the reviews conducted in this chapter, compelling evidence supporting the effectiveness of MFPH peptides in addressing hypertension and diabetes is brought to bear. The findings have the potential to improve cardiovascular health and control blood sugar levels for those battling with hypertension and diabetes. Additionally, researchers, industry professionals, and policymakers concerned about hypertension and diabetes can use this information to find ways and means to better exploit mesopelagic fish to produce protein hydrolysates that are valuable and applicable as functional ingredients for the development of new functional foods and nutraceuticals that could help manage the menace associated with hypertension and diabetes mellitus.

References

1. Scott, P., Global panel on agriculture and food systems for nutrition: Food systems and diets: Facing the challenges of the 21st century. *Food Secur. Sci. Sociol. Econ. Food Prod. Access Food*, 9, 653–654, 2017.
2. Ridoutt, B.G., Baird, D., Hendrie, G.A., Diets within planetary boundaries: what is the potential of dietary change alone? *Sustain. Prod. Consum.*, *28*, 802–810, 2021.
3. Kargwal, R., Kumar, A., Garg, M.K., Chanakaewsomboon, I., A review on global energy use patterns in major crop production systems. *Environ. Sci.: Adv.*, *1*, 5, 662–679, 2022.
4. Grote, U., Fasse, A., Nguyen, T.T., Erenstein, O., Food security and the dynamics of wheat and maize value chains in Africa and Asia. *Front. Sustain. Food Syst.*, *4*, 617009, 2021.
5. Henchion, M., Hayes, M., Mullen, A.M., Fenelon, M., Tiwari, B., Future protein supply and demand: strategies and factors influencing a sustainable equilibrium. *Foods*, *6*, 7, 53, 2017.
6. Shevkani, K. and Chourasia, S., Dietary Proteins: Functions, Health Benefits and Healthy Aging. *Nutr. Food Diet Ageing Longevity*, 14, 3–37, 2021.
7. Najafian, L., A review of bioactive peptides as functional food ingredients: mechanisms of action and their applications in active packaging and food quality improvement. *Food Funct.*, *14*, 13, 5835–5857, 2023.
8. Drazen, J.C. and Sutton, T.T., Dining in the deep: the feeding ecology of deep-sea fishes. *Annu. Rev. Mar. Sci.*, *9*, 337–366, 2017.
9. Proud, R., Handegard, N.O., Kloser, R.J., Cox, M.J., Brierley, A.S., From siphonophores to deep scattering layers: uncertainty ranges for the estimation of global mesopelagic fish biomass. *ICES J. Mar. Sci.*, *76*, 3, 718–733, 2019.
10. Sinclair, E.H., Pelland, N.A., Johnson, D.S., Community composition and spatial energetics of mesopelagic fishes and squids in the eastern Bering Sea as influenced by habitat variables. *Deep Sea Res. Part I*, *182*, 103704, 2022.
11. Olsen, R.E., Strand, E., Melle, W., Nørstebø, J.T., Lall, S.P., Ringø, E., Sprague, M., Can mesopelagic mixed layers be used as feed sources for salmon aquaculture? *Deep Sea Res. Part II*, *180*, 104722, 2020.
12. Raja, K., Suresh, K., Anbalagan, S., Ragini, Y.P., Kadirvel, V., Investigating the nutritional viability of marine-derived protein for sustainable future development. *Food Chem.*, *448*, 139087, 2024.
13. Leone, A., Lecci, R.M., Durante, M., Meli, F., Piraino, S., The bright side of gelatinous blooms: Nutraceutical value and antioxidant properties of three Mediterranean jellyfish (Scyphozoa). *Mar. Drugs*, *13*, 8, 4654–4681, 2015.
14. Dall'Olmo, G., Dingle, J., Polimene, L., Brewin, R.J.W., Claustre, H., Substantial energy input to the mesopelagic ecosystem from the seasonal mixed-layer pump. *Nat. Geosci.*, *9*, 11, 820–823, 2016.

15. Wan, P., Cai, B., Chen, H., Chen, D., Zhao, X., Yuan, H., Pan, J., Antidiabetic effects of protein hydrolysates from Trachinotus ovatus and identification and screening of peptides with α-amylase and DPP-IV inhibitory activities. *Curr. Res. Food Sci.*, *6*, 100446, 2023.
16. Rivero-Pino, F., Espejo-Carpio, F.J., Guadix, E.M., Production and identification of dipeptidyl peptidase IV (DPP-IV) inhibitory peptides from discarded Sardine pilchardus protein. *Food Chem.*, *328*, 127096, 2020.
17. Harnedy, P.A., Parthsarathy, V., McLaughlin, C.M., O'Keeffe, M.B., Allsopp, P.J., McSorley, E.M., FitzGerald, R.J., Atlantic salmon (Salmo salar) co-product-derived protein hydrolysates: A source of antidiabetic peptides. *Food Res. Int.*, *106*, 598–606, 2018.
18. Acosta, J., Roa, F., González-Chavarría, I., Astuya, A., Maura, R., Montesino, R., Valenzuela, A., *In vitro* immunomodulatory activities of peptides derived from Salmo salar NK-lysin and cathelicidin in fish cells. *Fish Shellfish Immunol.*, *88*, 587–594, 2019.
19. Narayanasamy, A., Balde, A., Raghavender, P., Shashanth, D., Abraham, J., Joshi, I., Nazeer, R.A., Isolation of marine crab (Charybdis natator) leg muscle peptide and its anti-inflammatory effects on macrophage cells. *Biocatal. Agric. Biotechnol.*, *25*, 101577, 2020.
20. Naik, A.S., Whitaker, R.D., Albrektsen, S., Solstad, R.G., Thoresen, L., Hayes, M., Mesopelagic fish protein hydrolysates and extracts: A source of novel anti-hypertensive and anti-diabetic peptides. *Front. Mar. Sci.*, *8*, 719608, 2021.
21. Qian, B., Zhao, X., Yang, Y., Tian, C., Antioxidant and anti-inflammatory peptide fraction from oyster soft tissue by enzymatic hydrolysis. *Food Sci. Nutr.*, *8*, 7, 3947–3956, 2020.
22. Hwang, D., Kang, M., Jo, M.J., Seo, Y.B., Park, N.G., Kim, G.-D., Anti-inflammatory activity of β-thymosin peptide derived from pacific oyster (Crassostrea gigas) on NO and PGE2 production by down-regulating NF-κB in LPS-induced RAW264. 7 macrophage cells. *Mar. Drugs*, *17*, 2, 129, 2019.
23. Park, S.Y., Fernando, I.P.S., Han, E.J., Kim, M.J., Jung, K., Kang, D.-S., Ahn, G., *In vivo* hepatoprotective effects of a peptide fraction from krill protein hydrolysates against alcohol-induced oxidative damage. *Mar. Drugs*, *17*, 12, 690, 2019.
24. Wang, Z., Ying, X., Gao, P., Wang, C., Wang, Y., Yu, X., Luo, H., Anti-inflammatory activity of a peptide from skipjack (Katsuwonus pelamis). *Mar. Drugs*, *17*, 10, 582, 2019.
25. Kumar, L.V., Shakila, R.J., Jeyasekaran, G., *In vitro* anti-cancer, anti-diabetic, anti-inflammation and wound healing properties of collagen peptides derived from unicorn leatherjacket (Aluterus monoceros) at different hydrolysis. *Turk. J. Fish. Aquat. Sci.*, *19*, 7, 551–560, 2019.
26. Irigoien, X., Klevjer, T.A., Røstad, A., Martinez, U., Boyra, G., Acuña, J.L., Hernandez-Leon, S., Large mesopelagic fishes biomass and trophic efficiency in the open ocean. *Nat. Commun.*, *5*, 1, 3271, 2014.

27. Eduardo, L.N., Bertrand, A., Mincarone, M.M., Santos, L.V., Fredou, T., Assunção, R.V., Le Loc'h, F., Hatchetfishes (Stomiiformes: Sternoptychidae) biodiversity, trophic ecology, vertical niche partitioning and functional roles in the western Tropical Atlantic. *Prog. Oceanogr.*, *187*, 102389, 2020.
28. Csirke, J., Review of the state of world fishery resources: Marine fisheries. *FAO Fish. Circ.,, 920*, 1, 173, 1997.
29. Gjøsaeter, J. and Kawaguchi, K., A review of the world resources of mesopelagic fish. *FAO Fish. Tech. Pap.*, 193, 1–151, 1980.
30. Standal, D. and Grimaldo, E., Institutional nuts and bolts for a mesopelagic fishery in Norway. *Mar. Policy*, *119*, 104043, 2020.
31. Fjeld, K., Tiller, R., Grimaldo, E., Grimsmo, L., Standal, I.-B., Mesopelagics–New gold rush or castle in the sky? *Mar. Policy*, *147*, 105359, 2023.
32. Hill, R.L., Hydrolysis of proteins. *Adv. Protein Chem.*, 20, 37–107, 1965.
33. Kristinsson, H.G. and Rasco, B.A., Fish protein hydrolysates: Production, biochemical, and functional properties, in: *Critical Reviews in Food Science and Nutrition*, vol. 40, 2000, https://doi.org/10.1080/10408690091189266.
34. Wisuthiphaet, N., Klinchan, S., Kongruang, S., Fish protein hydrolysate production by acid and enzymatic hydrolysis. *Appl. Sci. Eng. Prog.*, *9*, 4, 261–270, 2016.
35. Le Gouic, A.V., Harnedy, P.A., FitzGerald, R.J., Bioactive peptides from fish protein by-products, in: *Bioactive Molecules in Food*, J.M. Mérillon and K.G. Ramawat (eds.), Reference Series in Phytochemistry, Springer, Cham, pp. 1–35, 2019.
36. Wang, T., Wu, J., Yi, Y., Qi, J., Optimization of process conditions for infected animal tissues by alkaline hydrolysis technology. *Procedia Environ. Sci.*, *31*, 366–374, 2016.
37. Oliyaei, N., Ghorbani, M., Moosavi-Nasab, M., Sadeghimahoonak, A.R., Maghsoudloo, Y., Effect of temperature and alkaline pH on the physicochemical properties of the protein isolates extracted from the whole ungutted Lanternfish (Benthosema pterotum). *J. Aquat. Food Prod. Technol.*, *26*, 10, 1134–1143, 2017.
38. Goswami, B., Majumdar, S., Dutta, R., Bhowal, J., Optimization of enzymatic hydrolysis of Pleurotus ostreatus derived proteins through RSM and evaluation of nutritional and functional qualities of mushroom protein hydrolysates. *Braz. J. Food Technol.*, *25*, e2020186, 2022.
39. Ketemepi, H.K., Awang, M.A., Seelan, J.S.S., Noor, N.Q., II, Extraction process and applications of mushroom-derived protein hydrolysate: A comprehensive review. *Future Foods*, 9, 100359, 1–36, 2024.
40. Venkatesan, J., Anil, S., Kim, S.-K., Shim, M.S., Marine fish proteins and peptides for cosmeceuticals: A review. *Mar. Drugs*, *15*, 5, 143, 2017.
41. Ulug, S.K., Jahandideh, F., Wu, J., Novel technologies for the production of bioactive peptides. *Trends Food Sci. Technol.*, *108*, 27–39, 2021.

42. Aguilar, J.G., dos, S., Granato Cason, V., de Castro, R.J.S., Improving antioxidant activity of black bean protein by hydrolysis with protease combinations. *Int. J. Food Sci. Technol.*, *54*, 1, 34–41, 2019.
43. Goertzen, A.D., Nickerson, M.T., Tanaka, T., The improvement of the functional properties of a chickpea protein isolate through proteolysis with three proteases. *Cereal Chem.*, *98*, 3, 439–449, 2021.
44. Ding, Y., Chen, L., Shi, Y., Akhtar, M., Chen, J., Ettelaie, R., Emulsifying and emulsion stabilizing properties of soy protein hydrolysates, covalently bonded to polysaccharides: The impact of enzyme choice and the degree of hydrolysis. *Food Hydrocoll.*, *113*, 106519, 2021.
45. Pal, G.K. and Suresh, P.V., Sustainable valorisation of seafood by-products: Recovery of collagen and development of collagen-based novel functional food ingredients. *Innov. Food Sci. Emerg. Technol.*, *37*, 201–215, 2016.
46. Agyei, D. and Danquah, M.K., Industrial-scale manufacturing of pharmaceutical-grade bioactive peptides. *Biotechnol. Adv.*, *29*, 3, 272–277, 2011.
47. Rani, S., Pooja, K., Pal, G.K., Exploration of rice protein hydrolysates and peptides with special reference to antioxidant potential: Computational derived approaches for bio-activity determination. *Trends Food Sci. Technol.*, *80*, November 2017, 61–70, 2018. https://doi.org/10.1016/j.tifs.2018.07.013.
48. Saiz, L.C., Gorricho, J., Garjon, J., Celaya, M.C., Erviti, J., Leache, L., Blood pressure targets for the treatment of people with hypertension and cardiovascular disease. *Cochrane Database Syst. Rev.*, 11, 1–65, 2022.
49. Qi, Y., Han, X., Zhao, D., Wang, W., Wang, M., Sun, J., Hao, Y., Long-term cardiovascular risk associated with stage 1 hypertension defined by the 2017 ACC/AHA hypertension guideline. *J. Am. Coll. Cardiol.*, *72*, 11, 1201–1210, 2018.
50. Masenga, S.K. and Kirabo, A., Hypertensive heart disease: risk factors, complications and mechanisms. *Front. Cardiovasc. Med.*, *10*, 1205475, 2023.
51. Akmaljon o'g, M.A., Abdullajon o'g'li, M.S., Tolmasovich, T.R., Acute Disturbance of Blood Circulation in The Head. *West. Eur. J. Med. Med. Sci.*, *2*, 4, 27–31, 2024.
52. Wajngarten, M. and Silva, G.S., Hypertension and stroke: update on treatment. *Eur. Cardiol. Rev.*, *14*, 2, 111, 2019.
53. Ameer, O.Z., Hypertension in chronic kidney disease: What lies behind the scene. *Front. Pharmacol.*, *13*, 949260, 2022.
54. Thiagarajah, R., Kandasamy, R., Sellamuthu, P., Hypertensive Retinopathy and the Risk of Hemorrhagic Stroke. *J. Korean Neurosurg. Soc.*, *64*, 4, 543, 2021.
55. Soyoye, D.O., Abiodun, O.O., Ikem, R.T., Kolawole, B.A., Akintomide, A.O., Diabetes and peripheral artery disease: A review. *World J. Diabetes*, *12*, 6, 827, 2021.
56. Correia, M.C., Ogola, E.N., Kayima, J.K., Joshi, M.D., Silverstein, D.M., Kabinga, S.K., Erectile dysfunction in hypertensive males in Kenya: a tertiary referral hospital experience. *Afr. Health Sci.*, *22*, 2, 420–427, 2022.

57. Ungvari, Z., Toth, P., Tarantini, S., Prodan, C., II, Sorond, F., Merkely, B., Csiszar, A., Hypertension-induced cognitive impairment: from pathophysiology to public health. *Nat. Rev. Nephrol.*, *17*, 10, 639–654, 2021.
58. Li, G.-H., Le, G.-W., Shi, Y.-H., Shrestha, S., Angiotensin I–converting enzyme inhibitory peptides derived from food proteins and their physiological and pharmacological effects. *Nutr. Res.*, *24*, 7, 469–486, 2004.
59. Bora, S., Evren, V., Emek, S., Çakırlar, I., Agent-based modeling and simulation of blood vessels in the cardiovascular system. *Simulation*, *95*, 4, 297–312, 2019.
60. Bryan, N.S., Nitric oxide deficiency is a primary driver of hypertension. *Biochem. Pharmacol.*, *206*, 115325, 2022.
61. Chai, H.-J., Chan, Y.-L., Li, T.-L., Shiau, C.-Y., Wu, C.-J., Evaluation of lanternfish (Benthosema pterotum) hydrolysates as antioxidants against hydrogen peroxide induced oxidative injury. *Food Res. Int.*, *54*, 2, 1409–1418, 2013.
62. Coppola, D., Lauritano, C., Palma Esposito, F., Riccio, G., Rizzo, C., de Pascale, D., Fish waste: From problem to valuable resource. *Mar. Drugs*, *19*, 2, 116, 2021.
63. Hayes, M., Mora, L., Hussey, K., Aluko, R.E., Boarfish protein recovery using the pH-shift process and generation of protein hydrolysates with ACE-I and antihypertensive bioactivities in spontaneously hypertensive rats. *Innov. Food Sci. Emerg. Technol.*, *37*, 253–260, 2016.
64. Voronin, V.P., Artemenkov, D.V., Orlov, A.M., Murzina, S.A., Lipids and fatty acids in some mesopelagic fish species: General characteristics and peculiarities of adaptive response to deep-water habitat. *J. Mar. Sci. Eng.*, *10*, 7, 949, 2022.
65. Matić, J., Bøgwald, I., Tengstrand, E., Rønning, S.B., Afseth, N.K., Wubshet, S.G., Calanus finmarchicus as a novel source of health-promoting bioactive peptides: Enzymatic protein hydrolysis, characterization, and *in vitro* bioactivity. *Biocatal. Agric. Biotechnol.*, *52*, 102820, 2023.
66. Whitaker, R.D., Altintzoglou, T., Lian, K., Fernandez, E.N., Marine bioactive peptides in supplements and functional foods-A commercial perspective. *Curr. Pharm. Des.*, *27*, 11, 1353–1364, 2021.
67. Verma, S., Goand, U.K., Husain, A., Katekar, R.A., Garg, R., Gayen, J.R., Challenges of peptide and protein drug delivery by oral route: Current strategies to improve the bioavailability. *Drug Dev. Res.*, *82*, 7, 927–944, 2021.
68. Jia, L., Wang, L., Liu, C., Liang, Y., Lin, Q., Bioactive peptides from foods: Production, function, and application. *Food Funct.*, *12*, 16, 7108–7125, 2021.
69. Xu, Q., Hong, H., Wu, J., Yan, X., Bioavailability of bioactive peptides derived from food proteins across the intestinal epithelial membrane: A review. *Trends Food Sci. Technol.*, *86*, 399–411, 2019.
70. Abdelhedi, O. and Nasri, M., Basic and recent advances in marine antihypertensive peptides: Production, structure-activity relationship and bioavailability. *Trends Food Sci. Technol.*, *88*, 543–557, 2019.

71. Cunha, S.A. and Pintado, M.E., Bioactive peptides derived from marine sources: Biological and functional properties. *Trends Food Sci. Technol.*, *119*, 348–370, 2022.
72. Jogi, N., Yathisha, U.G., Bhat, I., Mamatha, B.S., Antihypertensive activity of orally consumed ACE-I inhibitory peptides. *Crit. Rev. Food Sci. Nutr.*, *62*, 32, 8986–8999, 2022.
73. Navaf, M., Sunooj, K.V., Aaliya, B., Sudheesh, C., Akhila, P.P., Mir, S.A., Khaneghah, A.M., Contemporary insights into the extraction, functional properties, and therapeutic applications of plant proteins. *J. Agric. Food Res.*, 14, 100861, 2023.
74. Šimat, V., Elabed, N., Kulawik, P., Ceylan, Z., Jamroz, E., Yazgan, H., Özogul, F., Recent advances in marine-based nutraceuticals and their health benefits. *Mar. Drugs*, *18*, 12, 627, 2020.
75. van der Koog, L., Gandek, T.B., Nagelkerke, A., Liposomes and extracellular vesicles as drug delivery systems: A comparison of composition, pharmacokinetics, and functionalization. *Adv. Healthcare Mater.*, *11*, 5, 2100639, 2022.
76. Mehrdadi, S., Lipid-based nanoparticles as oral drug delivery systems: Overcoming poor gastrointestinal absorption and enhancing bioavailability of peptide and protein therapeutics. *Adv. Pharm. Bull.*, *14*, 1, 48, 2024.
77. Awuchi, C.G., Echeta, C.K., Igwe, V.S., Diabetes and the nutrition and diets for its prevention and treatment: a systematic review and dietetic perspective. *Health Sci. Res.*, *6*, 1, 5–19, 2020.
78. Pearce, I., Simó, R., Lövestam-Adrian, M., Wong, D.T., Evans, M., Association between diabetic eye disease and other complications of diabetes: implications for care. A systematic review. *Diabetes Obes. Metab.*, *21*, 3, 467–478, 2019.
79. Eijaz, S., Lalpuria, S., Afzal, M., Yasmeen, A., Saleem, S., Begum, A., Prevalence of Complications Associated with Diabetes among Pakistani patients: a questionnaire-based Survey. *Curr. Diabetes Rev.*, *18*, 9, 87–95, 2022.
80. Umpierrez, G. and Korytkowski, M., Diabetic emergencies—ketoacidosis, hyperglycaemic hyperosmolar state and hypoglycaemia. *Nat. Rev. Endocrinol.*, *12*, 4, 222–232, 2016.
81. Mukhtar, Y., Galalain, A., Yunusa, U., A modern overview on diabetes mellitus: a chronic endocrine disorder. *Eur. J. Biol.*, *5*, 2, 1–14, 2020.
82. Nikpour, S., Mehrdad, N., Sanjari, M., Aalaa, M., Heshmat, R., Mafinejad, M.K., Ghezeljeh, T.N., Challenges of type 2 diabetes mellitus management from the perspective of patients: Conventional content analysis. *Interact. J. Med. Res.*, *11*, 2, e41933, 2022.
83. Chiang, J.L., Maahs, D.M., Garvey, K.C., Hood, K.K., Laffel, L.M., Weinzimer, S.A., Schatz, D., Type 1 diabetes in children and adolescents: a position statement by the American Diabetes Association. *Diabetes Care*, *41*, 9, 2026, 2018.

84. Green, A., Hede, S.M., Patterson, C.C., Wild, S.H., Imperatore, G., Roglic, G., Beran, D., Type 1 diabetes in 2017: global estimates of incident and prevalent cases in children and adults. *Diabetologia*, *64*, 2741–2750, 2021.
85. Fralick, M., Jenkins, A.J., Khunti, K., Mbanya, J.C., Mohan, V., Schmidt, M., II, Global accessibility of therapeutics for diabetes mellitus. *Nat. Rev. Endocrinol.*, *18*, 4, 199–204, 2022.
86. Roep, B.O., Thomaidou, S., van Tienhoven, R., Zaldumbide, A., Type 1 diabetes mellitus as a disease of the β-cell (do not blame the immune system?). *Nat. Rev. Endocrinol.*, *17*, 3, 150–161, 2021.
87. Taylor, R., Type 2 diabetes and remission: practical management guided by pathophysiology. *J. Intern. Med.*, *289*, 6, 754–770, 2021.
88. Misra, S. and Owen, K.R., Genetics of monogenic diabetes: present clinical challenges. *Curr. Diabetes Rep.*, *18*, 1–11, 2018.
89. McIntyre, H.D., Catalano, P., Zhang, C., Desoye, G., Mathiesen, E.R., Damm, P., Gestational diabetes mellitus. *Nat. Rev. Dis. Primers*, *5*, 1, 47, 2019.
90. Davies, M.J., Aroda, V.R., Collins, B.S., Gabbay, R.A., Green, J., Maruthur, N.M., Mingrone, G., Management of hyperglycaemia in type 2 diabetes, 2022. A consensus report by the American Diabetes Association (ADA) and the European Association for the Study of Diabetes (EASD). *Diabetologia*, *65*, 12, 1925–1966, 2022.
91. Lambrinou, E., Hansen, T.B., Beulens, J.W.J., Lifestyle factors, self-management and patient empowerment in diabetes care. *Eur. J. Prev. Cardiol.*, *26*, 2_suppl, 55–63, 2019.
92. Jiang, S., Liu, A., Ma, W., Liu, X., Luo, P., Zhan, M., Zhang, J., Lactobacillus gasseri CKCC1913 mediated modulation of the gut–liver axis alleviated insulin resistance and liver damage induced by type 2 diabetes. *Food Funct.*, *14*, 18, 8504–8520, 2023.
93. Ehrmann, D., Kulzer, B., Roos, T., Haak, T., Al-Khatib, M., Hermanns, N., Risk factors and prevention strategies for diabetic ketoacidosis in people with established type 1 diabetes. *Lancet Diabetes Endocrinol.*, *8*, 5, 436–446, 2020.
94. Karges, B., Tittel, S.R., Bey, A., Freiberg, C., Klinkert, C., Kordonouri, O., Holl, R.W., Continuous glucose monitoring versus blood glucose monitoring for risk of severe hypoglycaemia and diabetic ketoacidosis in children, adolescents, and young adults with type 1 diabetes: a population-based study. *Lancet Diabetes Endocrinol.*, *11*, 5, 314–323, 2023.
95. Parthsarathy, V., McLaughlin, C.M., Harnedy, P.A., Allsopp, P.J., Crowe, W., McSorley, E.M., O'Harte, F.P.M., Boarfish (Capros aper) protein hydrolysate has potent insulinotropic and GLP-1 secretory activity *in vitro* and acute glucose lowering effects in mice. *Int. J. Food Sci. Technol.*, *54*, 1, 271–281, 2019.
96. Callieri, C., Slabakova, V., Dzhembekova, N., Slabakova, N., Peneva, E., Cabello-Yeves, P.J., Corno, G., The mesopelagic anoxic Black Sea as an unexpected habitat for Synechococcus challenges our understanding of global "deep red fluorescence". *ISME J.*, *13*, 7, 1676–1687, 2019.

97. Elgaoud, I., Hamed, F., Lassoued, I., Chamkha, M., Oulahal, N., Degraeve, P., Barkia, A., Production of hydrolysates from swordfish (Xiphias gladius) head muscle as new protein source: evaluation of nutritional, antioxidant and functional properties. *Waste Biomass Valori.*, *15*, 2, 1065–1080, 2024.
98. Cavan, E.L., Belcher, A., Atkinson, A., Hill, S.L., Kawaguchi, S., McCormack, S., Schmidt, K., The importance of Antarctic krill in biogeochemical cycles. *Nat. Commun.*, *10*, 1, 4742, 2019.
99. Zheng, K., Wu, Y., Dai, Q., Yan, X., Liu, Y., Sun, D., Jiang, W., Extraction, identification, and molecular mechanisms of α-glucosidase inhibitory peptides from defatted Antarctic krill (Euphausia superba) powder hydrolysates. *Int. J. Biol. Macromol.*, *266*, 131126, 2024.

8

Mesopelagic Fishes: Potential Use for Anticancer and Antimicrobial Biotechnological Applications

Divya Kumari and Pracheta Janmeda*

Department of Bioscience and Biotechnology, Banasthali Vidyapith, Rajasthan, India

Abstract

The mesopelagic zone of the ocean, with its profusion of fish and zooplankton, offers chances for sustainable new product creation by serving as a source of unique raw materials. There is diversity in the bio-composition of these species due to the unique light and pressure circumstances in this ecological zone as well as their specific place in the marine food chain. Massive biomass aggregations made up of mesopelagic creatures support numerous top predators as well as significant pelagic trophic webs. Rich in proteins and omega-3 fatty acids, mesopelagic fish are an underappreciated resource that offer potential for the development of novel feed, food, and functional food components and products. On the other hand, care must be taken to prevent overfishing of this resource and to maximize the catch while adhering to sustainability objectives. Even though there are a few studies on the biology, ecology, and occurrence of these species, relatively little research has been done on how they might be used in biotechnological applications for antibacterial and anticancer purposes. Mesopelagic fishes, with their distinct biochemical and physiological adaptations, are a viable source of antibacterial and anticancer drugs. Research on their possible uses in biotechnology and medicine is still quite interesting. The use of several mesopelagic species for possible antibacterial and anticancer biotechnological applications is going to be examined in this chapter.

Keywords: Biotechnology, cancer, mesopelagic fishes, microbiology, omega-3 fatty acids

**Corresponding author*: pracheta@banasthali.in

Shaju S. S., Vipin. P. M. and Visakh. P.M. (eds.) Marine Biotechnology: Studies and Applications, (239–260) © 2025 Scrivener Publishing LLC

8.1 Introduction

The word "mesopelagic zone" describes the intermediate zone of the ocean, which is located between 200 and 1,000 m below the surface. These fish have evolved to withstand the intense pressure and low light levels of their surroundings. Due to their unusual adaptations and biochemical characteristics, which may have antibacterial and anticancer effects, they have attracted attention for possible health advantages. Due to the presence of omega-3 fatty acids, such as eicosapentaenoic acid (EPA) and docosahexaenoic acid (DHA), certain zooplankton species, including krill, are currently harvested and processed from this zone. Indeed, krill-derived omega-3 supplements are widely recognized for their associated health benefits and may be easily found in health stores. While mesopelagic fish are not appropriate for traditional culinary preparations due to their small size and bony shape, unlike krill, they may be useful as functional food ingredients. Furthermore, researchers find these understudied protein sources worth investigating further in accordance with policy directives, especially those stated by the UN, due to the worldwide demand for new, sustainable protein to feed the expanding population. *Benthosema glaciale*, also known as the Glacier lantern fish, *Maurolicus muelleri*, also known as the Silvery light fish, Northern Krill, and myctophids are the principal species of mesopelagic fish. According to some research done so far, these fish are an excellent source of minerals, bioactives, lipids, proteins, and lipids [1–3]. The mesopelagic species *Micromesistius poutassou*, also known as blue whiting, was previously reported to have an average protein level of 16.1 g/100 g of fish, while *M. muelleri* and *B. glaciale* were found to have average protein values of 12.3 and 14 g/100 g, respectively. According to earlier reports, total fat values for *M. muelleri* varied from 17.8 g/100 g of fish to 3.9 g/100 g of fish for *M. poutassou*.

Processing problems arise, nevertheless, because fish hauls for mesopelagic species frequently comprise mixed species. For instance, it was discovered in earlier research that amphipod mixtures found in fish hauls for mesopelagic species in open waters off the coast of Norway exceeded the maximum permissible level of cadmium in feedstuffs when the recovered biomass was processed into fish feed [4]. Thus, if the resource is managed properly, processing and refining mesopelagic fish species for bioactive compounds, like peptides and amino acids, may represent an economical and nutritionally sustainable approach to using this resource, in addition to improving catch and processing methods. There are an estimated 6–10 billion metric tons of mesopelagic fish in the world [5]. Numerous international projects are still assessing the estimating techniques and biomass. Because marine proteins contain essential amino acids, such as leucine, which is known to promote muscle growth and may be used in the development of products to prevent sarcopenia, they have excellent nutritional value and protein quality [6, 7]. One established technique for producing highly digestible and accessible proteins, peptides, and amino acids from fish and fishery by-products is enzymatic hydrolysis using enzymes such as pepsin, alcalase R, protamex R, and others. The final composition of the generated hydrolysate in terms of peptide and amino acid profiles is significantly influenced by the kind of enzyme, rate of hydrolysis, and beginning content of raw materials. Fish-derived bioactive peptides with hydrolysates are thought to be helpful in preventing conditions including hypertension, type 2 diabetes, and other disorders linked to the metabolic syndrome, such as inflammation, high cholesterol, and aging [8].

Although mesopelagic fish species are abundant, there are not many papers about using them to create protein-rich, bioactive hydrolysates that may be able to prevent common ailments in humans and animals. The Mediterranean krill *Meganyctiphanes norvegica* and the lanternfish *Myctophum punctatum* were found to be sources of antimicrobial and anticancer fractions rich in EPA, DHA, and eicosatetraenoic acid (ETA) in a study conducted by Lauritano *et al.* [3]. To our knowledge, mesopelagic animals have not, however, ever been shown to be effective in treating diseases that affect humans. It is critical to discover novel sources of bioactive chemicals, particularly for diseases that are resistant to drugs. The ongoing emergence and development of novel antibiotic-resistant organisms serves as a powerful driving force behind the hunt for novel bioactive chemicals from understudied habitats. This chapter's goal was to look at the possible applications of different mesopelagic species for biotechnological antibacterial and anticancer applications.

8.2 Mesopelagic Fishes

The mesopelagic zone, a unique layer of the ocean with particular environmental characteristics, is home to mesopelagic fishes. This is a thorough examination of their habitat:

Mesopelagic Zone

The depth of the mesopelagic zone is between 200 and 1,000 m below the ocean's surface. This zone is located above the bathypelagic (deep-sea) zone and below the epipelagic (or sunshine) zone. The mesopelagic zone obtains very little light; most of it comes from downwelling sunlight and bioluminescent creatures. Because of the extremely low light levels, the zone is sometimes known as the twilight zone. Natural light declines with depth. The mesopelagic zone has somewhat steady water temperatures; however, they can go very cold at 4°C to 10°C (39°F to 50°F). Ocean currents and geographic location have a small impact on temperature. The mesopelagic zone has pressures between 20 and 100 atm (2,000 and 10,000 kPa). Mesopelagic fish have evolved special adaptations to withstand the high pressure at these depths, including flexible bodies and enzymes that can withstand pressure. Despite occasionally being lower than in the ocean's top layers, the mesopelagic zone has enough oxygen to support a variety of living forms. According to Sutton *et al.* [9], the zone is frequently a transitional zone where oxygen levels drop but do not drop as much as they do in the deeper bathypelagic zone.

Habitat Characteristics

Fish that are mesopelagic have evolved a number of adaptations to deal with low light levels. Many have huge eyes to capture as much light as possible, and some have bioluminescent organs to aid in communication and predation. An essential tier of the ocean's food chain is the mesopelagic zone. Many species that feed on tiny zooplankton and other invertebrates use it as a home. To graze on phytoplankton and zooplankton in the higher layers throughout the night, mesopelagic fish frequently travel upward. Vertical migration is a common feature of mesopelagic fish, which travels up to the epipelagic zone at night to feed and then descends to deeper waters during the day. They can evade predators and seize eating opportunities thanks to this behavior. There are different types of predator–prey partnerships in the mesopelagic zone. Larger fish, squid, and marine mammals feed on mesopelagic fishes, which in turn feed on smaller zooplankton and invertebrates [10–13].

Environmental Conditions and Notable Adaptations

Similar to surface waters, the salinity in the mesopelagic zone is normally steady but can fluctuate slightly depending on local factors and ocean currents. The flow of organic material, including dead plankton and other detritus, from the higher layers downward can make the zone nutrient rich. This helps mesopelagic fishes have access to food [14]. Bioluminescent organs are found in many mesopelagic fish species, including lanternfishes and dragonfishes, which aid in their ability to conceal, attract prey, and communicate. These fish, which have big eyes to enhance their sense of light and movement, are designed to live in low light. Long, fang-like teeth are used by some species, such as dragonfishes, to catch prey in the murky depths [15].

8.3 Mesopelagic Fishes: Potential Use for Anticancer Applications

Fish that are mesopelagic create a variety of bioactive substances that may be anticancer. These comprise peptides, proteins, and lipids that have the ability to stop the proliferation of cancer cells or cause cancer cells to undergo programmed cell death or apoptosis. These substances have the ability to directly kill cancer cells or alter the signaling pathways that advance the disease (Figure 8.1). A number of fish peptides have been found to have the ability to either stimulate cancer cell death or prevent it from proliferating. Because of its capacity to promote tissue regeneration and regulate immunological responses, collagen produced from mesopelagic fish has shown promise and may have anticancer effects. The family Myctophidae includes a large number of mesopelagic fish that are high in omega-3 fatty acids, including EPA and DHA. Numerous studies have been conducted on the anticancer potential of omega-3 fatty acids. They can affect the metabolism of cancer cells, lessen inflammation, and stop the formation of tumors. The anticancer effects of these omega-3 fatty acids,

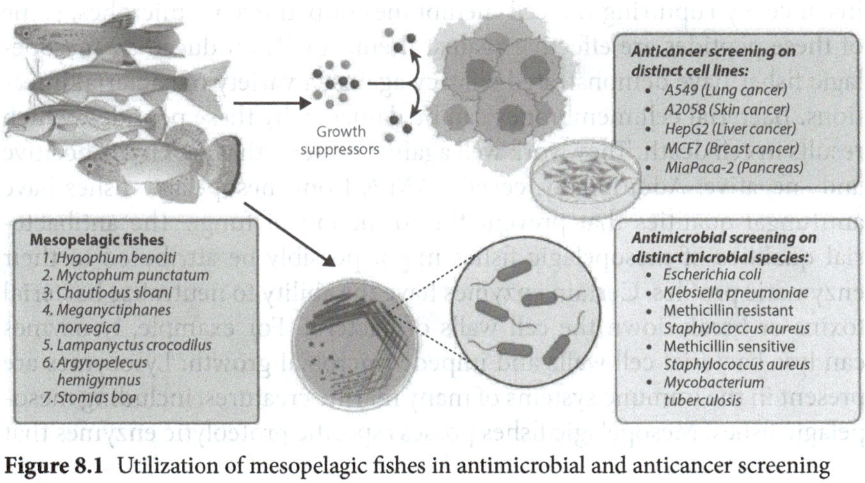

Figure 8.1 Utilization of mesopelagic fishes in antimicrobial and anticancer screening tests.

which are prevalent in mesopelagic fishes, have been well researched. EPA and DHA have the ability to stop the formation of cancer cells, encourage apoptosis, and lessen inflammation, which is a major cause of cancer. They might also make traditional cancer treatments more successful.

Antioxidants, like astaxanthin, which can counteract free radicals and lessen oxidative stress, a component that leads to the development of cancer, may also be present in high concentrations in these fish. Some mesopelagic fishes contain astaxanthin, a strong antioxidant that has been shown to lower oxidative stress and neutralize free radicals, both of which are linked to cancer. Astaxanthin may be able to shield DNA from deterioration and stop the growth of cancer cells. Additional antioxidant chemicals that guard against cellular damage and support anticancer action may be present in mesopelagic fishes. Certain substances derived from mesopelagic fishes have the ability to impede angiogenesis, the process by which tumors grow blood vessels. This may aid in halting the spread and growth of tumors. These fishes' bioactive chemicals have the potential to alter a number of cellular signaling pathways, including those that control the cell cycle, apoptosis, and metastasis [16].

8.4 Mesopelagic Fishes: Potential Use for Antimicrobial Applications

Antimicrobial peptides, or AMPs, are frequently produced by mesopelagic fishes as a component of their immune defense systems. These peptides have the ability to stop viruses, fungus, and bacteria from growing. For instance, by rupturing the cell membranes of pathogenic microbes, some of these peptides are effective against them. AMPs produced by mesopelagic fishes have demonstrated efficacy against a variety of bacterial infections. Bacterial cell membranes can be damaged by these peptides, which results in cell death. They work well against bacteria that are Gram-positive and -negative. Additionally, certain AMPs from mesopelagic fishes have antifungal qualities that prevent the formation of fungi. The antibacterial qualities of mesopelagic fishes might possibly be attributed to their enzymatic profiles. Certain enzymes have the ability to neutralize bacterial toxins or break down the cell walls of bacteria. For example, lysozymes can lyse bacterial cell walls and impede microbial growth. Lysozymes are present in the immune systems of many marine creatures, including mesopelagic fishes. Mesopelagic fishes possess specific proteolytic enzymes that have the ability to break down bacterial toxins or microbial proteins, hence enhancing their antibacterial properties.

These fishes may have developed strong antimicrobial compounds to defend themselves against microbial dangers in their habitat because of the harsh environmental factors found in the mesopelagic zone, such as high pressure and cold temperatures. Because of their particular habitat, mesopelagic fishes may produce novel antibacterial chemicals that are not present in other marine or terrestrial animals. The potential of these chemicals to treat resistant varieties of bacteria and other infections makes them remarkable (shown in Figure 8.1).

The extracts of the lanternfish *Myctophum punctatum* and the Mediterranean krill *Meganyctiphanes norvegica* were found to hinder the growth of methycillin-resistant *Staphylococcus aureus* (MRSA), methicillin-sensitive *Staphylococcus aureus* (MSSA), and *M. tuberculosis* in a recent study that screened mesopelagic species as a source of potential antimicrobial compounds. Both marine species' extracts were fractionated, and LC-UV analysis helped to clarify their composition. The most prevalent acids in the lipid components of both marine species were EPA, 4,7,10,13,16, and 19-docosahexaenoate (DHA, 22:6 n−3), and 8,11,14, and 17-eicosatetraenoic (ETA, 20:4 n−3). The gorgonian coral *Eunicea succinea* included the 14-methyl-5,9-pentadecadienoic FA, which was extracted and identified from the phospholipid pool. It showed efficacy against both *S. aureus* and *E. faecalis* [17].

8.5 Notable Mesopelagic Fishes with Anticancer and Antimicrobial Properties

- **Lanternfish (Family Myctophidae)**

Fish of the Myctophidae family that are mesopelagic are called myctophids. A total of 33 genera and over 250 species comprise their representation. Known as "Lanternfishes," they live in every ocean save the Arctic. They are widely recognized for their diel vertical migration between the meso- and epipelagic regions, as well as their adaptations to oxygen minimum zones (OMZs) in the upper 2,000 m. True to their name, lanternfishes have a dazzling appearance because of the photophores that are placed in a systematic manner throughout their bodies. This is only one of the many distinctive ecological characteristics of these fish. A traditional technique for capturing these fish is mid-water trawling, which typically accounts for biomass measured in million tons, as in the case of the Arabian Sea (20–100 million) or the Southern Ocean (70–200 million).

In terms of ecology, myctophids connect top predators, including squids, whales, and penguins, with primary consumers such as copepods and euphausiids. During migration, lanternfishes, together with other species including euphausiids, medusae, and fish juveniles, constitute a significant component of deep scattering layers (DSL). In their natural environments, myctophids are vulnerable to parasites, such as nematode larvae and siphonostomatoid copepods, similar to any other marine organism. Through their waste products and quickly sinking excrement, they contribute much to the organic carbon found on ocean floors. They are a cost-effective supply of minerals, lipids, and proteins that are utilized as fishmeal in agricultural fertilizers and as fodder for animals and poultry.

A small number of species are thought to be edible, although myctophids cannot be used as food for humans due to increased processing challenges. Myctophids have low fertility rates (100–2,000 eggs each spawn) and a life span of approximately 1–5 years. Additional examples include the spotted lanternfish, *Myctophum punctatum*; the winged lanternfish, *Lampanyctus alatus*; the deep-sea lanternfish, *Benthosema glaciale*; and the Risso's lanternfish, *Electrona rissoi*. Omega-3 fatty acids, like EPA and DHA, and other bioactive peptides, which have shown anticancer potential in preclinical research, are abundant in these fish and are well recognized for them. Different antimicrobial peptides and proteins are produced by lanternfishes. Their antibacterial effect may be attributed to their distinct biochemical makeup, which includes peptides and omega-3 fatty acids. *Benthosema glaciale* and *Myctophum punctatum* are two examples.

- **Dragonfishes (Family Stomiidae)**

Fish with ray fins found in deep waters, such as barbeled dragonfish, belong to the family Stomiidae. They are typically only 15 to 26 cm in size. These fish, which are apex predators, have massive jaws that are full with teeth that resemble fangs. Additionally, they have the ability to hinge the upper jaw system and neurocranium, which allows the jaw to open to a maximum of 100°. Because of this capacity, they can eat very enormous prey that is frequently 50% longer than they are normally. There are members of the Stomiidae family in every ocean. Depending on the optimal conditions for feeding and breeding in the water, they can also be found at a variety of depths within the bathypelagic zone ranging from thousands of meters below the surface.

This particular kind of deep-sea fish is capable of generating light on its own *via* a chemical process called bioluminescence. The photophore, a unique organ, contributes to the production of this light. The enormous heads and numerous sharp, fang-like teeth in their jaws are characteristics of deep-sea dragonfishes. They are chin-attached to a long, string-like structure called a barbel, which has a light-producing photophore at its tip. Additionally, photophores are affixed to the sides of their bodies. The *Chauliodus* is a particular species of Stomiidae that requires adrenaline to luminesce for longer than 30 min. On the other hand, it can produce light for several hours when adrenaline is present. They emit blue-green light, which has the longest wavelengths in the water.

To entice prey and possible mates, the deep-sea dragonfish creates flashing lights and waves its barbel back and forth. Fish in this family include the bioluminescent *Stomias boa* species; the black dragonfish, or *Malacosteus niger*, which is well-known for its bioluminescence; and the other bioluminescent dragonfish species, *Astronesthes niger*. Dragonfishes, a species of fish distinguished by their bioluminescence, are being researched for their potential anticancer properties due to the presence of several bioactive compounds, such as fatty acids and peptides. Bioactive substances and antimicrobial peptides are produced by dragonfish. Their capacity to create these substances is a response to the challenges posed by microbes in their deep-sea habitat. *Malacosteus niger* and *Stomias boa* are two examples [18, 19].

- **Bristlemouths (Family Gonostomatidae)**

Fish of this family include *Gonostoma elongatum*, which is distinguished by its long, fang-like teeth and bristle-like features; *Stomias boa*, which is recognized for its bioluminescent organs and elongated body; and *Gonostoma gracile*, which is also referred to as the slender bristlemouth. Fish with bristlemouths emit light. Bristlemouths use their bioluminescence to achieve various effects. While some use it to locate prey, others utilize it to evade predators. Since bristlemouths are protandrous, they are hermaphrodites that are first male. Some of them transition from being male to female in the beginning of their lives. Compared to female bristlemouths, males are smaller. With their powerful jaws, bristlemouths may take down larger prey than they are. Seventy percent of the length of the skull is made up of the lower jaw of the bristlemouth species *S. glarisianus*. It is impossible for bristlemouths to masticate their prey using their lower jaw. It is consequently assumed that they ingest the tail of their prey first. Bristlemouths are incredibly little, with an average size of 7.5 cm (3.0 in).

Bristlemouths feature big mouths, large jaws, short snouts, small eyes, and elongated bodies. Beginning in line with the anal fin is where the dorsal fin is located. The size of their photophores and the intensity of their pigmentation distinguish different species of bristlemouths. The morphology stays the same for most of the species. Although bristlemouths typically have dark coloring, they occasionally exhibit translucency. A pineal organ found in bristlemouths is used to sense gradually changing ambient light. As a result, the bristlemouth can regulate its seasonal behavior and circadian rhythm. The fish's small size makes them a simple meal for dragonfish and fangtooths. Omega-3 fatty acids and bioactive peptides found in bristlemouths have been studied for their ability to stop the growth of cancer cells and trigger apoptosis. The ability of these fishes to produce AMPs that are effective against a variety of bacteria and fungus has been examined, as seen in *Gonostoma elongatum* and *Stomias boa*, for instance [20].

• ***Anglerfish* (Family Lophiidae)**
Lophius piscatorius, a fish in this family that is mostly bathypelagic but also includes mesopelagic relatives, is one of the species. These are bony fish that get their name from their distinctive way of hunting, which involves luring other fish with a modified luminous fin ray called an esca or illicium. The symbiotic bacteria that live in and around the sea and are assumed to have received their luminescence from seawater are the source of the light. Omega-3 fatty acids and bioactive peptides produced by anglerfish are being investigated for their anticancer properties [21].

• **Silversides and Deep-Sea Smelts (Family Opisthoproctidae)**
Fish belonging to this family include the *Mytophum* species and *Opisthoproctus* species, which are distinguished by their transparent heads and bioluminescence, respectively. Because of their distinct biochemical makeup, which includes proteins with potential antibacterial and anticancer effects and omega-3 fatty acids, these fish are recognized for their potential.

8.5.1 Research and Biotechnological Applications

Mesopelagic fishes, which inhabit the twilight zone of the ocean (200–1,000 meters deep), have several important applications (shown in Figure 8.2). They are a key source of food in various marine ecosystems and are harvested for human consumption, particularly in Asia. Additionally, these fishes are used in the production of fishmeal for aquaculture and livestock feed. Their biochemical compounds, such as omega-3 fatty acids and antioxidants, are also of interest for the pharmaceutical and nutraceutical industries. Furthermore, studying mesopelagic fishes helps scientists understand deep-sea biodiversity and ecosystem functioning.

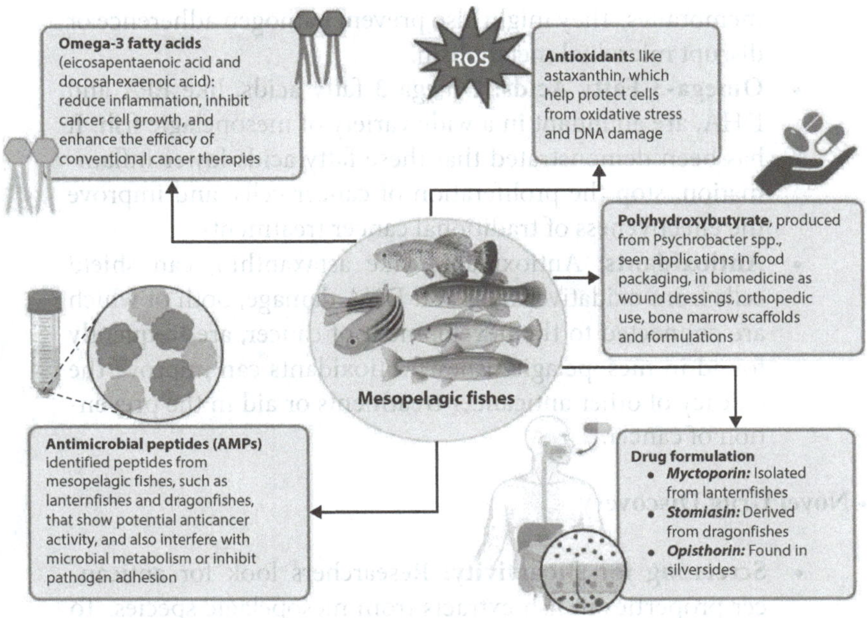

Figure 8.2 Major applications of mesopelagic fishes.

- **Unique Bioactive Compounds**
  - **Peptides and Proteins:** Fish that are mesopelagic produce a range of bioactive proteins and peptides that may have anticancer effects. Among these are peptides that have the ability to stop tumor development or cause cancer cells to undergo apoptosis, or programmed cell death. For instance, certain peptides may interfere with biological functions necessary for the growth of cancer cells. Peptides from mesopelagic fishes, like lanternfishes and dragonfishes, have been found in studies to have possible anticancer properties. These peptides have the potential to function through a number of pathways, such as apoptosis induction or tumor growth inhibition. As a component of their innate immunological defense, many mesopelagic fishes generate proteins and AMPs. These AMPs work well against a wide range of pathogens, including as viruses, fungus, and bacteria. Peptides from dragonfishes and lanternfishes, for example, have demonstrated antibacterial action. AMPs usually cause cell death in microorganisms by rupturing their cell

membranes. They might also prevent pathogen adherence or disrupt microbial metabolism.

- **Omega-3 Fatty Acids:** Omega-3 fatty acids, like EPA and DHA, are abundant in a wide variety of mesopelagic fish. It has been demonstrated that these fatty acids lower inflammation, stop the proliferation of cancer cells, and improve the effectiveness of traditional cancer treatments.
- **Antioxidants:** Antioxidants, like astaxanthin, can shield cells from oxidative stress and DNA damage, both of which are connected to the development of cancer, are frequently found in mesopelagic fishes. Antioxidants can improve the efficacy of other anticancer treatments or aid in the prevention of cancer.

- **Novel Drug Discovery**

    - **Screening for Bioactivity:** Researchers look for anticancer properties in fish extracts from mesopelagic species. To evaluate a compound's possible therapeutic effects, it is isolated and tested against cancer cell lines. The mechanisms of action, effectiveness, and safety of the chosen candidates are next investigated in further detail. The process of looking for new compounds in marine creatures is known as marine bio-prospecting. Because of their particular habitat, mesopelagic fishes create substances that are uncommon in terrestrial creatures. These new substances may be suitable candidates for the creation of antibiotic drugs.
    - **Marine Natural Products:** Mesopelagic fishes' distinct habitat stimulates the synthesis of new chemicals that are absent from terrestrial species. These novel substances may have distinct modes of action, which could make them promising candidates for brand-new cancer medications.

- **Marine Biotechnology**

    - **Genetic and Metabolic Studies:** The production of bioactive chemicals by mesopelagic fishes can be revealed by an analysis of their genetic makeup and metabolic pathways. By synthesizing or engineering similar molecules in the lab using this information, new medications may be developed.

- **Bio-Prospecting:** The process of looking for and identifying novel biological substances from marine creatures is known as marine bio-prospecting. Fish that are mesopelagic are crucial to this endeavor because they offer sources of special peptides, fatty acids, and other substances that may have anticancer effects.

- **Pharmacological Research**

    - **Mechanisms of Action:** Scientists look into the molecular effects of substances from mesopelagic fishes on cancer cells. Comprehending the working mechanisms of these substances facilitates the identification of their potential as therapeutic agents and the optimization of their use.
    - **Synergistic Effects:** Combining compounds from mesopelagic fishes with already available treatments may have synergistic benefits. Omega-3 fatty acids, for instance, may improve the efficacy of radiation or chemotherapy resulting in more efficient treatment plans.

- **Clinical Development**

    - **Preclinical Studies:** Preclinical models are used to assess the safety, effectiveness, and pharmacokinetics of promising medicines. Clinical studies are conducted after successful preclinical outcomes to evaluate the medications' potential as anticancer treatments for humans.
    - **Drug Formulation:** Based on their bioactivity and efficacy against cancer, compounds from mesopelagic fishes can be exploited to produce new medicine formulations, including injectable or oral therapy.

      Myctoporin: Derived from lanternfishes, this compound exhibits antibacterial properties against several microorganisms. Research is being done on its potential for medication development.

      Stomiasin: Stomiasin, which is derived from dragonfish, has antibacterial properties, including the ability to combat harmful fungi and bacteria.

      Opisthorin: This substance, which is present in silversides, has shown antibacterial qualities making it a promising option for additional study and development [22, 23].

8.5.2 Mesopelagic Jellyfish (*Periphylla periphylla*)

The jellyfish helmet in the mesopelagic zone, *Periphylla periphylla*, is a significant member of the zooplankton community. Their surfaces have a mucoidal layer coating that makes them an excellent carbon source that bacteria can easily use as an energy source. As a result, they provide an excellent environment for bacteria to live. Cnidarian jellyfish were once thought to be pathogen vectors because they were known to carry bacteria that are important for aquaculture diseases such tenacibaculosis in fish [24–26]. Nonetheless, jellyfish have the ability to actively or passively attract members of their microbial community underscoring the intricate and dynamic character of their relationship with microbes. The chemical defense theory helps to explain it in part; according to Yue *et al.* [27], jellyfish and/or some related microbes create antimicrobial toxins that shape the community of microbes to a limited number of taxa.

For instance, the peptide aurelin, which has antibacterial activity against both Gram-positive and -negative bacteria, is produced by *Aurelia aurita* [28].

As defense mechanisms to shield jellyfish from diseases and other microbes, jellyfish-associated bacteria have also developed antimicrobial chemicals, such as the neurotoxin tetrodotoxin and the polyketides paecilocin A–D [29].

The microbiota associated with jellyfish exhibited a varied chemistry encompassing primary and secondary metabolites, as demonstrated by the comparative untargeted metabolomics technique. All species had the same primary metabolites, which included phosphoethanolamines, purines, fatty acids, and ornithine lipids. Gram-negative bacteria are the only bacteria that create ornithine lipids (OL), which are more unique to them and are produced in greater quantities in response to phosphate being scarce [30]. *Bacillus megaterium* and *Escherichia coli* were the targets of action demonstrated by OL isolated from *Burkholderia gladioli* pv. *agaricicola* [31].

Since OL has been linked to a variety of other antimicrobial activities, these substances may have contributed to the antimicrobial activity seen in this investigation. A family of sphingosine lipids, known as sulfobacins, was initially identified from the Gram-negative *Chryseobacterium* sp.; nevertheless, no antibacterial action was observed. According to the current investigation, a *Psychrobacter* sp. isolated from a sponge named *Stelleta* sp. was likewise found to have derivatives of bile acid. According to Li *et al.* [32], seven bile acid derivatives were identified, given nuclear magnetic resonance (NMR) characterizations, and shown to have modest anticancer efficacy in addition to selective inhibitory action against a number of human diseases. According to Ma *et al.* [33], the putative polyhydroxybutyrate (PHB) class of chemicals was only generated by *Psychrobacter* spp. These polyhydroxyalkanoates have unique mechanical and thermal properties that make them excellent candidates for the production of biodegradable plastics. These properties are accumulated by a variety of microorganisms, including *Pseudohalocynthiibacter* sp., *Pseudomonas* sp., *Bacillus* sp., *Burkholderia* sp., and *Cupriavidus* sp. [34]. They have been used in the production of slow-release medications, bone marrow scaffolds, orthopedic use, food packaging, biomedicine as wound dressings, and more [35]. In a *Lemna minor* experiment, the low molecular weight PHB pentolide—which was also annotated in this study—exhibited phytotoxic action [36].

Antimicrobial properties have also been demonstrated by synthesized and modified PHB. *Shewanella* sp. SU126 was the main dominator of the diketopiperazine cluster. It is estimated that diketopiperazines are produced by 90% of Gram-negative bacteria. Additionally, Gram-positive bacteria, fungi, and higher organisms have been found to contain these cyclic dipeptides, which exhibit a variety of biological properties including cytotoxic, antifungal, and antibacterial properties.

8.5.3 Novel Bioactive Peptides Derived from Mesopelagic Fishes

In a study, mesopelagic fishes *M. muelleri* and *Benthosema* sp., hydrolyzed using the enzyme alcalase, were used to produce novel bioactive peptides with distinct amino acid sequences. For hydrolysates produced under specific conditions, such as an enzyme-to-substrate ratio of 1:100 (v/w) and a hydrolysis time of 3 h, peptides having cyclooxygenase (COX-1, COX-2) and monoacylglycerol lipase (MAGL) inhibitory activity were discovered. With an MAGL IC_{50} value of 6.35 μM, only one peptide—peptide NVGEVVCIFLTAALGLPEALI—derived from Irish M. muelleri, hydrolyzed with alcalase, inhibited the enzyme. With nitric oxide (NO) IC_{50} values of 572.096 and 674.435 mg/ml, respectively, that were reported for previously identified anti-inflammatory peptides, such as LREMLSTMCTARGA, AVGPGPRG, and VPWGPWPKG from a bromelain hydrolysate of sea cucumber, the COX and MAGL inhibitory potential of the identified peptides compares favorably with those of those previously identified [37].

Previous research has shown that anti-inflammatory peptides derived from pure amaranth containing the peptide sequences HGSEPFGPR and RPRYPWRYT can decrease the expression of LOX-1. The IC_{50} value of peptides were found to be similar to those found for mesopelagic-derived peptides against COX-1 and COX-2 [38]. QCPLHRPWAL had a selectivity index value of 10.81 and inhibited COX-1 by 82.90% (±0.54) and COX-2 by 53.84%. *In vivo* trials examining this peptide and the peptide NVGEVVCIFLTALGLPEALI are recommended due to their similarity, selectivity, and IC_{50} values. When tested at concentrations of 1 mg/ml, the whole hydrolysates produced from mesopelagic fish species separately did not exhibit COX-1 or COX-2 inhibitory activity more than 50%. Nevertheless, PreAIP-generated hydrolysates appear to be anti-inflammatory as well, according to *in silico* analysis studies. This effect could be amplified by producing molecular weight cut-off fractions to enrich the bioactive peptides. The only hydrolysate to inhibit MAGL by 62% at a dose of 1 mg/ml was the *Maurolicus muelleri* alcalase hydrolysate. It is true that a number of the bioactive peptides found in the mesopelagic hydrolysates that have been shown to have anti-inflammatory and antidiabetic properties are also expected to contribute umami flavors. The predicted values for the various peptides range from 1.68 mmol/L for peptide GFAGDDAPR to 37.67 mmol/L for peptide LACNCNLHARRCRFNM. The fifth basic taste sense, umami, is taken from the Japanese language and means "delicious savory taste" [39].

8.5.4 Anticancer and Antimicrobial Potential of Mediterranean Mesopelagic Fish Species

To our knowledge, mesopelagic animals have not, however, ever been shown to be effective in treating diseases that affect humans. It is critical to discover novel sources of bioactive chemicals, particularly for diseases that are resistant to drugs. The ongoing emergence and development of novel antibiotic-resistant organisms serves as a powerful driving force behind the hunt for novel bioactive chemicals from understudied habitats. A study was conducted to screen extracts of seven mesopelagic species for potential antibacterial (such as Gram-positive bacteria methicillin-resistant/sensitive *Staphylococcus aureus* and *Mycobacterium tuberculosis*) and anticancer (such as Gram-negative bacteria *Escherichia coli* and *Klebsiella pneumoniae*, skin cell line A2058, liver cell line HepG2, breast cell line MCF7, and pancreas cell line MiaPaca-2) properties.

One mesopelagic crustacean, *Meganyctiphanes norvegica* (Family Euphausiidae), and six mesopelagic fish, including *Hygophum benoiti*, *Lampanyctus crocodilus*, and *Myctophum punctatum* (Family Myctophidae),

Argyropelecus hemigymnus (Family Sternoptychidae), *Chauliodus sloani*, and *Stomias boa* (Family Stomiidae), were screened as part of the study. Out of all the mesopelagic organisms that were evaluated, the fish *M. punctatum* and the krill *M. norvegica* showed potent antibacterial and anticancer properties. *M. punctatum* showed greater activity against A549 (lung) and MCF7 (breast) cells in the anticancer assay, but *M. norvegica* demonstrated greater activity against HepG2 (liver) cells.

Depending on the investigated concentrations, MRSA, MSSA, and *Mycobacterium tuberculosis* growth was inhibited by extracts from *M. punctatum* and *M. norvegica* in antibacterial experiments. At extract concentrations ranging from 40 to 320 μg/ml, *M. punctatum* extract was able to completely inhibit MRSA viability. Lower concentrations resulted in less activity. Only at the highest tested concentrations (320 and 160 μg/ml) did the *M. punctatum* extract show activity against MSSA and *M. tuberculosis*. The *M. norvegica* extract showed activity against MRSA within the 80- to 320-μg/ml range, while only at the highest concentrations did it demonstrate activity against the other species. These activities represent the first time that these mesopelagic species have been identified as novel, rich sources of chemicals beneficial to human health.

8.6 Conclusions

It is imperative to acquire novel bioactive chemicals due to the growing need for innovative antimicrobial therapies to overcome antimicrobial resistance. The diversity of marine life and their biomolecules is influenced by a wide range of elements within the vast maritime environment. It was proposed that mesopelagic fish could serve as both a reservoir for the production of new, bioactive peptides and a supply of nutrition. Omega-3 fatty acids, antioxidants, and bioactive peptides found in the general biochemical profiles of mesopelagic fish species indicate that these fish may have anticancer potential, despite the paucity of particular studies on the subject. These fish are providing novel sources of antimicrobial compounds for research through bioprospecting, screening, and biotechnological applications. Research concentrating on the identification and extraction of particular bioactive substances from these fishes may open the door to novel medical interventions and prophylactic actions. To identify and assess these chemicals' potential for cancer prevention and treatment, more research is required.

References

1. Alvheim, A.R., Kjellevold, M., Strand, E., Sanden, M., Wiech, M., Mesopelagic species and their potential contribution to food and feed security-a case study from Norway. *Foods*, 9, 344–363, 2020. doi: 10.3390/foods9030344.
2. Wang, F., Wu, Y., Cui, Y., Chen, Z., Li, Z., Zhang, J., *et al.*, Factors affecting the fatty acid composition of mesopelagic fish of the continental slope in the South China Sea. *PeerJ [Preprints]*, 6, e26498v1, 2018. doi: 10.7287/peerj.preprints.26498v1.
3. Lauritano, C., Martínez, K.A., Battaglia, P., Granata, A., de la Cruz, M., Cautain, B., *et al.*, First evidence of anticancer and antimicrobial activity in Mediterranean mesopelagic species. *Sci. Rep.*, 10, 4929, 2020. doi: 10.1038/s41598-020- 61515-z.
4. Olsen, R.E., Strand, E., Melle, W., Nørstebø, J.T., Lall, S.P., Ringø, E., *et al.*, Can mesopelagic mixed layers be used as feed sources for salmon aquaculture? *Deep Sea Res. Part II Top. Stud. Oceanogr.*, 180, 104722–104732, 2020. doi: 10.1016/j.dsr2.2019.104722.
5. Irigoien, X., Klevjer, T.A., Røstad, A., Martinez, A., Boyra, G., Acuna, J.L., *et al.*, Large mesopelagic fishes biomass and trophic efficiency in the open ocean. *Nat. Commun.*, 5, 3271–3282, 2014.
6. Gigliotti, J.C., Determination of the Nutritional Value, Protein Quality and Safety of Krill Protein Concentrate Isolated Using an Isolelectric Solubilization/Precipitation Technique. *Graduate Theses, Dissertations, and Problem Reports,* West Virginia University, USA, p. 2555, 2007, Available online at: https://researchrepository.wvu.edu/etd/2555.
7. Breen, L. and Churchward-Venne, T.A., Leucine: a nutrient 'trigger' for muscle anabolism, but what more? *J. Physiol.*, 590, 2065–2066, 2012. doi: 10.1113/jphysiol.2012.230631.
8. Naik, A.S., Whitaker, R.D., Albrektsen, S., Solstad, R.G., Thoresen, L., Hayes, M., Mesopelagic Fish Protein Hydrolysates and Extracts: A Source of Novel Anti-Hypertensive and Anti-Diabetic Peptides. *Front. Mar. Sci.*, 8, 719608, 2021. doi: 10.3389/fmars.2021.719608.
9. Sutton, T.T., Clark, M.R., Dunn, D.C., Halpin, P.N., Rogers, A.D., Guinotte, J., *et al.*, A global biogeographic classification of the mesopelagic zone. *Deep-Sea Res. Pt. I Oceanogr. Res. Pap.*, 126, 85–102, 2017. http://doi.org/10.1016/j.dsr.2017.05.006.
10. Gjøsaeter, J. and Kawaguchi, K., *A Review of the World Resources of Mesopelagic Fish*, FAO, Rome, 1980.
11. Robinson, C., Steinberg, D.K., Anderson, T.R., Arístegui, J., Carlson, C.A., Frost, J.R., Ghiglione, J.F., Hernández-León, S., Jackson, G.A., Koppelmann, R., Quéguiner, B., Ragueneau, O., Rassoulzadegan, F., Robison, B.H., Tamburini, C., Tanaka, T., Wishner, K., Zhang, J., Mesopelagic zone ecology and biogeochemistry- a synthesis. *Deep-sea Res. Part II-topical Stud. Oceanogr.*, *57*, 1504–1518, 2010.

12. Kawabata, T., Lindsay, D.J., Kitamura, M., *et al.*, Evaluation of the bioactivities of water-soluble extracts from twelve deep-sea jellyfish species. *Fish. Sci.*, 79, 487–494, 2013. https://doi.org/10.1007/s12562-013-0612-y.
13. Tracey, D., Sophie, F., Saunders Ryan, A., Genner Martin, J., Large mesopelagic fish biomass in the Southern Ocean resolved by acoustic properties. *Proc. R. Soc. B.*, 289, 2021, 20211781, 2022. http://doi.org/10.1098/rspb.2021.1781.
14. Longo, F., Malara, D., Stipa, M.G., Consoli, P., Romeo, T., Sanfilippo, M., Abbate, F., Andaloro, F., Battaglia, P., Age, Growth and Otolith Microstructure of the Spotted Lanternfish Myctophum punctatum Rafinesque 1810. *J. Mar. Sci. Eng.*, 9, 801, 2021. https://doi.org/10.3390/jmse9080801.
15. Fjeld, K., Tiller, R.G., Grimaldo, E., Grimsmo, L., Standal, I.B., Mesopelagics–New gold rush or castle in the sky? *Mar. Policy*, 144, Article 105330, 2023.
16. Russo, P., Del Bufalo, A., Fini, M., Deep sea as a source of novel-anticancer drugs: update on discovery and preclinical/clinical evaluation in a systems medicine perspective. *Excli. J.*, *14*, 228–236, 2015. https://doi.org/10.17179/excli2015-632.
17. Alves, E., Dias, M., Lopes, D., Almeida, A., Domingues, M.d.R.; Rey, F. Antimicrobial Lipids from Plants and Marine Organisms: An Overview of the Current State-of-the-Art and Future Prospects. *Antibiotics*, *9*, 441, 2020. https://doi.org/10.3390/antibiotics9080441.
18. Christopher, P., Exploring Feeding Behavior In Deep-sea Dragonfishes (Teleostei: Stomiidae): Jaw Biomechanics and Functional Significance of a Loosejaw. *Biol. J. Linn. Soc.*, 106, 1, 224–240, 2012. http://doi.org/10.1111/j.1095-8312.2012.01854.x.
19. Mallefet, J., Duchatelet, L., Hermans, C., Baguet, F., Luminescence control of Stomiidae photophores. *Acta Histochem.*, 121, 1, 7–15, 2019-01-01. http://doi.org/10.1016/j.acthis.2018.10.001.
20. Přikryl, T., Prokofiev, A.M., Krzemiński, W., Feeding habits of the Oligocene bristlemouth fish Scopeloides glarisianus (Teleostei: Stomiiformes: Gonostomatidae). *Geobios*, 45, 4, 377–386, 2012. http://doi.org/10.1016/j.geobios.2011.10.012. ISSN: 0016-6995.
21. Nelson, J.S., Grande, T.C., Wilson, M.V.H., *Fishes of the World*, 5th, pp. 508–518, John Wiley & Sons, Hoboken, NJ, 2016, http://doi.org/10.1002/9781119174844.
22. St. John, M.A., Borja, A., Chust, G., Heath, M., Grigorov, I., Mariani, P., Martin, A.P., Santos, R.S., A Dark Hole in Our Understanding of Marine Ecosystems and Their Services: Perspectives from the Mesopelagic Community. *Front. Mar. Sci.*, 3, Article 31, 2016. doi: 10.3389/fmars.2016.00031.
23. Hayes, M., Naik, A., Mora, L., Iñarra, B., Ibarruri, J., Bald, C., Cariou, T., Reid, D., Gallagher, M., Dragøy, R., *et al.*, Generation, Characterisation and Identification of Bioactive Peptides from Mesopelagic Fish Protein Hydrolysates Using *In Silico* and *In Vitro* Approaches. *Mar. Drugs*, 22, 297, 2024. https://doi.org/ 10.3390/md22070297.

24. Ferguson, H.W., Christian, M.J.D., Hay, S., Nicolson, J., Sutherland, D., Crumlish, M., Jellyfish as vectors of bacterial disease for farmed salmon (Salmo Salar). *J. Vet. Diagn. Investig.*, 22, 376–382, 2010.
25. Delannoy, C.M.J., Houghton, J.D.R., Fleming, N.E.C., Ferguson, H.W., Mauve stingers (Pelagia noctiluca) as carriers of the bacterial fish pathogen Tenacibaculum maritimum. *Aquaculture*, 311, 255–257, 2011.
26. Småge, S.B., Brevik, OJ., Frisch, K., Watanabe, K., Duesund, H., Nylund, A., Concurrent jellyfish blooms and tenacibaculosis outbreaks in Northern Norwegian Atlantic salmon (Salmo salar) farms. *PLoS One*, 12, e0187476, 2017.
27. Yue, Y., Yu, H., Li, R., Xing, R., Liu, S., Li, P., Exploring the antibacterial and antifungal potential of jellyfish-associated marine fungi by cultivation-dependent approaches. *PLoS One*, 10, e0144394, 2015.
28. Ovchinnikova, T.V., Balandin, S.V., Aleshina, G.M., Tagaev, A.A., Leonova, Y.F., Krasnodembsky, E.D., Men'shenin, A.V., Kokryakov, V.N., Aurelin, a novel antimicrobial peptide from jellyfish Aurelia aurita with structural features of defensins and channel-blocking toxins. *Biochem. Biophys. Res. Commun.*, 348, 514–523, 2006.
29. Oppong-Danquah, E., Miranda, M., Blümel, M., Tasdemir, D., Bioactivity Profiling and Untargeted Metabolomics of Microbiota Associated with Mesopelagic Jellyfish Periphylla periphylla. *Mar. Drugs*, 21, 129, 2023. https://doi.org/10.3390/ md21020129.
30. Kim, S.K., Park, S.J., Li, X.H., Choi, Y.S., Im, D.S., Lee, J.H., Bacterial ornithine lipid, a surrogate membrane lipid under phosphate limiting conditions, plays important roles in bacterial persistence and interaction with host. *Environ. Microbiol.*, 20, 3992–4008, 2018.
31. Elshafie, H.S., Viggiani, L., Mostafa, M.S., El-Hashash, M.A., Camele, I., Bufo, S.A., Biological activity and chemical identification of ornithine lipid produced by Burkholderia gladioli pv. agaricicola ICMP 11096 using LC-MS and NMR analyses. *J. Biol. Res.*, 90, 1–8, 2017.
32. Li, H., Shinde, P.B., Lee, H.J., Yoo, E.S., Lee, C.-O., Hong, J., Choi, S.H., Jung, J.H., Bile acid derivatives from a sponge-associated bacterium Psychrobacter sp. *Arch. Pharmacal Res.*, 32, 857–862, 2009.
33. Ma, L., Zhang, Z., Li, J., Yang, X., Fei, B., Leung, P.H., Tao, X., A new antimicrobial agent: Poly (3-hydroxybutyric acid) oligomer. *Macromol. Biosci.*, 19, 1800432, 2019.
34. Esposito, F.P., Vecchiato, V., Buonocore, C., Tedesco, P., Noble, B., Basnett, P., de Pascale, D., Enhanced production of biobased, biodegradable, Poly (3-hydroxybutyrate) using an unexplored marine bacterium Pseudohalocynthiibacter aestuariivivens, isolated from highly polluted coastal environment. *Bioresour. Technol.*, 368, 128287, 2023.
35. Abdelwahab, M.A., Ahmed, A., El-Said, K.S., El Naggar, S.A., ElKholy, H.M., Evaluation of antibacterial and anticancer properties of poly

(3-hydroxybutyrate) functionalized with different amino compounds. *Int. J. Biol. Macromol.*, 122, 793–805, 2019.

36. Petta, T., Raichardt, L., Melo, I.S., Moraes, L.A.B., Bioassay-guided isolation of a low molecular weight PHB from Burkholderia sp. with phytotoxic activity. *Appl. Biochem. Biotechnol.*, 170, 1689–1701, 2013.
37. Ghanbari, R., Ebahimpour, M., Zarei, A., Ismail, A., Abdul-Hamid, A., Saari, N., Purification and Characterisation of Nitric Oxide Inhibitory peptides from Actinopyga lecanora through enzymatic hydrolysis. *Food Biotechnol.*, 30, 263–277, 2016.
38. Grancieri, M., Martino, H.S.D., Gonzalez de Mejia, E., Digested total protein and protein fractions from chia seed (Salvia hispanica L.) had high scavenging capacity and inhibited 5-LOX, COX-1-2, and iNOS enzymes. *Food Chem.*, 289, 204–214, 2019.
39. Mei, S., Yao, S., Mo, J., Wang, Y., Tang, J., Li, W., Wu, T., Integration of cloud-based molecular networking and docking for enhanced umami peptide screening from Pixian douban. *Food Chem. X*, 21, 101098, 2023.

9

Marine-Derived Drugs: Recent Advances in Cancer Therapy and Immune Signaling

Dhanisha Sulekha Suresh

Department of Surgery, The University of Alabama at Birmingham, Birmingham, Alabama, USA

Abstract

The marine realm presents a remarkable opportunity to explore innovative anti-cancer products and serves as a vast resource for isolating diverse cell targets for clinical intervention. However, the development of marine-derived anticancer drugs faces notable constraints. To surmount these challenges and limitations effectively, it is essential to foster extensive multidisciplinary collaborative efforts for advancing marine pharmaceutical production. Moreover, the integration of analytical spectrometry coupled with computational genetics, experimental therapeutics, and gene mining is crucial.

Keywords: Marine realm, bioactive compounds, anticancer products, immune signaling, chitosan, innovative methodologies, nanocarriers, cGAS/STING

9.1 Introduction

Over 70% of the Earth's surface is enveloped by water making the oceans the largest and most remarkable expanse on our planet. They constitute a vital reservoir of marine and freshwater organisms boasting unparalleled natural and biochemical diversity [1]. However, while certain drugs approved by the FDA have originated from terrestrial sources, there is a growing recognition of many complexes, potential anticancer drug candidates, and significant metabolites derived from expansive marine sources. Comprising over 80% of the world's diverse plant and animal species,

Email: dhanishass@gmail.com

Shaju S. S., Vipin. P. M. and Visakh. P.M. (eds.) Marine Biotechnology: Studies and Applications, (261–286) © 2025 Scrivener Publishing LLC

oceans are rich reservoirs of marine life. Various organisms, including tunicates, sponges, soft corals, fishes, sea hares, nudibranchs, opisthobranch mollusks, bryozoans, echinoderms, shells, prawns, marine microorganisms, and sea slugs, serve as abundant sources for bioactive compounds such as oils and cosmetics. Approximately 30,000 combinations of natural resources derived from marine sources have been identified. Notably, since 2008, there has been an annual production of over 1,000 compounds. Furthermore, these compounds are regularly categorized using advancements in structural classification, diversity, and randomness [2, 3]. The utilization of marine sources is crucial in acquiring essential drugs for various types of cancers and related diseases. Persistently, marine organisms have developed the capacity to produce complex compounds that possess cytotoxic properties adapting to the challenging and hostile conditions prevalent in marine ecosystems. Moreover, the consequences of evolutionary changes have also enabled marine animals to acquire these synthesized compounds enabling them to survive and adapt without physiological barriers. Regularly, challenging aquatic conditions linked to marine environments encompass elevated levels of salinity and pressure, limited oxygen availability for cellular functions, and fluctuating degrees of marine temperature in highly dynamic settings [4]. Natural products derived from marine sources, along with anticancer drugs, find broad applications in the anticancer drug industry and various cancer-related diseases. Furthermore, crucial agents for treatment extracted from these drugs, such as chemically derived compounds or chemical complexes, and natural immune agents, can be revitalized by utilizing the MNP technique. Despite extensive research efforts, including innovative and medically significant clinical approaches employing novel drug agents, it is noted that mortality rates from cancer persist at high levels [5, 6]. The WHO reports that more than 80% of the global population, particularly in developing nations, relies on herbal-based natural medicines for healthcare. Globally, most markets in the field of anticancer pharmaceuticals are reliant on compounds derived from natural sources and metabolites produced through engineering techniques. This includes a variety of compounds obtained from organisms residing in marine environments [7].

The development of therapeutic drugs requires relentless dedication to create new treatments for a wide range of medical conditions, including cancer, fungal and viral diseases, various degenerative neural conditions, and inflammatory disorders [8]. Moreover, advancements in overcoming microbial resistance to antibiotics and cancer cell resistance to anticancer drugs necessitate the development of innovative drug agents.

Following this, nanoparticles derived from marine sources have shown remarkable potential for the development of anticancer drugs, yet their utilization remains underexplored [9]. Furthermore, a significant drive for advancing new drug development has proven essential for future research into anticancer treatments. Presently, studies in nanotechnology offer a distinct approach to advancing drug discovery for anticancer therapy. Breakthroughs in nanomedicine have propelled innovative strategies involving nano-delivery systems and formulations, such as encapsulation techniques, to deliver anticancer drug treatments and synthetic compounds for precise targeting and eradication of diverse tumor types [10]. Numerous crucial anticancer medications sourced from marine origins have received approval from the United States (US) Food and Drug Administration (FDA). Among these, cytarabine (marketed as Cytosar U®) stands out as a chemotherapeutic agent utilized to combat malignant lymphocytic, acute myeloid and myelogenous leukemia, and non-Hodgkin's lymphoma. Trabectedin (marketed as Yondelis®) is a chemotherapeutic agent utilized to treat advanced soft tissue sarcoma exhibiting antitumor properties. Another therapeutic anticancer medication, marketed as Halaven®, is utilized in the treatment of both breast cancer and liposarcoma. Additionally, brentuximab vedotin, marketed as Adcetris®, represents an antibody–drug conjugate medication employed in treating anaplastic large-cell lymphoma and relapsed or refractory Hodgkin's lymphoma [11, 12].

Fundamentally, the innate immune pathways, including JAK/STAT signaling and cGAS-/STING-mediated immune responses, play an integral role in anticancer therapy. Earlier investigations have unveiled diverse applications of anticancer medications that exert substantial influence on immune pathways. Notably, the consistent utilization of innate immune responses induced by marine-derived drugs is noteworthy [13]. The extent of our comprehension regarding the influence of anticancer drugs sourced from marine origins on cell signaling and therapy is currently restricted. Therefore, we provide an in-depth exploration of the latest advancements in utilizing marine-derived anticancer drugs for disease management and overall health.

9.2 Marine-Derived Drugs

Marine natural products exhibit a diverse array of bioactive compounds relevant to pharmaceutical applications. Microbes dwelling in marine ecosystems, such as algae, corals, sponges, bacteria, and fungi, have been found to generate distinctive secondary metabolites, with diverse chemical compositions, playing a pivotal role in the development of anticancer medications. Anticancer drugs derived from different marine origins has been included in Table 9.1. Anticancer drugs derived from marine environments originate from a diverse range of organisms, including prokaryotes like marine bacteria such as *Noctiluca scintillans*, *Lactobacilli*, algae producing secondary anticancer metabolites, mangroves containing anticancer compounds, and other lesser-explored flora for potential anticancer properties. These marine organisms together constitute more than 70% of the Earth's biosphere [14].

Harnessing marine microorganisms to form distinct biological components holds the potential for discovering novel targets and medications.

Table 9.1 Bioactive compounds, taxonomy, and anticancer agents sourced from diverse marine origins.

Compounds	Category	Biological activity	Anticancer compounds	Marine source
Indoles, pyridoacrines, pyrroles, guadinines, isoquinolines, sterols, and aminoimidazoles	Alkaloids	Anticancer, cytotoxic, antimalarial, and antimicrobial	Tambjamine D, discorhabdines, topsentin	Tunicates, sponges, mollusks, anemones
Phenolic acids, flavonoids, lignin, anthocyanidins, tannins, epicatechin, gallic acid, epigallocatechin, catechin	Polyphenols	Anticancer, antioxidant, anti-inflammatory, antiviral, and metal chelators, inhibit human platelet aggregation	Phloroglucinal, scutellarein 4′-methyl ether, diecol, Ecol, caulerpenyne, usneoidols Z and E, 8,8′-biecol	Seagrass, seaweeds, and mangroves
Macrolides, polyols, polyethers, and aromatic compounds	Polyketides	Anticancer, antiparasitic, antibiotic, antifungal, and neurotoxic effects	SZ-685C, haloroquinone	Ascidians, sponges, bryozoans, soft corals, symbiotic or commensal bacteria
Monoterpenes, diterpenes, sesquiterpenes, tetraterpenes (carotenoids), and triterpenes (steroids)	Terpenes	Antiproliferative, cytotoxic, anticancer, antimicrobial, cytotoxic, and antifungal	Usneoidols Z and E, caulerpenyne	Sponges and soft corals

(*Continued*)

Table 9.1 Bioactive compounds, taxonomy, and anticancer agents sourced from diverse marine origins. (*Continued*)

Compounds	Category	Biological activity	Anticancer compounds	Marine source
Mirabamides, cyclic depsipeptides, celebesides, theopapuamides, callyaerins, callipeltins theonellamide, discodermin, pipecolidepsins	Peptides	Antiviral, cardiotonic, cardiotoxic, antimicrobial	Benzoxazolinone and brugine	Commensal or symbiotic bacteria, sponges, and fungi
Oligosaccharides, polysaccharides, aglycone, glycosides, gemcitabine analogs, nucleosides	Glycosides, carbohy-drates and others	Anticancer, drug delivery systems, tissue engineering, wound healing, hypocholeste-rolemic, antimutagenic, gel productions, glycosamino-glycans	Pentosan polysulfate, fucoidan, chondroitin 4 sulfate, heparin, LO A and B, chondroitin 6 sulfate	Tunicates and sponges

Pharmacologically significant anticancer compounds have been extracted from secondary metabolites synthesized by small marine organisms. Examples include bryostatins, discodermolide, eleutherobin, and sarcodictyin [15]. Similarly, the interaction with pathogenic organisms typically influences the utilization of probiotic microorganisms, like *lactobacilli* and *bifidobacteria*, altering the processing of anticancer compounds. The metabolite macrolactin-A has a notable inhibitory effect on B16F10 murine melanoma cells. Furthermore, *Streptomyces spinoverrucosal* dibohemamines B and C demonstrate promising potential in altering the cell lineage of non-small cell lung cancer [16, 17]. *Elysia rubefescens*, a mollusk, serves as a natural source of the depsipeptide kahalalide; its production is influenced by symbiotic microorganisms. Depsipeptide kahalalide demonstrates promising *in vitro* efficacy against solid tumors notably displaying precision targeting in prostate cancer cell lines [18, 19]. Moreover, the indole-based microbial anticancer agent trisindolal 38 has shown promising activity against melanoma (MEXF 462) and human breast (MAXF401) cancer cell lines [20]. Studies suggest that cyanobactin wewakazole B, extracted from cyanobacteria, exhibits cytotoxic effects against several human cancer cell lines [21]. The powerful cytotoxin called odoamide, which originates from Japanese Okeania, displays substantial anticancer activity against cervical HeLa S3 human tumor cells [22]. The highly effective anticancer drug Curacin-A is synthesized from the distinctive essence extracted from Curacao samples of *Lyngbya majuscule* [23]. Apratoxin A, a compound from algae *Lyngbya Apratoxin A*, demonstrates cytotoxicity against adenocarcinoma [24]. The extract from the red algae *Acanthophora spicifera* exhibits tumoricidal properties against Ehrlich's ascites malignancies [25]. Furthermore, microalgal-based nano-medication has been found to stimulate the interaction of p53 with pro-apoptotic factors leading to the subsequent modulation of caspase 3, which controls DNA degradation and ultimately leads to apoptosis (Figure 9.1).

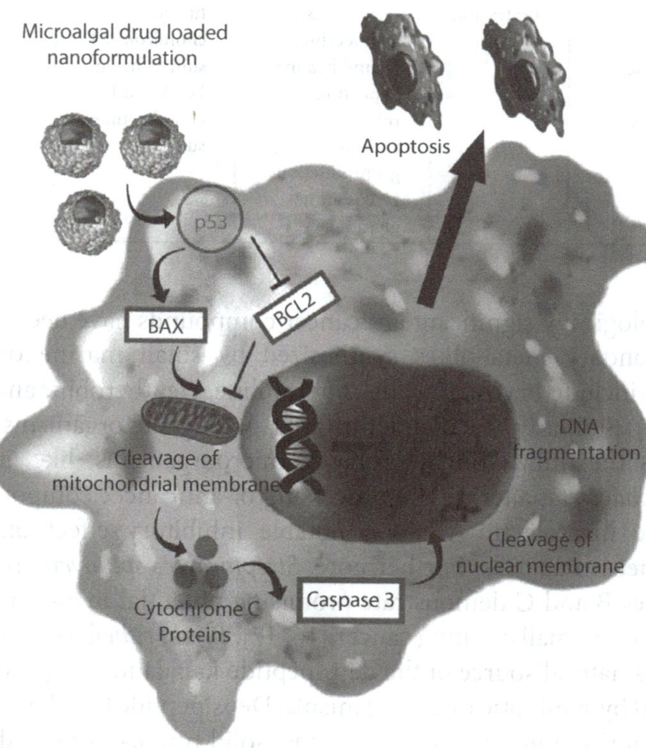

Figure 9.1 The apoptotic cascade is primarily initiated by mitochondrial-mediated molecular mechanisms. A nano-formulation primed with microalgal-derived compounds activates key regulatory pathways, including Bax, p53, and Bcl2, releasing cytochrome c from disrupted mitochondrial membranes, subsequently activating caspase 3. Once activated, caspase 3 facilitates the breakdown of nuclear membranes and initiates DNA fragmentation. Consequently, pro-apoptotic genes are upregulated, while anti-apoptotic genes are downregulated, collectively promoting apoptosis and cellular blebbing (zeiosis).

9.3 Marine-Derived Drugs: Recent Advances in Cancer Therapy

A diverse range of marine-derived products is a prominent source of bioactive constituents. Approved or clinically trialled anticancer drugs derived from marine origin has been included in Table 9.2. The FDA has approved four marine-derived medications, with a lead time of 20–30 years from discovery to market authorization [26]. Numerous marine-derived compounds exhibit significant anticancer potential and are currently undergoing various stages of clinical trials in the field of hematology and oncology, among this in 2011, FDA approved brentuximab vedotin 63 (Adcetris™) against anaplastic cell lymphoma and Hodgkin lymphoma [27]. Adcetris stands as the leading medication among seven major global markets, including France, the US, Germany, Spain, the United Kingdom, Japan, and Italy. Its significant impact on Hodgkin's lymphoma treatment has been crucial in recent years. Seattle Genetics (Nasdaq: SGEN), a US biotech firm, and Takeda (TYO: 4502), a Japanese pharmaceutical company, collaborate on both the production and marketing of Adcetris. In 1969, cytarabine (derived through the fermentation of *Streptomyces griseus*) received FDA approval for the treatment of cancer, including non-Hodkin's lymphoma, meningeal leukemia, and myeloid leukemia [28]. In 2010, the FDA approved Halaven® (eribulin mesylate) for the treatment of metastatic breast cancer. Eribulin (E7389), a pioneering macrocyclic ketone, was developed by researchers at the Eisai Research Institute. Halaven® combats malignancies by impeding the tubulin proteins crucial for the cytoskeleton required for the proliferation of tumor [29]. Tubulin is targeted by numerous chemotherapy treatments, including taxolin. Trabectedin, a groundbreaking antineoplastic alkaloid marketed as Yondelis®, is sourced from a Caribbean Sea squirt *Ecteinascidia turbinate.* Yondelis® seems to interact with DNA repair proteins and attach itself to a minor groove within the DNA structure [30]. PharmaMar, a pharmaceutical company, operates under joint ownership with Johnson and Johnson's drug research and innovation sector, and they have received approval for the treatment of soft tissue sarcoma. Since the approval of cytarabine over the past five decades, four medications derived from marine sources have gained approval [31].

Moreover, several other products are advancing through Phases I, II, and III clinical trials within the marine anticancer medicinal pipeline. Generally, clinical trials are comprised of four principal phases conducted before the drug market launch. The fourth phase commences after the pharmaceutical's first administration and extends throughout its lifespan.

Table 9.2 Approved or clinically trialed anticancer medications derived from marine sources.

Drug name	Class of drug	Sponsor	Marine source	Mechanism	Disease (marketed)	Status
Adcetris™ (brentuximab vedotin)	Anticancer	Seattle genetics	This mixture contains an antibody–drug conjugate consisting of an anti-CD30 antibody and MMAE (monomethyl auristatin E), which originates from the sea hare cyanobacteria or *Dollabella auricularia*	Triggers apoptosis in tumor cells by halting cell cycle progression from the G2 phase to the M phase, achieved through the disruption of the cytosolic microtubule pathway	Chronic large-cell anaplastic and Hodgkin's lymphoma	Approved by FDA (2018), Phases III and IV clinical trials are being conducted against Hodkin's lymphoma with chemotherapy drugs
Eribulin mesylate (Halaven®)	Macrolide	Eisai Co. Ltd.	Inhibitor of microtubule (derived from the sponge species *Halichondria okadai*)	Disrupts the growth cycle of cytoskeleton by preventing its assembly into functional structures affecting the cell cycle and sequestering tubulin	Metastasized breast cancer	Approved by FDA in 2016, treatment for metastatic liposarcoma

(*Continued*)

Table 9.2 Approved or clinically trialed anticancer medications derived from marine sources. (*Continued*)

Drug name	Class of drug	Sponsor	Marine source	Mechanism	Disease (marketed)	Status
Depoyt®, Cytosar-U (cytarabine, ara-C)	Anticancer	Jazz Pharmaceuticals	Pyrimidine nucleoside analog sourced from the sponge species *Cryptotheca crypta*	Inhibits DNA synthesis exerting its unique effects during the S phase of the cell cycle. It also possesses antiviral and immunosuppressive properties	Acute non-lymphoblastic leukemia	Approved by FDA in 2017, meningeal leukemia, acute myelogenous leukemia
Depatuxizumab mafodotin (Depatux-M)	Anticancer	Abbott Laboratories (AbbVie)	*Caldora penicillata* (cyanobacteria)	Inhibits tubulin polymerization process	Advanced solid tumors and adenocarcinoma	Phase III
Trabectedin® (Yondelis®)	Anticancer	PharmaMar and Johnson and Johnson	*Ecteinascidia turbinata*	Inhibits cell proliferation	Ovarian cancer, sarcoma	Approved by the FDA in 2015, liposarcoma and leiomyosarcoma
Enfortumab vedotin	Anticancer	Seattle Genetics, Agensys	*Caldora penicillata* (cyanobacterium)	Inhibits microtubular network	Urothelial cancers	Phase III
Marizomib	Anticancer	Nereus Pharmaceuticals	*Salinispora tropica* (actinomycete)	Inhibits proteasome and stimulates apoptosis	Glioblastoma and multiple myeloma	Phase III
Lurbinectedin (Zepsyre®)	Anticancer	PharmaMar	*Ecteinascidia turbinate* (tunicate)	RNA polymerase II inhibition	Small-cell lung cancer and ovarian cancer	Phase III
Polatuzumab vedotin	Anticancer	Genentech	*Caldora penicillata* (cyanobacteria)	Inhibits the polymerization of tubulin, apoptotic activator	Large B-cell lymphoma	Phase III

(*Continued*)

Table 9.2 Approved or clinically trialed anticancer medications derived from marine sources. (*Continued*)

Drug name	**Class of drug**	**Sponsor**	**Marine source**	**Mechanism**	**Disease (marketed)**	**Status**
Plinabulin	Anticancer	Beyond spring Pharmaceuticals	*Aspergillus* sp. (marine species)	Stimulant for guanine nucleotide exchange factors	Non-small-cell lung cancer and neutropenia	Phase III
Lestaurtinib	Anticancer	Kyowa Hakko	*Streptomyces staurosporeus* (bacteria)	Janus kinase 2 and tyrosine kinase 3 inhibitors	Acute myeloid leukemia	Phase III
Enzastaurin	Anticancer	Eli Lilly, Denovo Biopharma	*Streptomyces staurosporeus*	Inhibits protein kinase C beta, phosphatidyl inositol 3 and proto-oncogene protein c-akt	Glioblastoma, large B-cell lymphoma	Phase III
GSK2857916	Anticancer	GlaxoSmithKline	*Caldora pencillata* (cyanobacterium)	Inhibits tubulin polymerization, activates apoptosis	Multiple myeloma	Phase II
Ladiratuzumab vedotin	Anticancer	Merck & Co., Genentech, Genentech, Seattle Genetics, Quantum Leap	*Caldora penicillata* (cyanobacteria)	Activates apoptosis, inhibits the polymerization of tumor	Breast cancer	Phase II
Aplidine plitidepsin (Aplidin®)	Anticancer	PharmaMar	*Aplidium alpicans* (tunicate)	Activates apoptosis, Cell cycle inhibitor	Lymphoblastic leukemia-lymphoma, multiple myeloma	Phase II

(*Continued*)

Table 9.2 Approved or clinically trialed anticancer medications derived from marine sources. (*Continued*)

Drug name	Class of drug	Sponsor	Marine source	Mechanism	Disease (marketed)	Status
Tisotumab vedotin	Anticancer	Genmab, Seattle Genetics	*Caldora penicillata* (cyanobacteria)	Microtubule protein inhibitors	Cervical, bladder, esophageal, endometrial, prostate, and solid tumors	Phase II
PM060184	Anticancer	PharmaMar	*Lithoplocamia lithistoides* (sponge)	Inhibits tubulin polymerization	Colorectal and breast cancer	Phase II
Indusatumab vedotin	Anticancer	Takeda Oncology	*Caldora penicillata* (cyanobacteria)	Antagonists of guanylate cyclase C	Gastrointestinal and colorectal cancer	Phase II
Glembatumumab vedotin	Anticancer	National Cancer Institute, Celldex Therapeutics	*Caldora penicillata* (cyanobacterium)	Inhibitor of tubulin polymerization, activator of apoptosis	Breast, uveal melanoma, and brain cancer	Phase II
Denintuzumab mafodotin	Anticancer	Seattle Genetics	*Caldora penicillata* (cyanobacterium)	Immunomodulator, activates apoptosis, inhibits mitosis, and tubulin polymerization	Large B-cell lymphoma	Phase II

(*Continued*)

Table 9.2 Approved or clinically trialed anticancer medications derived from marine sources. (*Continued*)

Drug name	Class of drug	Sponsor	Marine source	Mechanism	Disease (marketed)	Status
PM00104 (Zalypsis®)	Anticancer	PharmaMar	*Jorunna funebris* (mollusk)	Cell cycle inhibitor	Endometrial cancer, Ewing's sarcoma, cervical cancer, Multiple myeloma, hematological malignancies, urogenital cancer, multiple myeloma	Phase II
Midostaurin (Rydapt®)	Anticancer	Novartis	*Streptomyces staurosporeus*	The antagonist of vascular endothelial growth factor receptor-2, platelet-derived growth alpha receptor, platelet-derived growth factor beta receptor, protein kinase C	Malignant melanoma, brain cancer, Hodgkin's lymphoma, systemic mastocytosis	Phase I

To ensure a successful drug trial, an extensive additional analysis is necessary to evaluate the candidate's market potential. The complex network of pharmaceutical advancement, involving the development and manufacturing of marine-derived anticancer drugs, opens up innovative avenues and resources to address the treatment challenges of patients with potentially resistant cancer types [32]. On the other hand, a plethora of drugs is currently undergoing clinical trials for the treatment of various cancers, which includes mafodotin, depatuxizumab, targeting advanced solid tumors and adenocarcinoma, enfortumab vedotin for urothelial cancers, marizomib for glioblastoma and multiple myeloma, lurbinectedin (zepsyre®) for small-cell lung cancer and ovarian cancer, polatuzumab vedotin, enzastaurin for large B-cell lymphoma, marizomib for glioblastoma and multiple myeloma, plinabulin for non-small-cell lung cancer and neutropenia, and lestaurtinib for acute myeloid leukemia [33–39]. In phase II clinical trials, various drugs are being investigated for their efficacy. This includes tisotumab vedotin for cervical cancer, bladder cancer, endometrial cancer, ovarian cancer, and esophageal cancer, GSK2857916 for multiple myeloma, plitidepsin (Aplidin®) and aplidine for lymphoblastic leukemia–lymphoma and multiple myeloma, PM060184 for colorectal and breast cancer, indusatumab vedotin for gastrointestinal, colorectal, and pancreatic cancer, and glembatumumab vedotin for breast cancer, squamous cell cancer, brain cancer, osteocarcinoma, malignant melanoma, and uveal cancer. Additionally, drugs in phase I clinical trials include midostaurin (Rydapt®) for systemic mastocytosis and myeloid leukemia [40–45] (Table 9.2).

9.4 Marine-Derived Drugs: Recent Advances in Immune Signaling

Marine flora exhibits intricate taxonomy, biological efficacy, and specialized chemical composition. It offers an exceptional platform, for the development of novel biopharmaceuticals, including medications targeting cancer and infectious disease, as well as developing immunomodulators. Despite the significant benefits of immunomodulators in managing immune-based conditions, the search for safe immunotherapies remains ongoing for many disorders. Numerous marine compounds exhibit the ability to mediate complex inhibitory effects on diverse cellular functions, including angiogenesis and apoptosis, and are also involved in intricate cellular signaling cascades [46].

9.4.1 cGAS/STING Signaling Pathways

The cGAS, a key cytosolic sensor that is responsible for detecting both endogenous/foreign DNA is involved in recognizing a broad range of intracellular bacterial/viral pathogens. This sensor triggers the production of interferons and various other cytokines responsible for different infections. Referred to as cyclic guanosine monophosphate (GMP)–adenosine monophosphate (AMP) synthase (cGAS), it serves as a pivotal component in innate immune responses. DNA serves as a messenger, inducing the generation of secondary messengers called cyclic GMP–AMP (cGAMP), which in turn triggers the production of 2′3′-cGAMP, acts as an endogenous ligand, and exhibits a robust affinity for the IFN stimulator genes (STING) [47]. The cGAS/STING pathway plays a crucial role in influencing the effects of anticancer and antiviral therapies, as well as cellular senescence. The cGAS sensor triggers the IFN mechanism, which regulates infection. The tumor DNA can also initiate this process. Moreover, the efficacy of antitumor response relies on the presentation of tumoral antigens to T lymphocytes especially CD8+ cells through antigen-presenting cells (APC). Additionally, cGAMP therapy not only enhances the antitumor effects of chemotherapy and ionizing irradiation but also boosts immune checkpoint blockade [47]. Correspondingly, the sustained activation of cGAS/STING-mediated immune activation triggers cell death, offering an additional protective barrier against tumor progression. The cGAS/STING innate immune pathway serves as a built-in defense mechanism against tumor progression functioning by linking DNA damage to a range of immune responses aimed at combating tumors. These responses encompass immune regulation, cellular senescence, and apoptosis collectively contributing to the suppression of tumor growth. As a result, this mechanism implies an enticing therapeutic aim of integrating these fundamental elements of antitumor properties for cancer treatment [47].

Recent findings have highlighted the immunoadjuvant potential of chitosan within the cGAS-activated immune system [45]. Chitosan is a chemical polymeric biomaterial derived from the exoskeletons of shellfish, including crabs and shrimp. Moreover, chitosan is characterized as a highly potent, large polycationic biopolymer with biodegradable, non-toxic, and biocompatible properties [48, 49]. Thus, it finds an extensive application in the field of biotechnology and agribusiness. Likewise, owing to its biodegradable characteristics, chitosan holds promise as a viable option for drug delivery, potentially extending the stability of medications and improving their overall bioavailability [50]. An earlier study provides compelling evidence suggesting that chitosan exhibits significant antitumor effects, both in laboratory and animal studies, indicating favorable conditions for its use as an adjuvant and carrier for antitumor immunity. Chitosan-based nanoparticles have garnered considerable attention in cutting-edge pharmaceutical applications encompassing targeted therapies for cancer, antioxidant properties, vaccine delivery, and genetic interventions [50].

Cancer stands as one of the most lethal diseases globally and faces significant challenges that constrain the effectiveness of modern chemotherapy and immunotherapeutic approaches for anticancer treatment. Nanocarriers specifically deliver chemotherapeutic agents to their target site. Chitosan nanoparticles, modified with surfaces loaded with therapeutic agents, exhibit increased stability, bioactivity, and porosity facilitating widespread utilization of chitosan in delivering anticancer medications, including doxorubicin, trastuzumab, polylactic acid, gambogic acid, docetaxl, methotrexate, 5-fluorouracil, ginsenoside compound K, curcumin, gemcitabine, epigallocatechin-3-gallate, metformin hydrochloride, temozolomide, and hesperetin [51]. Chitosan impacts the immune response of NLRP3 (NLR Pyrin Domain Family Group) triggering inflammation in phagocytic cells and facilitating the entry of interleukin-13 [52]. Moreover, there is a notion that chitosan induces interference with cGAS/STING pathway by affecting mitochondrial DNA resulting in the release of interferons (Figure 9.2). Likewise, mitochondrial reactive oxygen species generate oxidized DNA in the cytosol, thereby initiating the production of cGAS-/STING-mediated interferon production. Consequently, IFN prompts the generation of dendritic cells, which in turn facilitate antigen production followed by the induction of Th1 (Type 1 helper) immunity [53].

9.4.2 JAK/STAT Signaling

Numerous anticancer treatments hinge on the meticulous management of immune-related cellular signals implicated in carcinogenesis, governed by various cellular immune components, in conjunction with pathways including JAK/STAT, PI3K/AKT/mTOR, and NF-kB signaling [54]. Bryostatin 1, an anticancer drug sourced from *Bugula neritina*, is employed in the treatment of chronic lymphocytic leukemia. LS-1, a marine cembrenolide diterpene known as (1S, 2S, 3E, 7E, 11E)-3,7,11,15-cembratetraen-17,2-olide, exhibits anticancer properties specifically targeting colon cancers [55]. Bryostatin 1 initiates MAPK activation and induces interferon-γ production, activating JAK/STAT, contributing to its effectiveness in combating colon cancer [56]. Moreover, LS-1 treatment induces apoptosis in HT-29 human colorectal adenocarcinoma cells by targeting the

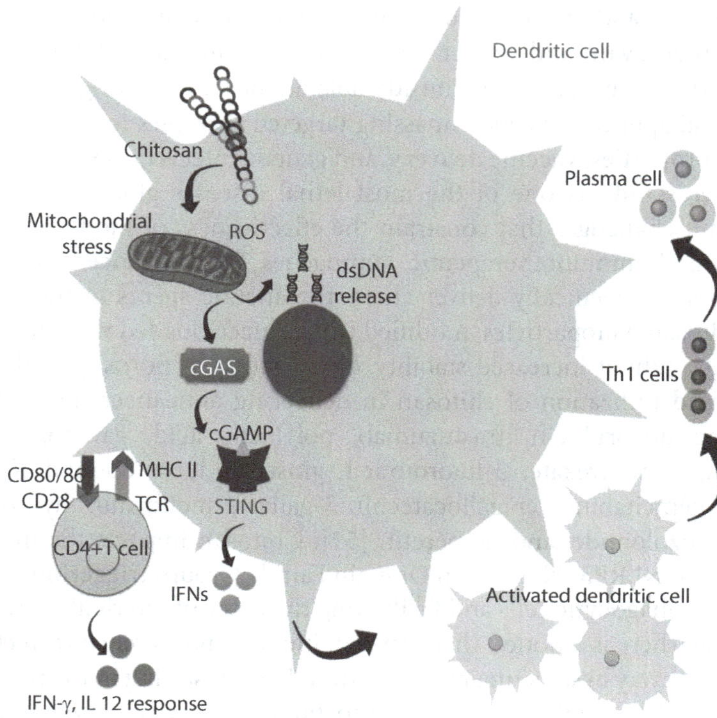

Figure 9.2 Chitosan exerts an immune-adjuvant effect by activating human DNA-mediated cGAS/STING pathway, promotes the transcription of interferons (IFNs), which in turn activate dendritic cells. The activated DCs enhance the immune response by stimulating Th1 cells and the differentiation of immunoglobulin-producing plasma cells.

intrinsic mitochondrial pathway enhancing resistance to anticancer effects and regulating STAT3 to limit its immunomodulatory function [57]. Sarcodiol, a semi-synthetic derivative of sarcophine, stands as a derivative merging both semi-engineered sarcophine and a marine natural product. Interestingly, sarcodiol modulates the activation potential of STAT-3 and cyclin D1, revitalizes the tumor-suppressor protein 53 (p53) within the cell, and initiates the series of events leading to caspase-dependent cell death [58]. Apratoxin A, a secondary metabolite originating from cyanobacteria, demonstrates cytotoxic characteristics as a marine natural product by promptly impeding STAT-3 phosphorylation. Apratoxin A disrupts JAK/STAT signaling triggered by IL6 and hinders the N-glycosylation of receptor tyrosine kinases involved in proteasomal degradation [59]. Moreover, GV-c9 and GV-c10, two unsaturated enone fatty acids derived from the marine red algae *Gracilaria verrucosa*, renowned for its antioxidant and anticancer properties, inhibit NF-kB activation by impeding the nuclear translocation of JAK/STAT (p-STAT1) and NF-kB signaling pathways [60].

9.5 Exploring Advancements and Hurdles in Marine Pharmaceutical Research

At present, there are a few FDA-approved drugs sourced from marine organisms, with a primary focus on antitumor medications available in the market. Concurrently, numerous others are undergoing clinical trials spanning phase I, I/II, or III. To overcome prevalent challenges in this endeavor, like availability and toxicity, the manufacturing of these medications has yielded valuable insights and crucial viewpoints. Over 20 compounds are undergoing clinical trials, with over 1,000 isolated compounds from marine organisms undergoing preclinical trials. Within the pharmaceutical production process, clinical trials aim to address two central inquiries: a) Does the drug demonstrate efficacy? and b) If so, is it safe for patent approval? Yet, in certain instances, finding definitive answers to these questions is challenging, particularly considering the expansive scope of most trials. Over the past five decades, over 30,000 novel compounds from marine environments have been discovered resulting in the approval of over 300 patents [61]. Aquatic organisms represent valuable reservoirs of intricate natural compounds exhibiting diverse pharmacological properties. The increasing number of patents granted in recent years strongly supports the recognition of these chemicals as valuable novel therapeutic agents. Given the critical supply challenges, researchers are exploring marine-derived resources as promising candidates for anticancer strategies. Numerous novel chemical entities with anticancer properties, derived from marine sources, have been patented due to their bioactive nature. Among the anticancer compounds derived from marine origins are *Streptomyces fulvissimus* ketone and *Streptomyces fulvissimus* acid (CN107619803; Zhejiang University) sourced from seaweed *Haliplanella lineata* [62]. Likewise, bacteria serve diverse novel chemical entities, such as cyclo proline-tyrosine (CN105199989) derived from *Streptomyces* sp. and phenaziterpene A and B (CN103360329 sourced from *Streptomyces niveus* [63]. Similarly, fungi contribute to the array of novel chemical entities featuring Gliocladine C, compounds 1 and 2, Luteoalbusin A and B (CN102911188), derived from *Acrostalagmus luteoalbus*, compound 3, pencilindole A and B isolated from *Xylocarpus granatum* (CN107827805), compounds I and II having octahydroperylene obtained from *Alternaria alternate* (CN106883113) [64].

Furthermore, the novel chemical entities sourced from sponges encompass Hippolide I (CN102617596) from *Hippospongia lachne*, Dysideanone D from *Dysidea avara* (CN104478688), Stelliferin A and Stelliferin B extracted from *Stelleta globostellata* [65, 66]. Likewise, novel chemical entities derived from plants comprise Salbige A and B sourced from *Asparagus* (CN103304622). Numerous pharmaceuticals derived from marine sources have been investigated as potential candidates for advanced cancer treatment. Trabectedin represents a novel DNA-alkylating drug utilized in the treatment of relapsed platinum-sensitive ovarian antiproliferative properties and presents a hopeful avenue for anticancer therapy. *Anthopleura anjunae*, an oligopeptide, demonstrates cytotoxic effects and holds significant promise for potential future therapy of prostate cancer [67–69].

9.6 Conclusions

The marine realm presents a remarkable opportunity to explore innovative anticancer products and serves as a vast resource for isolating diverse cell targets for clinical intervention. Remarkably, numerous human tumors are being leveraged to alleviate the scarcity of anticancer drugs. Moreover, despite 75% of the earth's surface being covered in water, research into the pharmacological properties of anticancer drugs from marine environments remains limited leaving a significant majority unexplored. Comparatively, there is a limited selection of anticancer therapeutic drugs, such as Adcetris™, Halaven®, Yondelis®, and Depocyt®, while only a few are undergoing late-stage clinical trials for various diseases. As a result, a significant gap exists in the distribution and utilization of anticancer drugs for clinical assessment. Chitosan, a remarkable marine biopolymer, offers significant potential for various drug applications. Consequently, it has been extensively employed in tumor treatment, as well as in the development of drug-delivery system nanoparticles and vectors. Moreover, a multitude of drugs, such as Apratoxin A, GV-c1, GV-c9, SD, and LS-1, have demonstrated pivotal anticancer attributes by impeding the JAK/STAT innate immune signaling pathway. Nevertheless, the signals regulated by cGAS/STING, which are crucial for anticancer effects, have not been extensively evaluated clinically with marine anticancer drugs. At present, the development of marine-derived anticancer drugs faces notable constraints. To surmount these challenges and limitations effectively, it is essential to foster extensive multidisciplinary collaborative efforts for advancing marine pharmaceutical production. Moreover, the integration of analytical spectrometry coupled with computational genetics, experimental therapeutics, gene mining, and other innovative methodologies is crucial in the toolkit for future efforts to explore novel constructs in marine natural product exploration.

References

1. Kiuru, P., D'Auria, M.V., Muller, C.D., Tammela, P., Vuorela, H., Yli-Kauhaluoma, J., Exploring marine resources for bioactive compounds. *Planta Med.*, *80*, 14, 1234–1246, 2014. https://doi.org/10.1055/s-0034-1383001.
2. Hu, Y., Chen, J., Hu, G., Yu, J., Zhu, X., Lin, Y., Chen, S., Yuan, J., Statistical research on the bioactivity of new marine natural products discovered during the 28 years from 1985 to 2012. *Mar. Drugs*, *13*, 1, 202–221, 2015. https://doi.org/10.3390/md13010202.
3. Hassan, S.S., Anjum, K., Abbas, S.Q., Akhter, N., Shagufta, B.I., Shah, S.A., Tasneem, U., Emerging biopharmaceuticals from marine actinobacteria. *Environ. Toxicol. Pharmacol.*, *49*, 34–47, 2017. https://doi.org/10.1016/j.etap.2016.11.015.
4. Gomes, N.G., Dasari, R., Chandra, S., Kiss, R., Kornienko, A., Marine Invertebrate Metabolites with Anticancer Activities: Solutions to the "Supply Problem". *Mar. Drugs*, *14*, 5, 98, 2016. https://doi.org/10.3390/md14050098.
5. Suarez-Jimenez, G.M., Burgos-Hernandez, A., Ezquerra-Brauer, J.M., Bioactive peptides and depsipeptides with anticancer potential: sources from marine animals. *Mar. Drugs*, *10*, 5, 963–986, 2012. https://doi.org/10.3390/md10050963.
6. Desbois, A.P., How might we increase success in marine-based drug discovery? *Expert Opin. Drug Discov.*, *9*, 9, 985–990, 2014. https://doi.org/10.1517/17460441.2014.927863.
7. Lee, J.Y., Orlikova, B., Diederich, M., Signal Transducers and Activators of Transcription (STAT) Regulatory Networks in Marine Organisms: From Physiological Observations towards Marine Drug Discovery. *Mar. Drugs*, *13*, 8, 4967–4984, 2015. https://doi.org/10.3390/md13084967.
8. Debela, D.T., Muzazu, S.G., Heraro, K.D., Ndalama, M.T., Mesele, B.W., Haile, D.C., Kitui, S.K., Manyazewal, T., New approaches and procedures for cancer treatment: Current perspectives. *SAGE Open Med.*, *9*, 20503121211034366, 2021. https://doi.org/10.1177/20503121211034366.
9. Jeong, G.J., Khan, S., Tabassum, N., Khan, F., Kim, Y.M., Marine-Bioinspired Nanoparticles as Potential Drugs for Multiple Biological Roles. *Mar. Drugs*, *20*, 8, 527, 2022. https://doi.org/10.3390/md20080527.
10. Dias, D.A., Urban, S., Roessner, U., A historical overview of natural products in drug discovery. *Metabolites*, *2*, 2, 303–336, 2012. https://doi.org/10.3390/metabo2020303.

11. van Andel, L., Rosing, H., Schellens, J.H., Beijnen, J.H., Review of Chromatographic Bioanalytical Assays for the Quantitative Determination of Marine-Derived Drugs for Cancer Treatment. *Mar. Drugs, 16*, 7, 246, 2018. https://doi.org/10.3390/md16070246.
12. Mayer, A.M., Glaser, K.B., Cuevas, C., Jacobs, R.S., Kem, W., Little, R.D., McIntosh, J.M., Newman, D.J., Potts, B.C., Shuster, D.E., The odyssey of marine pharmaceuticals: a current pipeline perspective. *Trends Pharmacol. Sci., 31*, 6, 255–265, 2010. https://doi.org/10.1016/j.tips.2010.02.005.
13. Song, X., Xiong, Y., Qi, X., Tang, W., Dai, J., Gu, Q., Li, J., Molecular Targets of Active Anticancer Compounds Derived from Marine Sources. *Mar. Drugs, 16*, 5, 175, 2018. https://doi.org/10.3390/md16050175.
14. Alves, C., Silva, J., Pinteus, S., Gaspar, H., Alpoim, M.C., Botana, L.M., Pedrosa, R., From Marine Origin to Therapeutics: The Antitumor Potential of Marine Algae-Derived Compounds. *Front. Pharmacol.*, *9*, 777, 2018. https://doi.org/10.3389/fphar.2018.00777.
15. Khalifa, S.A.M., Elias, N., Farag, M.A., Chen, L., Saeed, A., Hegazy, M.F., Moustafa, M.S., Abd El-Wahed, A., Al-Mousawi, S.M., Musharraf, S.G., Chang, F.R., Iwasaki, A., Suenaga, K., Alajlani, M., Göransson, U., El-Seedi, H.R., Marine Natural Products: A Source of Novel Anticancer Drugs. *Mar. Drugs, 17*, 9, 491, 2019. https://doi.org/10.3390/md17090491.
16. Wollowski, I., Rechkemmer, G., Pool-Zobel, B.L., Protective role of probiotics and prebiotics in colon cancer. *Am. J. Clin. Nutr.*, *73*, 2 Suppl, 451S–455S, 2001. https://doi.org/10.1093/ajcn/73.2.451s.
17. Fu, P., Legako, A., La, S., MacMillan, J.B., Discovery, Characterization, and Analogue Synthesis of Bohemamine Dimers Generated by Non-enzymatic Biosynthesis. *Chem. (Weinheim an der Bergstrasse Germany)*, *22*, 10, 3491–3495, 2016. https://doi.org/10.1002/chem.201600024.
18. Rafieezadeh, D. and Esfandyari, G. Marine bioactive peptides with anticancer potential, a narrative review. *Int. J. Biochem. Mol. Biol.*, *15*, 4, 118–126, 2024.
19. Sithranga Boopathy, N. and Kathiresan, K., Anticancer drugs from marine flora: an overview. *J. Oncol.*, *2010*, 214186, 2010. https://doi.org/10.1155/2010/214186.
20. Dhiman, A., Sharma, R., Singh, R.K., Target-based anticancer indole derivatives and insight into structure-activity relationship: A mechanistic review update (2018-2021). *Acta Pharm. Sinica. B*, *12*, 7, 3006–3027, 2022. https://doi.org/10.1016/j.apsb.2022.03.021.
21. Lopez, J.A., Al-Lihaibi, S.S., Alarif, W.M., Abdel-Lateff, A., Nogata, Y., Washio, K., Morikawa, M., Okino, T., Wewakazole B, a Cytotoxic Cyanobactin from the Cyanobacterium Moorea producens Collected in the Red Sea. *J. Nat. Prod.*, *79*, 4, 1213–1218, 2016. https://doi.org/10.1021/acs.jnatprod.6b00051.
22. Kaneda, M., Kawaguchi, S., Fujii, N., Ohno, H., Oishi, S., Structure-Activity Relationship Study on Odoamide: Insights into the Bioactivities of Aurilide-Family Hybrid Peptide-Polyketides. *ACS Med. Chem. Lett.*, *9*, 4, 365–369, 2018. https://doi.org/10.1021/acsmedchemlett.8b00028.

23. Wipf, P., Reeves, J.T., Day, B.W., Chemistry and biology of curacin A. *Curr. Pharm. Des.*, *10*, 12, 1417–1437, 2004. https://doi.org/10.2174/1381612043384853.
24. Khalifa, S.A.M., Shedid, E.S., Saied, E.M., Jassbi, A.R., Jamebozorgi, F.H., Rateb, M.E., Du, M., Abdel-Daim, M.M., Kai, G.Y., Al-Hammady, M.A.M., Xiao, J., Guo, Z., El-Seedi, H.R., Cyanobacteria-From the Oceans to the Potential Biotechnological and Biomedical Applications. *Mar. Drugs*, *19*, 5, 241, 2021. https://doi.org/10.3390/md19050241.
25. Saeed, A.F.U.H., Su, J., Ouyang, S., Marine-derived drugs: Recent advances in cancer therapy and immune signaling. *Biomed. Pharmacother. = Biomedecine Pharmacotherapie*, *134*, 111091, 2021. https://doi.org/10.1016/j.biopha.2020.111091.
26. Ahmed, I., Asgher, M., Sher, F., Hussain, S.M., Nazish, N., Joshi, N., Sharma, A., Parra-Saldívar, R., Bilal, M., Iqbal, H.M.N., Exploring Marine as a Rich Source of Bioactive Peptides: Challenges and Opportunities from Marine Pharmacology. *Mar. Drugs*, *20*, 3, 208, 2022. https://doi.org/10.3390/md20030208.
27. FDA approves brentuximab vedotin for Hodgkin lymphoma and systemic anaplastic large-cell lymphoma. *Oncol. (Williston Park N.Y.)*, *25*, 10, 904, 2011.
28. Dyshlovoy, S.A. and Honecker, F., Marine Compounds and Cancer: The First Two Decades of XXI Century. *Mar. Drugs*, *18*, 1, 20, 2019. https://doi.org/10.3390/md18010020.
29. Eribulin mesylate (Halaven) for breast cancer. *Med. Letter Drugs Ther.*, *53*, 1362, 30–31, 2011.
30. Monk, B.J., Dalton, H., Benjamin, I., Tanović, A., Trabectedin as a new chemotherapy option in the treatment of relapsed platinum sensitive ovarian cancer. *Curr. Pharm. Des.*, *18*, 25, 3754–3769, 2012. https://doi.org/10.2174/138161212802002814.
31. Cappello, E. and Nieri, P., From Life in the Sea to the Clinic: The Marine Drugs Approved and under Clinical Trial. *Life (Basel Switzerland)*, *11*, 12, 1390, 2021. https://doi.org/10.3390/life11121390.
32. Malve, H., Exploring the ocean for new drug developments: Marine pharmacology. *J. Pharm. Bioallied Sci.*, *8*, 2, 83–91, 2016. https://doi.org/10.4103/0975-7406.171700.
33. Yu, J., Fang, T., Yun, C., Liu, X., Cai, X., Antibody-Drug Conjugates Targeting the Human Epidermal Growth Factor Receptor Family in Cancers. *Front. Mol. Biosci.*, *9*, 847835, 2022. https://doi.org/10.3389/fmolb.2022.847835.
34. Powles, T., Rosenberg, J.E., Sonpavde, G.P., Loriot, Y., Durán, I., Lee, J.L., Matsubara, N., Vulsteke, C., Castellano, D., Wu, C., Campbell, M., Matsangou, M., Petrylak, D.P., Enfortumab Vedotin in Previously Treated Advanced Urothelial Carcinoma. *New Engl. J. Med.*, *384*, 12, 1125–1135, 2021. https://doi.org/10.1056/NEJMoa2035807.

35. Bota, D.A., Mason, W., Kesari, S., Magge, R., Winograd, B., Elias, I., Reich, S.D., Levin, N., Trikha, M., Desjardins, A., Marizomib alone or in combination with bevacizumab in patients with recurrent glioblastoma: Phase I/II clinical trial data. *Neuro-oncology Adv.*, *3*, 1, vdab142, 2021.
36. Patel, S., Petty, W.J., Sands, J.M., An overview of lurbinectedin as a new second-line treatment option for small cell lung cancer. *Ther. Adv. Med. Oncol.*, *13*, 17588359211020529, 2021. https://doi.org/10.1177/17588359211020529.
37. Davis, J.A., Shockley, A., Herbst, A., Hendrickson, L., Polatuzumab Vedotin for the Front-Line Treatment of Diffuse Large B-Cell Lymphoma: A New Standard of Care? *J. Adv. Practitioner Oncol.*, *14*, 1, 67–72, 2023. https://doi.org/10.6004/jadpro.2023.14.1.6.
38. Di, K., Lloyd, G.K., Abraham, V., MacLaren, A., Burrows, F.J., Desjardins, A., Trikha, M., Bota, D.A., Marizomib activity as a single agent in malignant gliomas: ability to cross the blood-brain barrier. *Neuro-oncology*, *18*, 6, 840–848, 2016. https://doi.org/10.1093/neuonc/nov299.
39. Hexner, E.O., Mascarenhas, J., Prchal, J., Roboz, G.J., Baer, M.R., Ritchie, E.K., Leibowitz, D., Demakos, E.P., Miller, C., Siuty, J., Kleczko, J., Price, L., Jeschke, G., Weinberg, R., Basu, T., Pahl, H.L., Orazi, A., Najfeld, V., Marchioli, R., Goldberg, J.D., Hoffman, R., Phase I dose escalation study of lestaurtinib in patients with myelofibrosis. *Leukemia Lymphoma*, *56*, 9, 2543–2551, 2015. https://doi.org/10.3109/10428194.2014.1001986.
40. Hong, D.S., Concin, N., Vergote, I., de Bono, J.S., Slomovitz, B.M., Drew, Y., Arkenau, H.T., Machiels, J.P., Spicer, J.F., Jones, R., Forster, M.D., Cornez, N., Gennigens, C., Johnson, M.L., Thistlethwaite, F.C., Rangwala, R.A., Ghatta, S., Windfeld, K., Harris, J.R., Lassen, U.N., Coleman, R.L., Tisotumab Vedotin in Previously Treated Recurrent or Metastatic Cervical Cancer. *Clin. Cancer Research an Off. J. Am. Assoc. Cancer Res.*, *26*, 6, 277, 2020. 1220–1228https://doi.org/10.1158/1078-0432.CCR-19-2962.
41. Trudel, S., Lendvai, N., Popat, R., Voorhees, P.M., Reeves, B., Libby, E.N., Richardson, P.G., Hoos, A., Gupta, I., Bragulat, V., He, Z., Opalinska, J.B., Cohen, A.D., Antibody-drug conjugate, GSK2857916, in relapsed/refractory multiple myeloma: an update on safety and efficacy from dose expansion phase I study. *Blood Cancer J.*, *9*, 4, 37, 2019. https://doi.org/10.1038/s41408-019-0196-6.
42. Martínez-Díez, M., Guillén-Navarro, M.J., Pera, B., Bouchet, B.P., Martínez-Leal, J.F., Barasoain, I., Cuevas, C., Andreu, J.M., García-Fernández, L.F., Díaz, J.F., Avilés, P., Galmarini, C.M., PM060184, a new tubulin binding agent with potent antitumor activity including P-glycoprotein over-expressing tumors. *Biochem. Pharmacol.*, *88*, 3, 291–302, 2014. https://doi.org/10.1016/j.bcp.2014.01.026.
43. Singh, D., Dheer, D., Samykutty, A., Shankar, R., Antibody drug conjugates in gastrointestinal cancer: From lab to clinical development. *J. Controlled Release: Off. J. Controlled Release Soc.*, *340*, 1–34, 2021. https://doi.org/10.1016/j.jconrel.2021.10.006.

44. Vahdat, L.T., Schmid, P., Forero-Torres, A., Blackwell, K., Telli, M.L., Melisko, M., Möbus, V., Cortes, J., Montero, A.J., Ma, C., Nanda, R., Wright, G.S., He, Y., Hawthorne, T., Bagley, R.G., Halim, A.B., Turner, C.D., Yardley, D.A., Glembatumumab vedotin for patients with metastatic, gpNMB overexpressing, triple-negative breast cancer ("METRIC"): a randomized multicenter study. *npj Breast Cancer*, *7*, 1, 57, 2021. https://doi.org/10.1038/s41523-021-00244-6.
45. Stone, R.M., Manley, P.W., Larson, R.A., Capdeville, R., Midostaurin: its odyssey from discovery to approval for treating acute myeloid leukemia and advanced systemic mastocytosis. *Blood Adv.*, *2*, 4, 444–453, 2018. https://doi.org/10.1182/bloodadvances.2017011080.
46. Karthikeyan, A., Joseph, A., Nair, B.G., Promising bioactive compounds from the marine environment and their potential effects on various diseases. *J. Genet. Eng. Biotechnol.*, *20*, 1, 14, 2022. https://doi.org/10.1186/s43141-021-00290-4.
47. Dhanisha, S.S. and Guruvayoorappan, C., Potential role of cGAS/STING pathway in regulating cancer progression. *Critical reviews in oncology/hematology. 178*, 103780, 277, 2022. https://doi.org/10.1016/j.critrevonc.2022.103780.
48. Gong, X., Gao, Y., Shu, J., Zhang, C., Zhao, K., Chitosan-Based Nanomaterial as Immune Adjuvant and Delivery Carrier for Vaccines. *Vaccines*, *10*, 11, 1906, 2022. https://doi.org/10.3390/vaccines10111906.
49. Jiménez-Gómez, C.P. and Cecilia, J.A., Chitosan: A Natural Biopolymer with a Wide and Varied Range of Applications. *Mol. (Basel Switzerland)*, *25*, 17, 3981, 2020. https://doi.org/10.3390/molecules25173981.
50. Aranaz, I., Alcántara, A.R., Civera, M.C., Arias, C., Elorza, B., Heras Caballero, A., Acosta, N., Chitosan: An Overview of Its Properties and Applications. *Polymers*, *13*, 19, 3256, 2021. https://doi.org/10.3390/polym13193256.
51. Bashir, S.M., Ahmed Rather, G., Patrício, A., Haq, Z., Sheikh, A.A., Shah, M.Z.U.H., Singh, H., Khan, A.A., Imtiyaz, S., Ahmad, S.B., Nabi, S., Rakhshan, R., Hassan, S., Fonte, P., Chitosan Nanoparticles: A Versatile Platform for Biomedical Applications. *Mater. (Basel Switzerland)*, *15*, 19, 6521, 2022. https://doi.org/10.3390/ma15196521.
51. Vasconcelos, D.P., de Torre-Minguela, C., Gomez, A.I., Águas, A.P., Barbosa, M.A., Pelegrín, P., Barbosa, J.N., 3D chitosan scaffolds impair NLRP3 inflammasome response in macrophages. *Acta Biomater.*, *91*, 123–134, 2019. https://doi.org/10.1016/j.actbio.2019.04.035.
52. Xian, H., Watari, K., Sanchez-Lopez, E., Offenberger, J., Onyuru, J., Sampath, H., Ying, W., Hoffman, H.M., Shadel, G.S., Karin, M., Oxidized DNA fragments exit mitochondria *via* mPTP- and VDAC-dependent channels to activate NLRP3 inflammasome and interferon signaling. *Immunity*, *55*, 8, 1370–1385.e8, 2022. https://doi.org/10.1016/j.immuni.2022.06.007.
53. Peterson, C., Denlinger, N., Yang, Y., Recent Advances and Challenges in Cancer Immunotherapy. *Cancers*, *14*, 16, 3972, 2022. https://doi.org/10.3390/cancers14163972.

54. Raghuvanshi, R. and Bharate, S.B., Preclinical and Clinical Studies on Bryostatins, A Class of Marine-Derived Protein Kinase C Modulators: A Mini-Review. *Curr. Top. Med. Chem.*, *20*, 12, 1124–1135, 2020. https://doi.org/10.2174/1568026620666200325110444.
55. Sun, M.K. and Alkon, D.L., Bryostatin-1: pharmacology and therapeutic potential as a CNS drug. *CNS Drug Rev.*, *12*, 1, 1–8, 2006. https://doi.org/10.1111/j.1527-3458.2006.00001.x.
56. Lee, H.S., Cho, H.J., Yu, R., Lee, K.W., Chun, H.S., Park, J.H., Mechanisms underlying apoptosis-inducing effects of Kaempferol in HT-29 human colon cancer cells. *Int. J. Mol. Sci.*, *15*, 2, 2722–2737, 2014. https://doi.org/10.3390/ijms15022722.
57. Mangialardi, P., Harper, R., Albertson, T.E., The pharmacotherapeutics of sarcoidosis. *Expert Rev. Clin. Pharmacol.*, *15*, 1, 51–64, 2022. https://doi.org/10.1080/17512433.2022.2032657.
58. Liu, Y., Law, B.K., Luesch, H., Apratoxin a reversibly inhibits the secretory pathway by preventing cotranslational translocation. *Mol. Pharmacol.*, *76*, 1, 91–104, 2009. https://doi.org/10.1124/mol.109.056085.
59. Pradhan, B., Nayak, R., Patra, S., Jit, B.P., Ragusa, A., Jena, M., Bioactive Metabolites from Marine Algae as Potent Pharmacophores against Oxidative Stress-Associated Human Diseases: A Comprehensive Review. *Mol. (Basel Switzerland)*, *26*, 1, 37, 2020. https://doi.org/10.3390/molecules26010037.
60. Srinivasan, R., Kannappan, A., Shi, C., Lin, X., Marine Bacterial Secondary Metabolites: A Treasure House for Structurally Unique and Effective Antimicrobial Compounds. *Mar. Drugs*, *19*, 10, 530, 2021. https://doi.org/10.3390/md19100530.
61. Alam, K., Mazumder, A., Sikdar, S., Zhao, Y.M., Hao, J., Song, C., Wang, Y., Sarkar, R., Islam, S., Zhang, Y., Li, A., *Streptomyces*: The biofactory of secondary metabolites. *Front. Microbiol.*, *13*, 968053, 2022. https://doi.org/10.3389/fmicb.2022.968053.
62. Sivalingam, P., Hong, K., Pote, J., Prabakar, K., Extreme Environment *Streptomyces*: Potential Sources for New Antibacterial and Anticancer Drug Leads? *Int. J. Microbiol.*, *2019*, 5283948, 2019. https://doi.org/10.1155/2019/5283948.
63. Liu, Z., Zhao, J.Y., Sun, S.F., Li, Y., Liu, Y.B., Fungi: outstanding source of novel chemical scaffolds. *J. Asian Nat. Prod. Res.*, *22*, 2, 99–120, 2020. https://doi.org/10.1080/10286020.2018.1488833.
64. Esposito, R., Ruocco, N., Viel, T., Federico, S., Zupo, V., Costantini, M., Sponges and Their Symbionts as a Source of Valuable Compounds in Cosmeceutical Field. *Mar. Drugs*, *19*, 8, 444, 2021. https://doi.org/10.3390/md19080444.

65. Kozhushnaya, A.B., Kolesnikova, S.A., Yurchenko, E.A., Lyakhova, E.G., Menshov, A.S., Kalinovsky, A.I., Popov, R.S., Dmitrenok, P.S., Ivanchina, N.V., Rhabdastrellosides A and B: Two New Isomalabaricane Glycosides from the Marine Sponge *Rhabdastrella globostellata*, and Their Cytotoxic and Cytoprotective Effects. *Mar. Drugs*, *21*, 11, 554, 2023. https://doi.org/10.3390/md21110554.
66. Bergonzi, M.C., Heard, C.M., Garcia-Pardo, J., Bioactive Molecules from Plants: Discovery and Pharmaceutical Applications. *Pharmaceutics*, *14*, 10, 2116, 2022. https://doi.org/10.3390/pharmaceutics14102116.
67. Katiyar, C., Gupta, A., Kanjilal, S., Katiyar, S., Drug discovery from plant sources: An integrated approach. *Ayu*, *33*, 1, 10–19, 2012. https://doi.org/10.4103/0974-8520.100295.
68. D'Incalci, M., Badri, N., Galmarini, C.M., Allavena, P., Trabectedin, a drug acting on both cancer cells and the tumour microenvironment. *Br. J. Cancer*, *111*, 4, 646–650, 2014. https://doi.org/10.1038/bjc.2014.149.
69. Wu, Z.Z., Ding, G.F., Huang, F.F., Yang, Z.S., Yu, F.M., Tang, Y.P., Jia, Y.L., Zheng, Y.Y., Chen, R., Anticancer Activity of Anthopleura anjunae Oligopeptides in Prostate Cancer DU-145 Cells. *Mar. Drugs*, *16*, 4, 125, 2018. https://doi.org/10.3390/md16040125.

10

Marine Bivalve-Derived Bioactive Compounds: Anticancer and Antimicrobial Effects

Suna Bektaş[1], Nahdhoit Ahamada Rachid[1], Önder Kılıç[2] and Serap Sancar[3]*

[1]Institute of Graduate Studies in Science, Istanbul University, Süleymaniye, Istanbul, Türkiye
[2]Section of General Biology, Biology Department, Faculty of Science, Istanbul University, Vezneciler, Istanbul, Türkiye
[3]Section of Molecular Biology, Biology Department, Faculty of Science, Istanbul University, Vezneciler, Istanbul, Türkiye

Abstract

Marine ecosystems possess a vast biodiversity of prokaryotic and eukaryotic organisms. Among all marine forms, mollusks play a crucial role in the food chain and ecosystem balance in those environments. Bivalve, a class of mollusks, harbors particularly in coastal regions and provides a sustainable food source with high-quality protein, omega-3 fatty acids, vitamin B12, and mineral contents. The extreme marine conditions, such as hypoxia, high temperature, acidification, and pathogenic microorganisms, lead to the evolution of novel bioactive compounds in marine bivalves like other marine invertebrate forms such as sponges, algae, crustaceans, and other mollusks. These compounds exhibit a diverse array of biochemical and therapeutic potentials. Therefore, in this work, we focused on the current literature regarding the bivalve-derived anticancer and antimicrobial bioactive compounds in peptide, polysaccharide, lipid, and lectin nature and their action mechanisms.

Keywords: Marine bivalves, anticancer activity, antimicrobial activity, bioactive peptides, bioactive polysaccharides, bioactive lipids, bioactive lectins

**Corresponding author*: ssancar@istanbul.edu.tr

Shaju S. S., Vipin. P. M. and Visakh. P.M. (eds.) Marine Biotechnology: Studies and Applications, (287–314) © 2025 Scrivener Publishing LLC

10.1 Introduction

The marine environments, encompassing vast biodiversity, have emerged as a promising source of such compounds related to supporting public health. Among the high biodiversity, marine invertebrates, including sponges, corals, bryozoans, mollusks, crustaceans, echinoderms, and tunicates, represent a significant portion. Their unique evolutionary pressures have led to the development of a rich array of chemically diverse compounds to defend against predators and competitors. Recent technological advancements have facilitated the discovery and isolation of marine compounds with novel mechanisms of action. These compounds often possess complex structures with high molecular weights and a distinctive halogen content due to the marine environment's parameters such as high salinity, low oxygen levels, and high pressure [1]. Notable chemical classes among these compounds include terpenoids, alkaloids, polysaccharides, peptides, and steroids [2]. The marine environment is a vast repository of potential medicinal compounds. Many marine-derived substances have

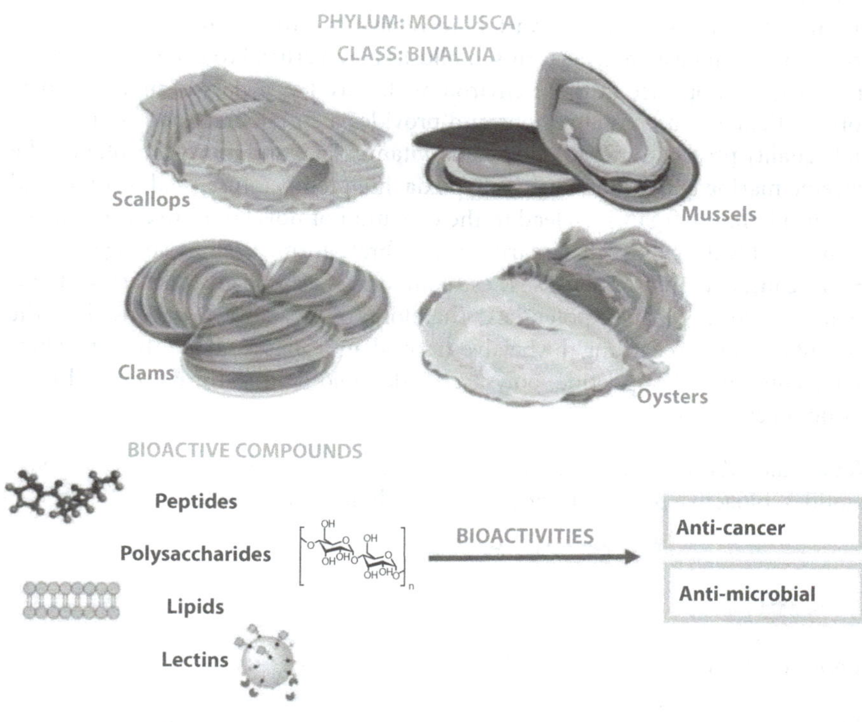

Figure 10.1 Bioactive molecules obtained from bivalves.

demonstrated the ability to modulate immune responses significantly offering opportunities for developing novel immunotherapies [3]. Proteins, peptides, and hydrolysates extracted from marine organisms exhibit a wide range of pharmacological properties, including antimicrobial, anticancer, antioxidant, antihypertensive, and anti-inflammatory activities [4].

The phylum Mollusca is one of the most expansive and diverse groups of marine animals. Within Mollusca, the class Bivalvia encompasses approximately 8,500 species [5] encompassing well-known invertebrates like clams, oysters, scallops, and mussels. Bivalves are found in various marine habitats from shallow coastal zones to deep environments [6]. Most of the bivalves are widely consumed by the public; therefore, they have commercial importance [7]. On the other hand, marine mollusks, particularly bivalves, offer a promising yet untapped resource for discovering novel bioactive compounds. In this section, we have mainly focused on the anticancer and antimicrobial activity of peptides, polysaccharides, lipids, and lectins derived from marine bivalves (Figure 10.1).

10.2 Bioactive Molecules from Marine Bivalves

10.2.1 Bioactive Peptides of Bivalves

Peptides, short chains of amino acids linked by covalent bonds, play a crucial role in regulating miscellaneous biological functions within cells and tissues. These molecular building blocks, typically composed of 10 to 100 amino acids, exert diverse effects, influencing processes ranging from cell signaling to metabolism. They can influence cell–cell communication and function as neurotransmitters/hormones. They also play innate immunity and may have antimicrobial activity. Marine bivalves live under extreme conditions, like polluted areas and exposure to pathogenic microorganisms, which leads to the production of unique peptides to survive in the environment [8].

10.2.1.1 Anticancer Peptides from Bivalves

Conventional chemotherapy is limited by drug resistance and severe adverse effects necessitating a search for novel anticancer agents. The anticancer peptides derived from marine organisms indicate great diversity in structure, including linear, cyclic, and basic peptides especially rich in lysine amino acids. The secondary structure of peptides could be α-helix, β-sheet, cyclic, and random coil [9]. Most of the bioactive molecules were defined in species that have been used for food and medicinal importance for years. One of these species is *Mytilus coruscus*, consumed as a valued food and used for the treatment of reproductive disorders and to strengthen the immune system [10]. Kim *et al.* [11] defined an anticancer peptide with AFNIHNRNLL amino acid sequence in the enzymatic (pepsin) hydrolysates of the mussels *M. coruscus* and observed that the peptide reduced cell viability and induced apoptosis in human cancer cells, including PC-3 prostate cancer cells, A549 lung cancer cell, and MDA-MB 231 breast cancer cells. Then, Kim *et al.* [12] isolated another peptide from an endemic clam of East Asian waters, Manila clam *Ruditapes philippinarum* (Adams and Reeve) by enzymatic (alpha-chymotrypsin) hydrolysis and investigated the anticancer activity on the same cell lines. The isolated peptide sequence with AVLVDKQCPD exhibited higher anticancer activity on the lung cancer cells. Both peptides represented selective anticancer activity with no effect on liver cells while inducing apoptosis in the cancer cells [11, 12]. The other peptide isolated from the latter species, *R. philippinarum* by enzymatic hydrolysis, is RPOI-1, which is a tetrapeptide sequence with Asp–Trp–Pro–His. RPOI-1 suppressed the viability of DU-145 human prostate cancer cells in a concentration- and time-dependent manner. RPOI-1 also induced apoptosis and G2 phase arrest [13]. In another study, the BCP-A peptide was defined in *Tegillarca granosa*, a clam species with significant ecological and economic value found in coastal and estuarine regions of Asia [14]. The peptide has three amino acids (Trp–Pro–Pro) and decreased cell viability in tested PC-3 and DU-145 human prostate cancer cells, H-1299 non-small cell lung carcinoma cells, and HeLa cervical cancer cells. In PC-3 cells, the peptide exhibited higher cytotoxicity and induced apoptotic cell death. Besides, the BCP-A peptide showed antioxidant properties [15].

Asiatic hard clam *Meretrix meretrix* (Linnaeus, 1758) is a historically significant marine species prized for both culinary and medicinal purposes in China. This species is widely distributed across the coastal regions of South and Southeast Asia [16]. Mere15 was defined in *M. meretrix* and selectively inhibited the viability of several human cancer cell lines, including HeLa, Aspc-1, A549, RKO, MCF-15, and BEL-7402 *in vitro*. The sensitivity of A549 human lung adenocarcinoma cells to Mere15 was higher than other cancer cells. The Mere15 led to cell cycle arrest and induced the intrinsic pathway of apoptosis by activating caspases, loss of mitochondrial membrane potential, and inducing DNA fragmentation. Besides, Mere15 inhibited the growth of A549 xenograft in nude mice in a dose-dependent manner [17]. Then, the same research group also observed that this peptide reduced cell adhesion and motility by decreasing the expression and secretion of matrix metalloproteinases [18]. In another study, Liu *et al.* [19] showed that Mere15 led to the inhibition of the proliferation of K562 human leukemia cells by disassembling microtubules, apoptosis, and cell cycle arrest. Moreover, Fan *et al.* [20] obtained a recombinant form of this protein and called it re-MM15. Re-MM15 pointed out the sequence similarity with tropomyosin of some marine invertebrates and cytotoxic effects on different cancer cell lines while having less cytotoxicity to NIH/3T3 fibroblast cells. Among the tested cancer cell lines, U87 human glioblastoma cells exhibited the highest sensitivity with an IC_{50} value of 2.12 μM, and re-MM15 led G2/M phase arrest in U87 cells by promoting tubulin polymerization and apoptosis. Furthermore, re-MM15 decreased U87 cell motility by inhibiting FAK/Akt/MMPs signaling and suppressed the tumor growth of U87 xenograft in nude mice [20]. Similarly, Xu *et al.* [21] found that Mere15 inhibited the proliferation of non-small-cell lung cancer and cell motility by decreasing the expression of MMP-2, MMP-9, and Snail and increasing the expression of E-cadherin. Mere-15 inhibited the PI3K/Akt/mTOR pathway in NSCLC cells and suppressed xenograft growth in nude mice bearing NCI-H460 cancer cells [21]. The other peptide isolated from *M. meretrix* is a coelomic fluid protein, MML, which has a 40-kDa molecular weight [22]. MML was reported to inhibit the cell proliferation of human hepatoma BEL-7402, human breast cancer MCF-7, and human colon cancer HCT116 cells without affecting normal cells, including human breast epithelial cell MCF-10A and murine fibroblast NIH3T3 cells. The inhibition was more significant in human hepatoma BEL-7402, and treated BEL-7402 cells showed membrane disruption and LDH leakage. MML was also demonstrated to inhibit microtubule polymerization and lead to G2/M arrest in the cell cycle [22].

Sinonovacula constricta (Lamarck, 1818), an economically and nutritionally important bivalve species, thrives in the intertidal and estuarine habitats of China, Japan, and Korea [23]. Huang *et al.* [24] isolated 12 peptides, SCH-P1–P12, by enzymatic hydrolysis from *S. constricta*. Two of these peptides, SCH-P9 and SCH-P10, inhibited the viability of human prostate cells DU-145 and PC-3 cells in a time- and dose-dependent manner more than the remaining 10 peptides. Although SCH-P10 increased the number of cells at G0/G1 and decreased the G2/M phase cells in both cell lines, SCH-P9 did not cause any significant change in the number of cells at different cell cycle stages. However, both peptides induced apoptotic cell death in DU-145 and PC-3 cells by quantification of phosphatidylserine externalization with Annexin V/FITC staining [24]. The other studied source organisms belong to the genus *Arca*. Song *et al.* [25] purified two proteins, namely, G-4-2 and G-6; the latter has a saccharide from

A. subcrenata (Lischke, 1869), which is consumed as a seafood in China and has been used for centuries in traditional Chinese medicine with its properties [26]. Two peptides reduced cell viability in human tumor cells including HeLa (cervical carcinoma cells), HL-60 (leukemia cells), and KB (oral epithelial cancer cells). Then, Hu *et al.* [27] studied a polypeptide fraction, P2, from *A. subcrenata* and observed that the fraction has both *in vitro* and *in vivo* antitumor effects. P2 inhibited cell viability in various tumor cell lines, especially HeLa and HT-29 with IC_{50} values of 11.43 and 13 μg/ml, respectively. P2 showed less cytotoxicity in normal liver cells, L-02. P2 administration was also well tolerated by mice and reduced the growth of hepatoma cells H-22 and sarcoma cells S-180 xenograft tumors in a dose-dependent manner [27]. The same research group also showed that P2 inhibited the secretion of inflammatory cytokines, IL-6, and TNF-α from human cervical cancer HeLa cells and nitric oxide production from activated RAW264.7 macrophage cells. P2 also downregulated the COX-2 and iNOS protein expression and gene expression of IL-6 cytokine and IL-8 chemokine in a dose-dependent manner in HeLa cells. Therefore, P2 has the inhibitory effects on cancer-related inflammation [28]. This fraction was renamed as PAS (polypeptide fraction from *A. subcrenata*), and Hu *et al.* [29] observed that it inhibited the cell viability in human colon cancer HT-29 cells in a concentration-dependent manner with no toxicity on normal colon cells, IEC-6. PAS was also indicated to lead to cell cycle arrest at the G2/M phase by lowering the expression of Cyclin B1 and Cdc 2 and induced apoptosis by cleavage of caspase 3, PARP, and decreasing Bcl-2/Bax ratio in HT-29 cells. These effects of PAS were not seen in IEC-6 cells. Moreover, PAS was reported to reduce the growth of HT-29 xenograft tumors in nude mice with little effect on normal mucosa. Additionally, PAS's inhibitory effects on HT-29 cells and tumor tissues were linked to the suppression of IGF-1R phosphorylation and Akt/mTOR pathway and a reduction in ATP production [29]. Chen *et al.* [30] isolated a novel antioxidant and anticancer peptide, H3, from the visceral mass of *A. subcrenata.* The antiproliferative activity of H3 was tested on cervix epithelioid carcinoma (HeLa), lung adenocarcinoma cell lines (A549, SPC-A-1), human hepatocarcinoma (HepG2), colon adenocarcinoma (HT-29), and normal human liver cells (L-02). H3 was shown to have selective toxicity to cancer cells without affecting the viability of L-02 cells. The most sensitive cells were reported to be HepG2, HT-29, and HeLa cells with IC_{50} values of 10.1, 10.5, and 10.8 μg/ml, respectively. The amino acid composition and antioxidant activity of H3 could play a role in anticancer activity [30].

The other marine species, *Arca inflata* (Reeve, 1844), is widely in the Asian coastal regions. Xu *et al.* [31] purified a novel antitumor peptide, J2-C3, from the visceral mass of *A. inflata*. J2-C3 protein is rich in lysine, and α-helix, β-sheet, and β-turn secondary structures were determined in the peptide. J2-C3 inhibited the cell viability of human lung adenocarcinoma cells A549, human colorectal cancer cells HT-29, and human hepatocarcinoma cells HepG2 *in vitro* in a dose- and time-dependent manner [31]. Another peptide is J2-C2, which was isolated from the mussels of *A. inflata* and has a similar secondary structure to J2-C3. J2-C2 reduced the viability of human lung adenocarcinoma cell lines (A549, SPC-A-1) and human hepatocarcinoma cell line (HepG2) [32]. Li *et al.* (2022) studied the antitumor mechanism of P6, which was a recently isolated peptide from the hemolymph of *A. inflata*, in colorectal carcinoma both *in vitro* and *in vivo*. P6 is rich in lysine amino acids and has a helical structure. P6 was demonstrated to inhibit the viability of colorectal cancer cells DLD-1, HT-29, HCT116, and SW620 in a concentration-dependent manner. DLD-1 and HT-29 had the lowest IC_{50} values of 2.14 ± 0.28 and 4.43 ± 0.15 μg/ml, respectively. P2 induced apoptotic cell death and cell cycle arrest at the S/G2 transition in a dose-dependent manner in both cell lines. P6 caused mitochondrial membrane potential changes and increased the levels of cleaved caspase-3, cleaved PARP, proapoptotic proteins, Bax, and cytochrome c. The phosphorylation of p38 was also increased with the P6 dose increase. P6 also led to the influx of Ca^{+2} levels and intracellular ROS in both colorectal cells. Moreover, P6 suppressed the growth of HT-29 tumors in xenograft nude mice by inducing apoptosis, and major organ damage was not observed in mice treated with P6 [33].

In Table 10.1, we present all the anticancer peptides mentioned above. Despite promising preclinical evidence, no bivalve-derived anticancer peptide has yet advanced to clinical trials. A limited knowledge of the structure–function relationship of these peptides hinders drug development efforts. However, the growing number of identified anticancer peptides from marine species, including bivalves, offers hope. These peptides often exhibit selective activity, low drug resistance, and biocompatibility making them attractive candidates for drug development. Furthermore, the studies suggest potential synergistic effects between bivalve anticancer peptides and conventional chemotherapeutic agents [34]. This underscores the potential of bivalve peptides, such as P2 from *Arca subcrenata*, P6 from *A. inflata*, and MM15 from *M. meretrix*, as promising candidates for future anticancer drug development.

Table 10.1 Anticancer peptides of Bivalves.

Peptide	Source organism	Experimental model	Action mechanisms	Reference
Anticancer peptide AFNIHNRNLL	*Mytilus coruscus*	Cytotoxic to human prostate PC-3, human breast cancer MDA-MB 231, and human lung cancer A549 cells	Apoptosis	[11]
Anticancer peptide AVLVDKQCPD	*Ruditapes philippinarum*	Cytotoxic to human prostate PC-3, human breast cancer MDA-MB 231, and human lung cancer A549 cells	Apoptosis	[12]
BCP-A peptide	*Tegillarca granosa*	Human prostate cancer PC-3 and DU-145 cells, non-small-cell lung carcinoma H-1299 cells, and cervical cancer HeLa cells	Decreased cell viability, induced apoptosis in PC-3 cells	[15]
G-4-2 and G6	*Arca subcrenata*	Human cervical carcinoma cells HeLA, human leukemia HL-60 cells, and human oral epithelial cancer KB cells	Decreased cell viability	[25]
H3	*Arca subcrenata*	Human cervix epithelioid carcinoma (HeLa), lung adenocarcinoma cell lines (A549, SPC-A-1), human hepatocarcinoma (HepG2), colon adenocarcinoma (HT-29), and normal human liver cells (L-02)	Selective toxicity to cancer cells by its antioxidant activity	[30]

(*Continued*)

Table 10.1 Anticancer peptides of Bivalves. (*Continued*)

Peptide	Source organism	Experimental model	Action mechanisms	Reference
J2-C2	*Arca inflata*	Human lung adenocarcinoma cell lines (A549, SPC-A-1) and human hepatocarcinoma cells HepG2	Decreased cell viability	[32]
J2-C3	*Arca inflata*	Human lung adenocarcinoma cells A549, human colorectal cancer cells HT-29, and human hepatocarcinoma cells HepG2	Decreased cell viability	[31]
Mere15	*Meretrix meretrix*	A549 lung adenocarcinoma cell line	Cell cycle arrest, apoptosis	[17]
Mere15	*Meretrix meretrix*	A549 cell xenograft tumor model in nude mice	Suppressed tumor growth	[18]
Mere15	*Meretrix meretrix*	Human K562 chronic myelogenous leukemia cells	Apoptosis induction, cell cycle arrest, and microtubule disassembly	[19]
MML	*Meretrix meretrix*	Human hepatoma BEL-7402, human breast cancer MCF-7, and human colon cancer HCT116 cells	Possibly an increase in membrane permeability and a decrease in tubulin polymerization	[22]
P2	*Arca subcrenata*	More cytotoxic to human colorectal cancer cells HT-29 and cervical cancer HeLa cells than other tested cancer cell lines Hepatoma cells H-22 and sarcoma cells S-180 xenograft mice model	Decreased cell viability Suppressed tumor growth	[27]

(*Continued*)

Table 10.1 Anticancer peptides of Bivalves. (*Continued*)

Peptide	Source organism	Experimental model	Action mechanisms	Reference
P2 (polypeptide fraction)	*Arca subcrenata*	Human cervical cancer HeLa cells	Reduced proinflammatory cytokine levels and immune-related protein expression	[28]
P6	*Arca inflata*	Human colorectal cancer cells HT-29, DLD-1, SW620, and HCT116 HT-29 cells xenograft tumor model in nude mice	Apoptosis, cell cycle arrest Suppressed tumor growth by apoptosis	[33]
PAS (polypeptide fraction)	*Arca subcrenata*	Human colorectal cancer HT-29 cells HT-29 cell xenograft tumor model in nude mice	Apoptosis and cell cycle arrest Suppressed tumor growth	[29]
Recombinant Mere15		NCl-H460 cell xenograft tumor model in nude mice	Suppressed tumor growth	[21]
Recombinant MM15		U87 glioma cells U87 cell xenograft tumor model in nude mice	Cell cycle arrest, apoptosis, inhibition of migration and invasion of U87 cells through downregulating FAK/Akt/MMPs signaling. Suppressed tumor growth	[20]
RPOI1	*Ruditapes philippinarum*	Human prostate cancer DU-145 cells	Inhibited proliferation of DU-145 cells, induced apoptosis, and cell cycle arrest	[13]
SCH-P9 and SCH-P10	*Sinonovacula constricta*	Human prostate cancer PC-3 and DU-145 cells	Decreased cell viability, apoptosis, cell cycle arrest	[24]

10.2.1.2 Antimicrobial Activities of Bivalve's Peptides

Producing new antimicrobial agents is important and urgent because of the regular multiplication of drug-resistant microorganisms. In this regard, exploiting new sources can allow antimicrobials with new characteristics and properties to be found. The marine environment houses a broad biodiversity, including plants, animals, and microorganisms of various features and living conditions. Due to its high ecological diversity, the marine world constitutes fascinating laboratories and industries for research and development. Peptides are the attractive components of marine organisms. They include most bioactive compounds [35], usually coming from the hydrolysis activities of enzymes, like trypsin, pepsin, proteinases, chymotrypsins, and alcalase, which are reported on the hydrolysis of the parent proteins [36–38].

Bivalves are known for their highly developed innate immune system and their less developed acquired one [39, 40]. Antimicrobial peptides (AMPs) are determined as major components of the innate immune systems of these animals. Most marine antimicrobial peptides are thought to be obtained from mussels [38, 41]. The lifestyle of these animals can largely contribute to the production of these agents. Briefly, marine mussels are known to attach themselves to rocks or each other, remain immobile, and take up nutrients through their siphons, which will be transported to the gill, and assembled in the mucus, then transferred to the mouth [39, 41]. Being immobile, these animals are exposed to various microorganisms and toxic materials; however, they exhibit less development of their acquired immune system. It is reported that marine mussels produce antimicrobial peptides of different groups, including defensins, mytimacin, myticusin, bigdefensin, mytimycin, mytilin, defensin, myticins, and myticalins [42, 43].

In general, these AMPs are constituted of short peptides classified according to their length, structure, and composition, especially the presence or absence of S2 bridges [38]. AMPs are also characterized by their cationic, hydrophobic, and amphipathic features [38, 44]. The AMPs in the organisms act as a primary defense against microbial infection in the organism's fluids before the action of the immune system's cells [39, 45, 46]. On the other hand, the AMPs can act outside the microbial cells or intracellularly depending on their membrane permeability property [41, 47]. Oh *et al.* [39] determined myticusin-beta as an immune-related AMP extracted at a high level from the mantle of the horse mussel *M. coruscus*. This AMP, composed of 104 amino acids, is observed to display antibacterial activity against both Gram-positive and -negative bacteria.

While the Gram-positive bacteria targeted by this AMP include *Bacillus cereus*, *Staphylococcus aureus*, and *Clostridium perfringens*, the targeted Gram-negative bacteria include strains of *E. coli*, *Klebsiella pneumoniae*, and *Pseudomonas aeruginosa*. All these targeted bacteria are known for their human pathogenicity. The authors suggested the efficiency of this AMP as an alternative antibiotic [39].

Clam is known to be another marine mollusk group in terms of AMP richness. Yang *et al.* [48] explored the antibacterial activity and mode of action of TGH1, a novel AMP from *Tegillarca granosa*. They observed that 12.5 μg/ml of the peptide constitutes the minimum inhibitor concentration against the foodborne pathogen *Vibrio parahaemolyticus*. The mode of action was noted on the pathogen cell surface using transmission electron microscopy and laser confocal microscopy. The peptide aggregates on the cell surface and causes perforation of the cell wall with an integral destruction of the cell membrane. In addition, the TGH1 showed an antibiofilm effect by preventing biofilm formation and an increase in the inner membrane permeability. This antiparallel 55.6% β-sheet peptide was also observed without cytotoxicity on human normal hepatocytes. Therefore, the authors concluded the high potential application of this AMP in food pathogen control as an antimicrobial agent [48].

Another study focused on the two macins of the clam species *Venerupis philippinanum* [49]. In this study, VpMacin1 and VpMacin2 are identified as two peptides synthesized in the animal's gills and hepatopancreas, respectively. These peptides have been studied for their antimicrobial activities against *Vibrio anguillarum*. Upon infection, the hemocytes of the animal induce the secretion of these AMPs. After 6 days of the infection, the mortality rate of the pathogen was observed to be 22% and 49% for VpMacin1 and VpMacin2, respectively. In addition, the recombinants of these two proteins were tested against *E. coli* cells. The recombinant rVpMacin2 showed broad-spectrum antimicrobial and antibiofilm activities. The permeability of the pathogenic cell membrane was increased after the treatment with the rAMPs. Furthermore, the treatment of rVpMacin-1 induced phagocytic and chemotactic effects. Therefore, these two macin AMPs are naturally secreted by the animal to enhance its defense system, and studying their potential against drug resistance can result in promising results [49].

10.2.2 Bioactive Polysaccharides of Bivalves

Polysaccharides are complex carbohydrates composed of multiple monosaccharide units linked together by glycosidic bonds [50]. Bioactive polysaccharides are known as sugar-based functional polysaccharides produced from living organisms and/or have biological effects on organisms [51]. Bioactive polysaccharides have become the focus of research as therapeutic agents and have several advantages, including biodegradability, non-toxicity, and biocompatibility making them promising candidates for treating various chronic diseases [52]. In addition, studies have revealed that polysaccharides exhibit various biological activities such as anti-inflammatory, anticancer, antioxidant, and antiviral [53–55]. Up until now, natural polysaccharide compounds have primarily been extracted from terrestrial plants or fungi. However, the aquatic environment (marine and freshwater) has much higher biodiversity and is therefore a unique source of a huge amount of polysaccharide compounds with different bioactivities [56]. The structures and biological activities of polysaccharides from aquatic organisms are quite different from those of polysaccharides from terrestrial ones [57]. Polysaccharides derived from aquatic environments represent a promising alternative to traditional carbohydrate-based therapeutics [58]. Bivalves are a major aquatic source of polysaccharide bioactive compounds. In this section, the anticancer and antimicrobial effects of polysaccharides, which are obtained from bivalves, were discussed.

10.2.2.1 Anticancer Effects of Bivalve's Polysaccharides

A study investigated the effects of polysaccharides obtained from five bivalve species, including *Perna viridis* and *various Donax clams*, on cell viability in human cancer cell lines. MTT assays revealed that these polysaccharides inhibited the growth of MDA-MB-231 (breast), HeLa (cervical), HepG2 (liver), and HT-29 (colon) cancer cells. According to MTT results, it was revealed that the polysaccharide that showed the strongest antiproliferative effect in all cell lines was *D. variabilis* polysaccharide. For this reason, the apoptotic effects of *D. variabilis* polysaccharide in these cell lines were analyzed by AO/EtBr staining. Based on the research, these polysaccharides have strong antiproliferative effects, and it has been pointed out that these polysaccharides may have promising potential as anticancer agents with further studies [59]. The same study pointed out that the effects of polysaccharides obtained from clam species (*Donax variabilis, Donax incarnatus, Donax cuneatus, Sunetta scripta*) and mussel (*Perna viridis*) were investigated in normal (Vero) and two lung cancer cell lines [lung adenocarcinoma (A549) and lung large-cell carcinoma (NCI-H460)]. Although all of the polysaccharides showed significant inhibitory effects on the growth of lung cancer cells, the polysaccharide obtained from *D. variabilis* exhibited the highest cytotoxicity [54]. Similarly, a previous study showed that polysaccharide obtained from *D. variabilis* induced reactive oxygen species (ROS)-induced mitochondrial-mediated apoptosis in lung cancer cells (A549) [60]. In another study, sulfated polysaccharides isolated from *Cerastoderma edule* were shown to cause antiproliferative effects on chronic myelogenous leukemia (K562) and relapsed acute lymphoblastic leukemia cell lines (MOLT4) [61]. Polysaccharides isolated from oysters, *Crassostrea gigas*, were reported to exhibit antitumor activity against HepG2 cells without toxicity to normal cells [62]. In the study conducted by Liao *et al.* [63], *in vitro* antioxidant and anticancer activities of polysaccharides (CFPS-2) isolated from the clam *Corbicula fluminea* were evaluated. CFPS-2 demonstrated a dose-dependent increase in antioxidant activity and significantly inhibited the growth of human gastric cancer cells (SGC7901) and ovarian carcinoma cells (SKOV3 and A2780). In a study conducted by Jiang *et al.* [64], anticancer effects were investigated by content analysis of three polysaccharide fractions (CSPS1, CSPS2, CSPS3) isolated from *Cyclina sinensis.* According to the analyses performed in this study, it was revealed that the composition and properties of CSPS-3, which has higher protein, uronic acid, sulfate content, and a more complex monosaccharide composition, were quite different from those of CSPS-1 and CSPS-2. In addition, CSPS-3 has antioxidant properties and a strong inhibitory effect on the growth of human gastric cancer BGC-823 cells. In a study conducted to investigate the anticancer activity of water-soluble polysaccharide (WSP) isolated from the mantle of Freshwater pearl mussel, *Hyriopsis cumingii*, MTT was performed on human hepatocarcinoma HepG2 cells. It was revealed that mRNA and protein levels of c-myc, bcl-2, and cyclin D1 decreased in the group to which this polysaccharide was applied in HepG2 cells. The results of the study showed that WSP exhibits strong anticancer activity [65]. Zhang *et al.* [66] isolated two water-soluble polysaccharides, PEF1 and PEF2, from *R. philippinarum.* According to the results, PEF1 showed higher cytotoxicity than PEF2 on human hepatoma SMMC-7721 cells. Another study showed that protein-bound polysaccharide, CFPS-1, isolated from *Corbicula fluminea*, suppressed the proliferation of human breast cancer MCF-7 and MDA-MB-231 cells in a concentration-dependent manner by inducing apoptosis and inhibiting the cell cycle in the S phase [67]. Polysaccharides BEP1, BEP2, and BEP3, isolated from *Bullacta exarata*, demonstrated dose-dependent antioxidant activities. Among these, BEP3 exhibited the strongest antioxidant properties [68]. It was also reported that BEP3 exhibited anticancer activity in breast cancer (Bcap37), pancreatic cancer (SW1990), and cervical cancer (HeLa) cells [68]. A sulfated polygalactan from the bivalve oyster *Crassostrea madrasensis* attenuated NF-κB signaling in LPS-activated macrophages leading to a reduction in cytokine production and reactive oxygen species [69].

10.2.2.2 Antimicrobial Potentials of Polysaccharides from Marine Bivalves

Antimicrobial activity from polysaccharides has received little attention compared to the bioactivity of polypeptides. This activity in polysaccharides derived from marine bivalves is also in its early study stage. The main antimicrobial activity registered from polysaccharides isolated from these animals is the antiviral activity. A glucosaminoglycan identified from a scallop was reported to have an HSV-1 inhibitory effect [70–72]. This polysaccharide acts to reduce the cytopathic effect of the virus and inhibit its replication. The experimentation of this study showed a positive correlation between the polysaccharide antiviral effectivity and the treatment period [71, 72]. In another study, Amornrut *et al.* [73] isolated a galactan sulfate from the marine clam species *Meretrix petechialis*. This sulfated polysaccharide was found to exhibit potential in antiviral drugs' development [72, 73]. It is composed of homogenous D-galactose entities with β-(1-3)-glycosidic linkage. The antiviral activity of the galactan was observed against HIV (human immunodeficiency virus) [73, 74].

Besides the antiviral activities reported from bivalve organisms, a recent study discovered the antibacterial activity of two polysaccharides extracted from the razor clam *Solen marginatus*. The SM-CP and SM-DP are sulfated polysaccharides with bioactive features such as anti-coagulation, antioxidant, and antibacterial activity. Their antibacterial activities were against both Gram-negative and -positive bacteria, including *S. aureus*, *L. monocytogenes*, *S. enterica*, and *E. Coli*, with the highest activity observed against *S. aureus* [75]. Chitosan is known as a common polysaccharide with higher bioactive potential. It is extracted from chitin, which is a main component of animal and insect shells [76]. Varma and Vasudevan [77] processed an extraction of chitosan from the horse mussel *Modiolus modiolus*. This chitosan was characterized by various functional groups of organic compounds such as C–OC– stretching and CH_2– stretching. In addition, this polysaccharide was found to be composed of higher sulfur and hydrogen, and the degree of acetylation was evaluated to be 57.43%. The authors explored the antibacterial activities of this extract, and they found that all tested bacteria were inhibited by the polysaccharide at a concentration of 200 μg/ml with the highest activity registered against *E. coli* and *B. subtilis* [77].

10.2.3 Bioactive Lipids of Bivalves

Lipids, which are essential metabolites in mussels, hold significant promise for commercialization as health-promoting functional foods or dietary supplements [6]. Omega-3 reduces the risk of cancer and cardiovascular diseases, and mollusk and crustacean species are important for nutritional sources of omega-3 (n−3) very long-chain polyunsaturated fatty acids (VLC-PUFAs), such as eicosapentaenoic (EPA) and docosahexaenoic acids (DHA) [78].

10.2.3.1 Anticancer Effects of Lipid from Bivalves

Studies on shellfish lipids have revealed promising anticancer properties. Polar lipid fractions from *Crassostrea gigas* demonstrated more inhibitory effects on cancer cell growth in HepG2, MDA-MB-231, and RD cells compared to neutral lipid fractions [79]. Lipid extracts from various shellfish species, including *Mytilus coruscus, Ruditapes philippinarum*, and *Perna canaliculus*, have exhibited cytotoxicity against multiple cancer cell lines. These extracts have shown potential to induce apoptosis and inhibit tumor growth *in vitro* [80–82]. For example, *Mytilus coruscus* extract, containing seven steroids and one polycyclic aromatic compound, exhibited strong cytotoxicity against AGS, DLD-1, and HeLa cells [82]. A phase I clinical trial of GLMLE, a lipid extract from *Perna canaliculus*, demonstrated good tolerability in patients with advanced breast and prostate cancer [83].

Furthermore, lipid extracts from various shellfish species, including the Pacific oyster, common Orient clam, Yesso scallop, Far Eastern mussel, and Japanese littleneck, have shown cytotoxicity against colon carcinoma, epithelial melanoma, glioblastoma multiforme, lung carcinoma, and adenocarcinoma cell lines [78]. These findings suggest that shellfish lipids, particularly those rich in omega-3 fatty acids, may hold promise as novel anticancer agents.

10.2.3.2 Antimicrobial Effects of Lipid from Bivalves

Lipids are one of the major molecular components of living beings' cells. Their function varies depending on the cells. They are classified into different classes such as free fatty acids, sterols, sterol-esters, phospholipids, and triglycerides [6]. Fatty acids are the most studied, and more beneficial attributes have been assessed regarding the antimicrobial potential of lipids.

Fatty acids are reported to show a broad spectrum of antimicrobial activities against different prokaryotic and eukaryotic cells such as Gram(−) and Gram(+) bacteria, fungi, archaea, and viruses [84]. In addition, free fatty acids are generally reported to act with low concentrations that cannot affect the non-target cells and tissues. Antimicrobial fatty acids are attributed to rare microbial resistance, reduced expression and dissemination of microbial virulence factors, are eco-friendly, act in synergy with other antimicrobial agents, have rare side effects, and are stable and non-corrosive [84]. The mechanisms of antimicrobial fatty acids depend on various factors such as the type of fatty acid, targeted microbes, and environmental conditions [84–86]. Antimicrobial lipids are reported in human, animal, plant, and microorganism cells. Their principal function is to ensure the protection of these organisms against external colonizing cells and toxins [84, 87].

Literature about the antimicrobial activities of bivalve lipids is seen to be less effective than that about antimicrobial peptides. This refers to studies with low quality conducted on this topic. However, the main reason might be difficulties in their extraction and characterization. Fatty acid (FA) extracted with ethyl acetate from the rock oyster *Saccostrea glomerata* was studied for its antibacterial, antifungal, and antiviral potentials. Results showed that the FA suppressed the growth of bacterial pathogens, including *P. aeruginosa*, *A. hydrophila*, *S. aureus*, *V. parahaemolyticus*, and *V. Harveyi*, for more than 15 mm of inhibition zone. In addition, the extract showed antifungal activities against some fungal species like *A. flavus*, *Fusarium* sp., *A. niger*, and *C. albucans*. Besides, the extract showed antiviral activity against the White Spot Syndrome Virus (WSSV) with a survival rate of 70% after the treatment of infected shrimp [88].

10.2.4 Bioactive Lectins from Bivalves

Lectins, glycan-binding proteins, recognize a diverse array of glycan structures enabling the deciphering of glycocodes that define oligosaccharide composition in glycoconjugates. Lectins encoded by bivalve genes have been observed to up- or downregulate in response to pathogenic microbial infections highlighting their role in responding to external stimuli [89]. The study on lectins from mollusks is particularly intriguing due to their reported antitumor, antiviral, and antimicrobial activities. Within the Phylum Mollusca, lectins isolated from species with economic or medical significance, such as mussels and oysters, hold special interest [90].

10.2.4.1 Anticancer Effects of Bivalve's Lectins

Galactose-binding lectins isolated from marine mussels have demonstrated significant anticancer potential. MytiLec, a lectin from *Mytilus galloprovincialis*, exhibited cytotoxicity against Burkitt's lymphoma Raji cells expressing Gb3, a cell surface glycosphingolipid [91]. Mechanistically, MytiLec triggers the activation of multiple cell death pathways, including the MAPK cascade, death receptor signaling, and mitochondria-mediated apoptosis, by interaction with glycan Gb3 [92]. Other lectins obtained from marine bivalve species, such as CGL from *Crenomytilus grayanus* and FL-96 from *Pinctada fucata*, have also shown anticancer activity. CGL-induced cytotoxicity, cell cycle arrest at G2/M phase, and apoptosis in Gb3-expressing Raji cells without any effects on erythroleukemia K562 cells [93] and cell death in breast cancer cells by recognizing Gb3. FL-96 inhibited the proliferation of HepG2 and C666-1 cancer cells [93]. SeviL, a lectin from *Mytilisepta virgata*, exhibited cytotoxicity against various cancer cell lines expressing a special oligosaccharide, asialo-GM1, at the cell surface. In HeLa cells, SeviL activated kinase MKK 3/6, p38 MAPK, and caspase-3/9 leading to apoptosis [94]. *In vivo* studies have demonstrated the anticancer potential of MytiLec-1. Intraperitoneal injection of MytiLec-1 in mice inhibited cancer cell growth in a dose-dependent manner. MytiLec-1, administered intraperitoneally to mice at doses of 1.0 and 2.0 mg/kg/day for five consecutive days, significantly suppressed cancer cell growth. *In vitro* studies using the MTT assay demonstrated the lectin's antiproliferative activity against U937 and HeLa cells. Further analysis of cell morphology and apoptosis-related gene expression (p53, Bax, Bcl-X, and NF-κB) suggested that MytiLec-1 induced apoptosis in these cells [95]. Jhou *et al.* [96] found that CGL could bind a range of glycans containing either α- or β-Gal/GalNAc residues, which suggests that CGL is a potential probe for carcinogenesis and metastatic ability caused by hyper-sialylation. CGL-biotin-SP successfully differentiated between HeLa cells and desialylated HeLa cells, and also lung and breast cancer cells with varying metastatic abilities [96]. Overall, these findings highlight the potential of marine lectins as promising candidates for novel anticancer therapies.

10.2.4.2 Antimicrobial Potentials of Bivalve's Lectins

Lectins are also known for their antimicrobial activities, and they are reported to be one of the large glycoproteins from bivalve mollusks [97, 98]. Hossain *et al.* [98] classified lectins identified from mollusks into F-type, S-type, C-type, P-type, S-type, R-type, ficolins- and chitinase-like lectins based on the similarity of their amino acids in the carbohydrate recognition domain. The *in vitro* study on the antifungal activity of lectin derived from mussel *Crenomytilus grayanus* revealed that this lectin is distributed in the tissues of the muscles, gills, hepatopancreas, and hemocytes of this species. Besides, the upregulation of this lectin and the inhibition of the fungal spore germination were observed after the animal infection by the yeast *Pichia pastoris* in this study. The study also confirmed this lectin's immuno-innate status and its overexpression upon fungal infection [97].

Another study explored the physical properties and antimicrobial activities of a recombinant lectin belonging to the mytilectin family of mussels. The observed results showed that the recombinant form of the lectin extracted from *Mytilus galloprovincialis* presents hemagglutination activity at a wide range of temperatures (20°C–40°C) and pH range between 7 and 9. The antimicrobial activity assay showed antibacterial, antibiofilm, and antifungal potentials. The antibacterial activities consisted of a bactericidal activity observed against *Pseudomonas aeruginosa* with a growth inhibition rate of 65% at 200 μg/ml. In addition, their antibacterial activity was observed with high inhibition zones around 50 and 100 μg/disk of lectin against *Bacillus cereus* and species of the genus *Shigella*. The antibiofilm potential was also observed in the inhibition of 15% of *P. aeruginosa* biofilm compared with the negative control. On the other hand, in this study, the antifungal activity demonstrated using the growth inhibition of *Aspergillus niger* and the agglutination of spores formed by *A. niger* and *Trichoderma reesei* [98].

Similarly, He *et al.* [99] investigated gene expression and antimicrobial properties of C-type lectin extracted from the pearl oyster species *Pinctada fucata martensii*. This study showed that the PmCTL-1 was a novel C-type lectin highly expressed in the mantle of the bivalve but also expressed at a low level in other animal compartments like the gill, hemocyte, gonad, adductor muscle, and hepatopancreas. The sequencing of the cDNA of this lectin showed a length of 636 bp encoding 149 amino acids. The gene expression responds positively to the bacterial stimulation showing this lectin's innate immune-response and the antibacterial potential [99].

10.3 Conclusion

The adverse effects of drugs, the emergence of antibiotic-resistant bacterial species, and the necessity for the development of drugs that do not show drug resistance by acting differently than traditional cancer drugs have led to an increase in interest in natural molecules produced by living beings.

Biotechnology and molecular biology developments allow the purification of natural products revealing their molecular structures and determining their modes of action. Developments in organic chemistry offer people the opportunity to synthesize these molecules. This allows the development of new drugs. There is a huge molluscan biodiversity in the marine ecosystems. Among the mollusks, bivalves have different forms, such as mussels, oysters, and scallops, and they are filter-feeding benthic organisms. These living beings are exposed to stress factors (hypoxia, high temperature, acidification, etc.) and pathogenic microorganisms due to changing environmental conditions. Those environmental factors lead them to synthesize a wide variety of bioactive molecules, and those molecules enable them to adapt to the changing environmental conditions by evolutionary mechanisms; thus, they can succeed in surviving in that environment. Bivalves, which do not have an adaptive immune system like many marine invertebrates, have achieved this thanks to their genes and evolution. Extracts obtained from bivalves have been used in traditional medicine for centuries to treat various diseases due to their anti-inflammatory, anticancer, antidiabetic, antioxidant, antimicrobial, and wound-healing effects. Overall, this section compiles studies on the anticancer and antimicrobial properties of bioactive peptides, polysaccharides, lipids, and lectins obtained from bivalves. However, considering the species diversity of bivalves and their significant potential for the development of bioactive compounds, we can conclude that many of the bivalves' bioactive molecules remain relatively unexplored. Therefore, the necessity of further studies on the purification of natural compounds from marine bivalves and the determination of their activities is important in identifying the new bioactive compounds.

References

1. Lindequist, U., Marine-derived pharmaceuticals–challenges and opportunities. *Biomol. Ther.*, 24, 6, 561, 2016.
2. López, Y., Cepas, V., Soto, S.M., The marine ecosystem as a source of antibiotics, in: *Grand Challenges in Marine Biotechnology*, P.H. Rampelotto and A. Trincone (eds.), Chapter 1, Springer, Cham, Switzerland, pp. 3–48, 2018.
3. Ahmad, B., Shah, M., Choi, S., Oceans as a source of immunotherapy. *Mar. Drugs*, 17, 5, 282, 2019.
4. Jo, C., Khan, F.F., Khan, M.I., Iqbal, J., Marine bioactive peptides: Types, structures, and physiological functions. *Food Rev. Int.*, 33, 1, 44–61, 2017.

5. Huber, M., Compendium of Bivalves 2. A full-color guide to the remaining seven families, in: *A systematic listing of 8'500 bivalve species and 10'500 synonyms*, ConchBooks, Harxheim, 2015.
6. Grienke, U., Silke, J., Tasdemir, D., Bioactive compounds from marine mussels and their effects on human health. *Food Chem.*, 142, 48–60, 2014.
7. Gyawali, P. and Hewitt, J., Faecal contamination in bivalve molluscan shellfish: can the application of the microbial source tracking method minimise public health risks? *Curr. Opin. Environ. Sci. Health*, 16, 14–21, 2020.
8. Kijjoa, A. and Sawangwong, P., Drugs and cosmetics from the sea. *Mar. Drugs*, 2, 2, 73–82, 2004.
9. Xie, M., Liu, D., Yang, Y., Anti-cancer peptides: classification, mechanism of action, reconstruction and modification. *Open Biol.*, 10, 7, 2020.
10. Xu, H., Guo, T., Guo, Y.F., Zhang, J.E., Li, Y., Feng, W., Jiao, B., Characterization and protection on acute liver injury of a polysaccharide MP-I from *Mytilus coruscus*. *Glycobiology*, 18, 1, 97–103, 2008.
11. Kim, E.K., Joung, H.J., Kim, Y.S., Hwang, J.W., Ahn, C.B., Jeon, Y.J., *et al.*, Purification of a novel anticancer peptide from enzymatic hydrolysate of *Mytilus coruscus*. *JMB*, 22, 10, 1381–1387, 2012.
12. Kim, E.K., Kim, Y.S., Hwang, J.W., Lee, J.S., Moon, S.H., Jeon, B.T., Park, P.J., Purification and characterization of a novel anticancer peptide derived from *Ruditapes philippinarum*. *Process Biochem.*, 48, 7, 1086–1090, 2013.
13. Yang, Z., Zhao, Y., Yan, H., Xu, L., Ding, G., Yu, D., Sun, Y., Isolation and purification of oligopeptides from *Ruditapes philippinarum* and its inhibition on the growth of DU145 cells *in vitro*. *Mol. Med. Rep.*, 11, 2, 1063–1068, 2015.
14. Shao, Y., Chai, X., Xiao, G., Zhang, J., Lin, Z., Liu, G., Population genetic structure of the blood clam, *Tegillarca granosa*, along the Pacific coast of Asia: isolation by distance in the sea. *Malacologia*, 59, 2, 303–312, 2016.
15. Chi, C.F., Hu, F.Y., Wang, B., Li, T., Ding, G.F., Antioxidant and anticancer peptides from the protein hydrolysate of blood clam (*Tegillarca granosa*) muscle. *J. Funct. Foods*, 15, 301–313, 2015.
16. Liu, B., Dong, B., Tang, B., Zhang, T., Xiang, J., Effect of stocking density on growth, settlement and survival of clam larvae, *Meretrix meretrix*. *Aquaculture*, 258, 1–4, 344–349, 2006.
17. Wang, C., Liu, M., Cheng, L., Wei, J., Wu, N., Zheng, L., Lin, X., A novel polypeptide from *Meretrix meretrix* Linnaeus inhibits the growth of human lung adenocarcinoma. *Exp. Biol. Med.*, 237, 442–450, 2012.
18. Wang, H., Wei, J., Wu, N., Liu, M., Wang, C., Zhang, Y., Wang, F., Liu, H., Lin, X., Mere15, a novel polypeptide from *Meretrix meretrix*, inhibits adhesion, migration and invasion of human lung cancer A549 cells *via* down-regulating MMPs. *Pharm. Biol.*, 51, 145–151, 2013.
19. Liu, M., Zhao, X., Zhao, J., Xiao, L., Liu, H., Wang, C., Cheng, L., Wu, N., Lin, X., Induction of Apoptosis, G0/G1 Phase Arrest and Microtubule

Disassembly in K562 Leukemia Cells by Mere15, a Novel Polypeptide from *Meretrix meretrix* Linnaeus. *Mar. Drugs*, 10, 11, 2596–2607, 2012.

20. Fan, Z., Xu, Q., Wang, C., Lin, X., Zhang, Q., Wu, N., A tropomyosin-like *Meretrix meretrix* Linnaeus polypeptide inhibits the proliferation and metastasis of glioma cells *via* microtubule polymerization and FAK/Akt/MMPs signaling. *Int. J. Biol. Macromol.*, 15, 145, 154–164, 2020.
21. Xu, B., Guo, M., Ma, L., Farooqi, A.A., Wang, L., Qiao, G., Liu, M., Zuo, L., Ye, H., Lin, X., *et al.*, Mere15, a novel polypeptide from *Meretrix meretrix*, inhibits proliferation and metastasis of human non-small cell lung cancer cells through regulating the PI3K/Akt/mTOR signaling pathway. *Neoplasma*, 68, 1181–1189, 2021.
22. Ning, X., Zhao, J., Zhang, Y., Cao, S., Liu, M., Ling, P., Lin, X., A novel antitumor protein extracted from *Meretrix meretrix* Linnaeus induces cell death by increasing cell permeability and inhibiting tubulin polymerization. *Int. J. Oncol.*, 35, 805–812, 2009.
23. Song, J., Li, L., Liu, Z., Li, Q., Ran, P., Sequence analysis and expression of a cDNA clone encoding tropomysin in *Sinonovacula constricta*. *Mol. Bio. Rep.*, 36, 315–321, 2009.
24. Huang, F., Ding, G., Yang, Z., Yu, F., Two novel peptides derived from *Sinonovacula constricta* inhibit the proliferation and induce apoptosis of human prostate cancer cells. *Mol. Med. Rep.*, 16, 5, 6697–6707, 2017.
25. Song, L., Ren, S., Yu, R., Yan, C., Li, T., Zhao, Y., Purification, Characterization and *in vitro* Anti-Tumor Activity of Proteins from *Arca subcrenata* Lischke. *Mar. Drugs*, 6, 3, 418–430, 2008.
26. Guo, X.S. and Li, Y., *Marine Chinese Traditional Medicines*, pp. 138–141, Sciences Press, Beijing, P.R. China, 2003.
27. Hu, X., Song, L., Huang, L., Zheng, Q., Yu, R., Antitumor effect of a polypeptide fraction from *Arca subcrenata in vitro* and *in vivo*. *Mar. Drugs*, 10, 2782–2794, 2012.
28. Wu, Y., Hu, X., Song, L., Zhu, J., Yu, R., The inhibitory effect of a novel polypeptide fraction from *Arca subcrenata* on cancer-related inflammation in human cervical cancer HeLa cells. *Sci. World J.*, 1, 768938, 2014.
29. Hu, X., Zheng, W., Luo, Y., Ou, X., Song, L., Zhang, S., *et al.*, *Arca subcrenata* polypeptides inhibit human colorectal cancer HT-29 cells growth *via* suppression of IGF-1R/Akt/mTOR signaling and ATP production. *Nut. Cancer*, 72, 2, 260–272, 2020.
30. Chen, L., Song, L., Li, T., Zhu, J., Xu, J., Zheng, Q., Yu, R., A New Antiproliferative and Antioxidant Peptide Isolated from *Arca subcrenata*. *Mar. Drugs*, 11, 6, 1800–1814, 2013.
31. Xu, J., Chen, Z., Song, L., Chen, L., Zhu, J., Lv, S., Yu, R., A New *in Vitro* Anti-Tumor Polypeptide Isolated from *Arca inflata*. *Mar. Drugs*, 11, 12, 4773–4787, 2013.

32. Zhu, J., Xu, J., Wang, Y., Li, C., Chen, Z., Song, L., *et al.*, Purification and structural characterization of a novel anti-tumor protein from *Arca inflata*. *Int. J. Biol. Macromol.*, 105, 103–110, 2017.
33. Li, C., Zhang, S., Zhu, J., Huang, W., Luo, Y., Shi, H., Yu, D., Chen, L., Song, L., Yu, R.A., Novel Peptide Derived from *Arca inflata* Induces Apoptosis in Colorectal Cancer Cells through Mitochondria and the p38 MAPK Pathway. *Mar. Drugs*, 20, 110, 2022.
34. Chinnadurai, R.K., Khan, N., Meghwanshi, G.K., Ponne, S., Althobiti, M., Kumar, R., Current research status of anti-cancer peptides: Mechanism of action, production, and clinical applications. *Biomed. Pharmacother.*, 164, 114996, 2023.
35. Hu, Y., Chen, J., Hu, G., Yu, J., Zhu, X., Lin, Y., Chen, S., Yuan, J., Statistical research on the bioactivity of new marine natural products discovered during the 28 years from 1985 to 2012. *Mar. Drugs*, 13, 202–221, 2015.
36. Najafian, L. and Babji, A.S., A reviews of fish-derived antioxidant and antimicrobial peptides: their production, assessment, and applications. *Peptides*, 33, 178–185, 2012.
37. Senevirathne, M. and Kim, S.K., Development of bioactive peptides from fish proteins and their health promoting ability. *Adv. Food Nutr. Res.*, 65, 235–248, 2012.
38. Shanmugam, S., Shankar, K., Ramachandiran, S., Naidu, K., Kalimuthu, K., Muthuvel, A., *In Vitro* Studies and Characterization of Tissue Protein from Green Mussel, *Perna viridis* (Linnaeus, 1758) for Antioxidant and Antibacterial Potential. *Int. J. Pept. Res. Ther.*, 26, 159–169, 2020.
39. Oh, R., Lee, M.J., Kim, Y.O., Nam, B.H., Kong, H.J., Kim, J.W., Park, J.Y., Seo, K.J., Kim, D.G., Myticusin-beta, antimicrobial peptide from the marine bivalve, *Mytilus coruscus*. *Fish Shellfish Immunol.*, 99, 342–352, 2020.
40. Tincu, J.A. and Taylor, S.W., Antimicrobial peptides from marine invertebrates. *Antimicrob. Agents Chemother.*, 48, 10, 3645–3654, 2004.
41. Cunha, S.A. and Pintado, M.E., Bioactive peptides derived from marine sources: Biological and functional properties. *Trends Food Sci. Technol.*, 119, 348–370, 2022.
42. Qin, C.L., Huang, W., Zhou, S.Q., Wang, X.C., Liu, H.H., Fan, M.H., Wang, R.X., Gao, P., Liao, Z., Characterization of a novel antimicrobial peptide with chiting-biding domain from *Mytilus coruscus*. *Fish Shellfish Immunol.*, 41, 2, 362–370, 2014.
43. Leoni, G., De Poli, A., Mardirossian, M., Gambato, S., Florian, F., Venier, P., Wilson, D.N., Tossi, A., Pallavicini, A., Gerdol, M., Myticalins: a novel multigenic family of linear, cationic antimicrobial peptides from marine mussels (*Mytilus spp.*). *Mar. Drugs*, 15, 8, 261, 2017.
44. Costa, R., Silva, N., Sarmento, B., Pintado, M., Delivery systems for antimicrobial peptides and proteins: towards optimization of bioavailability and targeting. *Curr. Pharm. Biotechnol.*, 18, 2, 108–120, 2017.

45. Mayer, A.M., Rodriguez, A.D., Berlinck, R.G., Fusetani, N., Marine pharmacology in 2007–8: marine compounds with antibacterial, anticoagulant, antifungal, anti-inflammatory, antimalarial, antiprotozoal, antituberculosis, and antiviral activities; affecting the immune and nervous system, and other miscellaneous mechanisms of action. *Comp. Biochem. Physiol. C*, 153, 191–222, 2011.
46. Wang, L., Qiu, L., Zhou, Z., Song, L., Research progress on the mollusc immunity in China. *Dev. Comp. Immunol.*, 39, 1-2, 2–10, 2013.
47. Semreen, M.H., El-Gamal, M.I., Abdin, S., Alkhazraji, H., Kamal, L., Hammad, S., Al-Awady, F., Waleed, D., Kourbaj, L., Recent updates of marine antimicrobial peptides. *Saudi Pharm. J.*, 26, 3, 396–409, 2018.
48. Yang, S., Dong, Y., Aweya, J.J., Xie, T., Zeng, B., Zhang, Y., Liu, G.M., Antimicrobial activity and acting mechanism of *Tegillarca granosa* hemoglobin-derived peptide (TGH1) against Vibrio parahaemolyticus. *Microb. Pathog.*, 147, 104302, 2020.
49. Xue, R., Han, Y., Li, F., Chen, L., Yang, D., Zhao, J., Identification, antibacterial activities and action mode of two macins from manila clam *Venerupis philippinarum*. *Fish Shellfish Immunol.*, 118, 411–420, 2021.
50. Guo, M.Q., Hu, X., Wang, C., Ai, L., Polysaccharides: structure and solubility. *Solubility Polysaccharides*, 2, 8–21, 2017.
51. Ullah, S., Khalil, A.A., Shaukat, F., Song, Y., Sources, extraction and biomedical properties of polysaccharides. *Foods*, 8, 8, 304, 2019.
52. Colegate, S.M. and Molyneux, R.J. (Eds.), *Bioactive natural products: detection, isolation, and structural determination*, CRC Press, Boca Raton, 2007.
53. Liu, Z., Zhang, Y., Ai, C., Tian, W., Wen, C., Song, S., Zhu, B., An acidic polysaccharide from *Patinopecten yessoensis* skirt prevents obesity and improves gut microbiota and metabolism of mice induced by high-fat diet. *Food Res. Int.*, 154, 110980, 2022.
54. Padmanaban, D., Samuel, A., Xavier, I.A., Ulaganathan, I., Sahayanathan, G.J., Raja, K., Chinnasamy, A., *In vitro* Anticancer and Antioxidant Activity of Partially Purified Poly-saccharides from Marine Bivalves Against Human Lung Cancer Cell Lines (A549 and NCI-H460). *J. Biol. Act. Prod. Nat.*, 12, 2, 94–110, 2022.
55. Xiang, X.W., Wang, R., Chen, H., Chen, Y.F., Shen, G.X., Liu, S.L., *et al.*, Structural characterization of a novel marine polysaccharide from mussel and its antioxidant activity in RAW264. 7 cells induced by H2O2. *Food Biosci.*, 47, 101659, 2022.
56. Tan, K., Lu, S.Y., Tan, K., Ransangan, J., Cai, X., Cheong, K.L., Bioactivity of polysaccharides derived from bivalves. *Int. J. Biol. Macromol.*, 126096, 2023.
57. Farhadi, A., Lv, L., Song, J., Zhang, Y., Ye, S., Zhang, N., *et al.*, Whole-transcriptome RNA sequencing revealed the roles of chitin-related genes in the eyestalk abnormality of a novel mud crab hybrid (*Scylla serrata*♀× *S. paramamosain*♂). *Int. J. Biol. Macromol.*, 208, 611–626, 2022.

58. Laurienzo, P., Marine polysaccharides in pharmaceutical applications: an overview. *Mar. Drugs*, 8, 9, 2435–2465, 2010.
59. Padmanaban, D., Samuel, A., Sahayanathan, G.J., Raja, K., Chinnasamy, A., Anticancer effect of marine bivalves derived polysaccharides against human cancer cells. *ISBAB*, 39, 102240, 2022.
60. Sahayanathan, G.J., Padmanaban, D., Raja, K., Chinnasamy, A., Anticancer effect of purified polysaccharide from marine clam *Donax variabilis* on A549 cells. *J. Food Biochem.*, 44, 11, e13486, 2020.
61. Aldairi, A.F., Ogundipe, O.D., Pye, D.A., Antiproliferative activity of glycosaminoglycan-like polysaccharides derived from marine molluscs. *Mar. Drugs*, 16, 2, 63, 2018.
62. Cheong, K.L., Xia, L.X., Liu, Y., Isolation and characterization of polysaccharides from oysters (*Crassostrea gigas*) with anti-tumor activities using an aqueous two-phase system. *Mar. Drugs*, 15, 11, 338, 2017.
63. Liao, N., Chen, S., Ye, X., Zhong, J., Wu, N., Dong, S., *et al.*, Antioxidant and anti-tumor activity of a polysaccharide from freshwater clam, *Corbicula fluminea. Food Funct.*, 4, 4, 539–548, 2013.
64. Jiang, C., Wang, M., Liu, J., Gan, D., Zeng, X., Extraction, preliminary characterization, antioxidant and anticancer activities *in vitro* of polysaccharides from *Cyclina sinensis. Carbohydr. Polym.*, 84, 3, 851–857, 2011.
65. Qiu, S., Huang, S., Huang, J., Pan, J., Zhang, W., Antitumor activity of the water-soluble polysaccharide from *Hyriopsis cumingii* in vitro. *Toxicol. Ind. Health*, 26, 3, 151–161, 2010.
66. Zhang, L., Liu, W., Han, B., Sun, J., Wang, D., Isolation and characterization of antitumor polysaccharides from the marine mollusk *Ruditapes philippinarum. Eur. Food Res. Technol.*, 227, 103–110, 2008.
67. Liao, N., Zhong, J., Zhang, R., Ye, X., Zhang, Y., Wang, W., *et al.*, Protein-bound polysaccharide from *Corbicula fluminea* inhibits cell growth in MCF-7 and MDA-MB-231 human breast cancer cells. *PLoS One*, 11, 12, e0167889, 2016.
68. Zhang, D., Wu, H., Xia, Z., Wang, C., Cai, J., Huang, Z., *et al.*, Partial characterization, antioxidant and antitumor activities of three sulfated polysaccharides purified from *Bullacta exarata. J. Funct. Foods*, 4, 4, 784–792, 2012.
69. Chakraborty, K., Krishnan, S., Joy, M., Polygalactan from bivalve *Crassostrea madrasensis* attenuates nuclear factor-κB activation and cytokine production in lipopolysaccharide-activated macrophage. *Carbohydr. Polym.*, 249, 116817, 2020.
70. Yu, N., Liu, S., Han, J.J., Sun, F.S., The depressive effect of glycosaminoglycan from scallop on type-I herpes simplex virus. *Acta Acad. Med. Qingdao Univ.*, 2, 111–114, 2008.
71. Wang, W., Wang, S.X., Guan, H.S., The antiviral activities and mechanisms of marine polysaccharides: an overview. *Mar. Drugs*, 10, 12, 2795–2816, 2012.

72. Wang, L.C., Di, L.Q., Li, J.S., Hu, L.H., Cheng, J.M., Wu, H., Elaboration in type, primary structure, and bioactivity of polysaccharides derived from mollusks. *Crit. Rev. Food Sci. Nutr.*, 59, 7, 1091–1114, 2017.
73. Amornrut, C., Toida, T., Imanari, T., Woo, E.R., Linhardt, R., Wu, S.J., Kim, Y.S., A new sulfated β-galactan from clams with anti-HIV activity. *Carbohydr. Res.*, 321, 1–2, 121–127, 1999.
74. Woo, E.R., Kim, W.S., Kim, Y.S., Virus-cell fusion inhibitory activity for the polysaccharides from various Korean edible clams. *Arch. Pharm. Res.*, 24, 514–517, 2001.
75. Souissi, N., Boughriba, S., Abdelhedi, O., Hamdi, M., Jridi, M., Li, S., Nasri, M., Extraction, structural characterization, and thermal and biomedical properties of sulfated polysaccharides from razor clam *Solen marginatus. RSC Adv.*, 9, 20, 11538–11551, 2019.
76. Rinaudo, M., Chitin and chitosan: Properties and applications. *Prog. Polym. Sci.*, 31, 603–632, 2006.
77. Varma, R. and Vasudevan, S., Extraction, characterization, and antimicrobial activity of chitosan from horse mussel *Modiolus modiolus. ACS Omega*, 32), 20224–20230, 2020.
78. Shin, J., Song, M.H., Yu, J.W., Ko, E.Y., Shang, X., Oh, J.W., *et al.*, Anticancer potential of lipophilic constituents of eleven shellfish species commonly consumed in Korea. *Antioxidants*, 10, 10, 1629, 2021.
79. Nguyen, M.V., Kakooza, D., Do, T.H.T., Tran, A.P., Nguyen, H.T., Tran, N.Q., Fatty Acid Composition and Anticancer Activity of Neutral and Polar Lipids of Pacific Oyster (*Crassostrea gigas*) Cultured in Khanh Hoa Coast in Vietnam. *Pol. J. Food Nutr. Sci.*, 74, 2, 169–176, 2024.
80. Kim, E.K., Kim, Y.S., Lee, S.J., Jeon, Y.J., Ahn, C.B., Kim, Y.T., *et al.*, Effect of partially purified lipid from the mussel *Mytilus coruscus* on apoptosis in cancer cells. *J. Korean Soc. Appl. Biol. Chem.*, 54, 59–65, 2011.
81. Kim, E.K., Kim, Y.S., Hwang, J.W., Moon, S.H., Jeon, B.T., Park, P.J., A partially purified lipid extracted from *Ruditapes philippinarum* suppresses cancer cell proliferation. *Food Sci. Biotechnol.*, 24, 2177–2183, 2015.
82. Wikarta, J.M. and Kim, S.M., Nitric oxide synthesis inhibition and cytotoxicity of Korean horse mussel *Modiolus modiolus* extracts on cancer cells in culture. *Cytotechnology*, 68, 879–890, 2016.
83. Sukumaran, S., Pittman, K.B., Patterson, W.K., Dickson, J., Yeend, S., Townsend, A., *et al.*, A phase I study to determine the safety, tolerability and maximum tolerated dose of green-lipped mussel (*Perna canaliculus*) lipid extract, in patients with advanced prostate and breast cancer. *Ann. Oncol.*, 2, 5, 1089–1093, 2010.
84. Desbois, A., Potential applications of antimicrobial fatty acids in medicine, agriculture and other industries. *Recent Pat. Anticancer Drug Discov.*, 7, 2, 111–122, 2012.

85. Yang., D., Pornpattananangkul., D., Nakatsuji., T., Chan, M., Carson, D., Huang, C.M., Zhang, L., The antimicrobial activity of liposomal lauric acids against Propionibacterium acnes. *Biomater*, 30, 6035–40, 2009.
86. Mil-Homens, D., Bernardes, N., Fialho, A.M., The antibacterial properties of docosahexaenoic omega-3 fatty acid against the cystic fibrosis multiresistant pathogen *Burkholderia cenocepacia. FEMS Microbiol. Lett.*, 328, 61–9, 2012.
87. Alves, E., Dias, M., Lopes, D., Almeida, A., Domingues, MR., Rey, F., Antimicrobial Lipids from Plants and Marine Organisms: An Overview of the Current State-of-the-Art and Future Prospects. *Antibiotics*, 9, 441, 2020.
88. Karthikeyan, S.C., Velmurugan, S., Donio, M.B.S., Michaelbabu, M., Citarasu, T., Studies on the antimicrobial potential and structural characterization of fatty acids extracted from Sydney rock oyster *Saccostrea glomerata. Ann. Clin. Microbiol. Antimicrob.*, 13, 332, 2014.
89. Venier, P., Varotto, L., Rosani, U., Millino, C., Celegato, B., Bernante, F., *et al.*, Insights into the innate immunity of the Mediterranean mussel *Mytilus galloprovincialis. BMC Genomics*, 12, 1–19, 2011.
90. Chernikov, O., Kuzmich, A., Chikalovets, I., Molchanova, V., Hua, K.F., Lectin CGL from the sea mussel *Crenomytilus grayanus* induces Burkitt's lymphoma cells death *via* interaction with surface glycan. *Int. J. Biol. Macromol.*, 104, 508–514, 2017.
91. Fujii, Y., Dohmae, N., Takio, K., Kawsar, S.M., Matsumoto, R., Hasan, I., *et al.*, A lectin from the mussel *Mytilus galloprovincialis* has a highly novel primary structure and induces glycan-mediated cytotoxicity of globotriaosylceramide-expressing lymphoma cells. *JBC*, 287, 53, 44772–44783, 2012.
92. Hasan, Sugawara, S., Fujii, Y., Koide, Y., Terada, D., Iimura, N., *et al.*, MytiLec, a mussel R-type lectin, interacts with surface glycan Gb3 on Burkitt's lymphoma cells to trigger apoptosis through multiple pathways. *I.Mar. Drugs*, 13, 12, 7377–7389, 2015.
93. Liu, P., Li, W., Peng, Y., Han, S., Liang, Z., Cen, Y., *et al.*, Molecular cloning, expression, and functional analysis of a putative lectin from the pearl oyster (*Pinctada fucata*, Gould 1850). *Fish Shellfish Immunol.*, 143, 109215, 2023.
94. Fujii, Y., Gerdol, M., Kawsar, S.M., Hasan, I., Spazzali, F., Yoshida, T., *et al.*, A GM1b/asialo-GM1 oligosaccharide-binding R-type lectin from purplish bifurcate mussels *Mytilisepta virgata* and its effect on MAP kinases. *FEBS J.*, 287, 12, 2612–2630, 2020.
95. Hasan, I., Asaduzzaman, A.K.M., Swarna, R.R., Fujii, Y., Ozeki, Y., Uddin, M.B., Kabir, S.R., MytiLec-1 shows glycan-dependent toxicity against brine shrimp Artemia and induces apoptotic death of Ehrlich ascites carcinoma cells *in vivo*. 17, 9, 502, 2019.AU: Please provide journal title.
96. Jhou, Y.F., Yang, F.L., Liao, K.S., Wu, C.Y., Lee, I.M., A lectin from *Crenomytilus grayanus* as a probe for the detection of widespread cancerous and metastatic cells. *Glycobiology*, 33, 5, 423–431, 2023.
97. Chikalovets, I.V., Chernikov, O.V., Pivkin, M.V., Molchanova, V.I., Litovchenko, A.P., Li, W., Lukyanov, P.A., A lectin with antifungal activity

from the mussel *Crenomytilus grayanus*. *Fish Shellfish Immunol.*, 42, 2, 503–507, 2015.

98. Hossain, M.M., Rajia, S., Ohkawa, M., Yoshimoto, S., Fujii, Y., Kawsar, S.M., Ozeki, Y., Hasan, I., Physicochemical properties and antimicrobial activities of MytiLec-1, a member from the mytilectin family of mussels. *Int. J. Biol. Macromol.*, 253, 127628, 2023.
99. He, J., Shen, C., Liang, H., Fang, X., Lu, J., Antimicrobial properties and immune-related gene expression of a C-type lectin isolated from *Pinctada fucata* martensii. *Fish Shellfish Immunol.*, 105, 330–340, 2020.

Index

Printed in the USA/Agawam, MA
September 16, 2026

910661.086